Nanocatalysis
Applications and Technologies

Editors
Vanesa Calvino-Casilda
Departamento de Ingeniería Eléctrica, Electrónica, Control
Telemática y Química Aplicada a la Ingeniería
ETS de Ingenieros Industriales
Universidad Nacional de Educación a Distancia (UNED)
Madrid, Spain

Antonio José López-Peinado
Departamento de Química Inorgánica
y Química Técnica. Facultad de Ciencias
Universidad Nacional de Educación a Distancia (UNED)
Madrid, Spain

Rosa María Martín-Aranda
Departamento de Química
Inorgánica y Química Técnica
Facultad de Ciencias
Universidad Nacional de Educación a Distancia (UNED)
Madrid, Spain

Elena Pérez-Mayoral
Departamento de Química Inorgánica
y Química Técnica, Facultad de Ciencias
Universidad Nacional de Educación a Distancia (UNED)
Madrid, Spain

CRC Press
Taylor & Francis Group
Boca Raton London New York

CRC Press is an imprint of the
Taylor & Francis Group, an **informa** business
A SCIENCE PUBLISHERS BOOK

Cover illustration reproduced by kind courtesy of F.G. Caballero (CENIM, CSIC) and M.L. Miller (ORNL, USA)

CRC Press
Taylor & Francis Group
6000 Broken Sound Parkway NW, Suite 300
Boca Raton, FL 33487-2742

First issued in paperback 2020

© 2019 by Taylor & Francis Group, LLC
CRC Press is an imprint of Taylor & Francis Group, an Informa business

No claim to original U.S. Government works

ISBN-13: 978-1-138-70379-7 (hbk)
ISBN-13: 978-0-367-78025-8 (pbk)

Library of Congress Cataloging-in-Publication Data

Names: Calvino-Casilda, Vanesa, editor.
Title: Nanocatalysis : applications and technologies / editors, Vanesa
    Calvino-Casilda (Departamento de Ingenieria Electrica, Electronica,
    Control, Telemâatica y Quimica Aplicada a la Ingenieria, Universidad
    Nacional de Educacion a Distancia (UNED), ETS de Ingenieros Industriales,
    Madrid, Spain) [and three others].
Description: Boca Raton, FL : CRC Press, Taylor & Francis Group, 2019. | "A
    science publishers book." | Includes bibliographical references and index.
Identifiers: LCCN 2018051046 | ISBN 9781138703797 (hardback)
Subjects: LCSH: Catalysts. | Nanostructured materials.
Classification: LCC TP159.C3 N345 2019 | DDC 660/.2995--dc23
LC record available at https://lccn.loc.gov/2018051046

Visit the Taylor & Francis Web site at
http://www.taylorandfrancis.com

and the CRC Press Web site at
http://www.crcpress.com

Preface

Catalysis is an important area where nanotechnology offers interesting new opportunities. The use of nanocatalysts has increased the opening of new fields for the synthesis of industrially important products. The special properties of nanomaterials are mainly due to their increased relative surface area, and quantum effects, that can modify or enhance their reactivity.

Nanocatalysts are promising materials for different chemical processes. However, they have potential health risks.

Catalytic technologies are critical to future energy, chemical process, and environmental industries (production of petrochemical and chemical products, fuels from coal and natural gas or conversion of crude oil).

Production of high value products from cheaper raw materials, and friendly chemical processes are the drivers for nanocatalysts development.

The main applications of nanocatalysts are found in medicine, water purification, energy storage, biodiesel production, photocatalytic activity, environment protection or production of hydrogen for fuel cell.

Different materials and elements, such as silica, clays, iron, aluminium or titanium, have been used as nanocatalysts. Nevertheless, a full investigation of their catalytic behaviour has still not been completely described.

In this sense, new investigations such as the use of magnetic supports to recover the catalyst from the catalytic media, or the preparation of metallic nanocatalysts include gold, ruthenium, rhodium, silver, palladium, iron, nickel and platinum nanoparticles supported on zeolites, silica, clays, or alumina have been recently developed. Nanocatalysis plays a central role in the industry. The industrial impact of nanocatalysis is clearly reflected by the increasing number of patents, technologies and products in the market. Thus, it is necessary to provide a better understanding of the effects of the size and shape of the nanoparticles and their interactions with support materials or stabilizing agents.

Hence, this book reflects the cooperation between international scientists of diverse disciplines including chemistry, material sciences, engineering and toxicology.

Thus, we hope this book can provide information about nanocatalysts, their applications and technologies, inspiring new investigation in this field.

Vanesa Calvino-Casilda
Antonio José López-Peinado
Rosa María Martín-Aranda
Elena Pérez-Mayoral

Contents

Part I
Nanotoxicity

Chapter 1

Nanotechnology
Concepts of Nanotoxicity

Pablo Fernández-Rodríguez,[1,2] Jorge Hurtado de Mendoza,[2]
José Luis López-Colón,[2] Antonio José López-Peinado[1] and
Rosa María Martín-Aranda[1,]*

INTRODUCTION

Nanotechnology has experienced rapid development due to the variety of applications of nanomaterials (Gajanan and Tijare 2018; Yata et al. 2018; Yin and Talapin 2013). New materials for industrial applications have new and improved physical and chemical properties (Yan 2018; Tiwari and Dhoble 2018; Wang and Astruc 2018) that are different when compared with their micron-sized counterparts (Dilnawaz et al. 2018; Ovid'ko et al. 2018). However, it is inevitable the human exposure to these substances (Sardoiwala et al. 2018; Guggenheim et al. 2018). The life-cycle of these materials' (development, manufacture, consumer usage, final disposal) affects animal species, environment (water, air, soil) and consumers (Villaseñor et al. 2018; Deng et al. 2018). Recently, a range of toxic effects of the so-called engineered nanomaterials (ENM) have been confirmed, indicating that they are a risk for human health (De Matteis and Rinaldi 2018).

The toxicity of engineered nanomaterials is due to some of their physical and chemical properties such as charge, surface area and reactivity (Sukhanova 2018). It is necessary to study the relationship between the structure and properties of engineered nanomaterials and the interaction of these nanomaterials with living organisms for a better understanding of their nanotoxicity. Engineered nanomaterials

[1] Departamento de Química Inorgánica y Química Técnica. Facultad de Ciencias, UNED, Paseo Senda del Rey, 9, 28040-Madrid, Spain.
[2] Instituto de Toxicología de la Defensa, Área de Espectroscopía de Emisión Atómica. Gta. Del Ejército, 1, 28047-Madrid, Spain.
   Emails: pablofer.rodriguez@gmail.com; jhm@grilab.es; jlopcol@oc.mde.es; alopez@ccia.uned.es
[*] Corresponding author: rmartin@ccia.uned.es

can interact with living organisms (Kumar et al. 2018) and damage them, due to their small size. Unfortunately, the unique properties of engineered nanomaterials should be associated with some of their undesired health hazards (Yang et al. 2017). The study of toxicity of these new materials is a recent discipline called nanotoxicology (Boraschi et al. 2018).

Nanomaterials have important uses in a wide range of areas, such as catalysis (Sharma et al. 2015; Khalaj 2018), biotechnology (Karimi et al. 2018), energy (Levchenko et al. 2018), structural materials (Wu et al. 2018), electronic materials (Saha et al. 2018), biosensors (Fàbrega et al. 2018; Ly et al. 2018) medicine, environmental sciences (Singh 2018; Golchin et al. 2018), among others. In the last years, important advances have been made in the preparation and characterization of nanomaterials to generate nanoparticles with different size, shape, properties, and surface potential, which have totally changed the industry (Chen et al. 2018; Xu et al. 2018). Increasing nanotechnology applications have led to the increasing demand for procedures and safety protocols for workers, researcher and consumers. Hence, nanosafety is a priority for developing a sustainable environment (Tran et al. 2017).

With the increasing utilization of ENM, the exposure of workers and consumers to ENM increases significantly. The risks of exposure to human health, concerning the attractive characteristics of ENM (high surface, small size and high surface reactivity) has to be studied and monitored in different scenarios, and models have to be developed (Swierczewska et al. 2018).

There are no described standard protocols to control the toxicity of ENM, because their composition can change over time after their release into the environment and during their lifecycle (Oberdörster et al. 2007). There are very few Occupational Exposure Limits for nanomaterials (OELs). For instance, 1 $\mu g\ m^{-3}$ elemental carbon as Recommended Exposure Limit (REL) in the respirable fraction as an 8 h time weighted average, has been determined for all types of carbon nanotubes and carbon nanofibers. Several studies provided useful information to estimate the effects on consumers through the environment, but they do not provide quantitative exposure information (Karimi et al. 2018). In the same way, there is very little information on the potential toxicity for exposure to ENM containing products during recycling steps and disposal. So, further investigation is necessary for a better understanding of the toxicity and end-of-life processes of nanomaterials, and full characterization of ENMs is necessary to understand their properties (Chatterjee et al. 2018; He et al. 2018). It is also important to develop *in silico* methods to predict the biological effects of ENM as a complementary study to the experimental investigation. Due to the complex interactions between living organisms with ENM, and the increasing development of high throughput screening methods and statistical modelling, such as QSAR, quantitative structure-activity relationships or machine learning methods are the preferred methods of very recent studies (Baweja and Dhawan 2018).

Characterization of nanomaterials

During the production of nanomaterials, chemical requirements and toxicological information have to be controlled. For this reason, there is an increasing interest in investigating and developing appropriate methods and standard procedures for physical and chemical characterization of nanomaterials (Vlastou et al. 2018). Some of these methods are very specific for particular characteristics, and it is necessary to combine several specific techniques to understand the properties of nanomaterials and their applications (Pulido-Reyes et al. 2018). Moreover, we will describe how such information is key in understanding the possible toxicity of ENM (Laux et al. 2018). A variety of analytical techniques are used to generate information on textural properties and structural characteristics (Jo et al. 2018; Mackevica et al. 2018). Electron microscopy is the most important technique for the chemical and structural investigation of nanomaterials at high resolution, near to atomic levels (Hendriks et al. 2018; Najafi et al. 2018). Other non-destructive methods such as X-ray diffraction and gas adsorption isotherms are also useful. This type of data are important to understand the toxicity of the nanoparticles. Significant development has been made during the last decade, in the correlation of physical and chemical parameters with nanomaterials toxicity (Bundschuh et al. 2018). In this way, physicochemical properties, which describe the state of the nanomaterials during and after release and/ or exposure, and their reactivity have been investigated in the last years.

Many nanomaterials based on SiO_2, ZnO, TiO_2, Fe-oxide, carbon nanotubes, carbon black or nanoclays are used in a number of applications such as catalysis, energy, electronics, chemicals, coatings or medicine and have to be produced in very large amounts (Ibrahim et al. 2018). Other technologically important materials (silver and nanoporous materials) are produced in a smaller scale (Zaharescu et al. 2018) but, in general, their fabrication has to be controlled (Kumar et al. 2018; Saba et al. 2018; Arshad et al. 2018; Sánchez-Calvo et al. 2018; Sharma et al. 2018; AlZoubi et al. 2018).

Composition

The chemical composition is fundamental to classify nanomaterials and the atomic structure of materials is a key point for the identification of the toxicological effects of ENMs.

The classification can be in organic or inorganic groups, polymers, ceramic and carbon based materials. The different properties (chemical, physical and biological) of nanomaterials are influenced by their chemical composition. Several methods can be employed to determine their chemical composition. The most common methods to study nanomaterials are Inductively Coupled Plasma-Mass Spectrometer (ICP-MS), Nuclear Magnetic Resonance, Raman Spectroscopy, X-Ray photoelectron spectroscopy (XPS), X-Ray fluorescence spectroscopy (XRF), Energy Dispersive X-Ray Spectroscopy (EDS/EDS) and Electron Energy Loss Spectroscopy (EELS).

The biological response of many nanoparticles depends on the crystal phases. As an example, various forms of silica differ in their physicochemical and toxicological properties, or the inhalation of crystalline silica affords serious negative effects, while synthetic amorphous silica has not shown any negative effects due to their amorphous state. X-ray diffraction (XRD) is an important technique to identify the crystalline phase in nanomaterials. SAXS is also an X-ray diffraction-based technique employed for the same purpose. Both techniques, SAXS and XRD, exclusively give the information on the diffraction pattern for further interpretation of the crystal phases. On the other hand, transmission electron microscopy, TEM, atomic force microscopy, AFM and scanning electron microscopy, SEM, which are microscopy methods, generate bidimensional images, at the atomic scale of the nanoparticles. TEM images give amplitude and phase information of the crystal structure. TEM information is very important in the study of pore connectivity in the amorphous silica network and mesoporous silicas.

Surface modification

Surface modification of nanomaterials (doping or functionalization) provokes changes in physical and chemical properties with respect to their pristine structure. For example, methylation of the silica surface is an effective route to disperse silica nanoparticles in organic solvents. In general, modification of the surface of nanoparticles can be carried out by physical modifications and chemical treatments. Frequently, the combination of both chemical and physical methods is employed for the preparation of engineered nanomaterials. Moreover, additional techniques such as XRF, XPS, electron microscopy FT-IR, XRD, Raman, and mass spectrometry, can be used to determine the nature of doping and surface chemical modification of nanomaterials; thermal analysis is an easier technique for quantitative analysis.

Size and shape

The shape and size of nanoparticles show important toxicity effects in our organism (Tran et al. 2017). In nanomaterials, it is very important to determine the size and size-distribution. It was observed that, when the particle size decreases, the concentration of atoms on the particle surface increases and, in consequence, the chemical and physical properties can be altered. Particle size measurements afford the basis for the study of toxicity of nanoparticles. Different methods present different aspects for the analysis of the particle size and size distribution. In this sense, light diffraction methods or electron microscopy are important analytical techniques.

Nanoparticles present physical and chemical properties associated with hierarchical shapes. The surface energy considerations and thermodynamic, at the nanoscale, are not easy to study. Bottom-up preparation methods are employed to generate a large variety of nanoparticles. Different shapes such as tubes, wires, spheres, rods, spheroids or plates can be obtained, depending on the synthetic conditions and composition. The shape of nanoparticles is an important factor controlling the properties associated at nanoscale. For instance, the shape of nanoparticles is a key factor in medicine applications (the bio-distribution, drug delivery and circulation time, among others).

Electron microscopy such as SEM, TEM and SPM are used for shape identification of nanoparticles, while intrinsic properties, such as surface morphology and crystal structure, are studied by tomography combined with electron microscopy.

Agglomeration effects in nanomaterials are related with high surface energies and other phenomena. The surface energy can be classified into magnetic properties (ferromagnetism), physical interlock (entanglement), electrical properties (Van der Waals forces) and bridging (liquid film or greasy coatings). The dispersion energy for these phenomena is moderate. Nevertheless, others phenomena have a strong influence and interface contacts and high binding energies generate aggregates.

The surface characterization of nanomaterials and their properties are difficult to analyse when agglomerates and aggregates are formed. The understanding of agglomeration and aggregation are not clear for defining the particle state. The state of these nanoparticles in biological systems and the environment is still under study. It is very important for the understanding agglomeration and aggregation mechanisms. The techniques to determine the aggregation and agglomeration level of nanoparticles are the same as used for the nanoparticle size calculation.

Porosity and surface area

The accessible surface area is a singular characteristic of nanomaterials. Nanomaterials possess high surface area because of the high number of atoms on the surface with respect to the atoms inside the particles. For this reason, they are very attractive, and offers new properties with interesting applications demanding high surface such as chemical sensing, hydrogen storage and catalysts. A higher toxicity was observed from nanomaterials than from their larger counterparts. For instance, ultrafine nanoparticles of titanium oxide are more cytotoxic than the homologous fine sized particles.

Surface area can be calculated by the BET equation, which is one of the most frequently employed methods to identify nanomaterials. The total pore volume, and the pore size distribution of the nanoparticles are calculated from the adsorption branch of the isotherm and fitting various pore shapes (bottleneck, cylindrical or slit type pores). For the study of material powders, the specific surface area is normally calculated by the BET equation using nitrogen adsorption. Adsorption of gases is an important technique for surfaces characterization and porosity measurements. To determine the textural properties of solid material, different gases (N_2, Ar, He) at different pressures are adsorbed on a material, and information about the surface area, pore volume and pore size are easily obtained. Five types of adsorption isotherms have been described. The Type I isotherm is characteristic of microporous materials. Type II isotherm is employed to analyse macroporous and non-porous solids. Type III and Type V isotherms are characteristic of weak adsorbate-adsorbent interactions and are commonly associated with microporous or mesoporous materials, respectively, and Type IV isotherms are typical for mesopores where pore-filling occurs by pore condensation.

Stages of the life cycle of nanoproducts

During the last years, nanotechnology has been included in an important number of industrial processes and, in consequence, the increasing use of nanomaterials enhanced the potential risks for the environment, consumers and workers (Fojtů et al. 2017). In order to understand the life cycle of industrial nanoproducts, it is necessary to define the life cycle and exposure cases. Hence, we have described here the general processes at all the stages of the life cycle of the nanoproducts, from the synthesis of the nanomaterial to the end of the life of the product (Salieri et al.

2018). Identification of all the processes involved in the release is also important (Naumenko and Fakhrullin 2017). The type of release (isolated particles, aggregates, etc.) may be a determinant for the toxicity evaluation (Alsaleh and Brown 2018).

Industrial preparation of nanomaterials

New materials have allowed an important expansion of nanotechnology in the industry (Zhao et al. 2017). During the fabrication, the worker is in contact with the materials. In general, silica, carbon nanotubes, titanium dioxide, zinc, iron, cerium and aluminium oxides are the nanomaterials with highest production volumes. For this reason, the production of ENMs received high attention due to the potential risks for human health (Laux et al. 2018; Xu et al. 2018). On the contrary, the release of ENMs to the environment during the synthesis has been less studied (Giovannetti et al. 2017).

The synthetic methods for the preparation of nanomaterials are gas-, solid- or liquid-phase reactions, and top-down and bottom-up methods under dry or wet synthesis. The subsequent purification step depends on the material and on the synthetic method. More than one step with an extra purification path is needed in the case of coated particles (Thambiraj et al. 2017).

The addition of nano additive materials into products involves several steps. Most of the them are aggressive and induce the release of the nanomaterial. The release of nanomaterials from nanocomposites during the synthesis has received special attention in comparison to other steps during preparation and use. Most of the studies investigate, in detail, the polymeric nanocomposites on the matrix, due to their high production volume (Shin et al. 2018). Nevertheless, standard protocols have been developed to control the nanomaterial release during the fabrication.

Preparation of carbon based nanomaterials

Several advances have been done in the preparation of carbon based nanomaterials (Iwamura et al. 2018; Vogli et al. 2018; Kaymakci et al. 2018). Carbon based nanomaterials can be produced mainly by thermal and hydrothermal syntheses and plasma processes (Xu et al. 2018; Morère 2017). Arc Discharge and Laser Ablation are the most frequent plasma processes. In the first one, a current between both graphite electrodes, at low pressure, under hydrogen, helium or methane circulate in the presence of a catalyst containing a transition metal. During the process, graphite is vapourized and condensed over the cathode of the reactor. Carbon nanotubes can also be prepared by Laser Ablation, in a similar process to Arc Discharge. The laser vapourizes graphite and catalyst, and carbon nanomaterials grow over the formed nanocatalysts. Usually, the deposition of a carbon source (hydrocarbon over a transition metal catalyst) is observed. Carbon based nanomaterials can also be produced by thermal, hydrothermal or sonochemical methods. These methods consist of heating a mixture of hydrocarbon and water under pressure in the presence of a catalyst. Small amounts of carbon nanomaterials such as SWCNT, MWCNT or fullerenes, mixed with other forms of carbon, are obtained, being necessary for the purification of the final product. Different methods can be used to prepare graphene, graphene oxide and their corresponding derivatives. Graphene can be also produced by top-down approaches (Giovannetti et al. 2018), and the purest graphene can be synthetized by exfoliation of graphite oxide. The bottom-up approaches are employed for the preparation of carbon based nanomaterials such as bidimensional carbon layers over a support. Other strategies including acylation, arylation or carbene or nitrene addition (Ravi et al. 2018) can be employed to modify the surface of carbon nanomaterials to improve their dispersibility, their compatibility with a matrix or to functionalize them with quantum dots or gold nanoparticles, among others.

Carbon nanofibers and carbon nanotubes production is continuously increasing and show high toxicity during the synthesis processes (De Marchi et al. 2018). Although the synthesis takes place in a closed reactor, the exposure occurs during maintenance operations and the material recovery (Kim et al. 2016; Dong and Liu 2016).

Preparation of metallic and oxide based nanomaterials

The exposure to nanometal oxides and metallic based nanomaterials has not been thoroughly investigated (Sánchez et al. 2014). Moreover the release of metallic nanomaterials to the environment has to be controlled.

There are many approaches to the preparation of metallic containing nanomaterials (Ding et al. 2018; Blanco-Flores et al. 2018). Metallic nanomaterials are conventionally prepared by the reduction of a precursor in the presence of a stabilizer. Considering the occupational exposure, the wet synthesis is considered to be lower risk due to the lower aerosol formation compared to the use of powders. Nanoparticles of noble metals, Au, Ag, Pt, are generally synthesized by the reduction of a precursor salt in water (Chen et al. 2017; Song et al. 2018). The colloids so synthesized, they can be directly functionalized *in situ* or in an organic solvent for subsequent surface functionalization. The reaction conditions determine the size and morphology of the resulting material (Ramos-Delgado et al. 2016). Other methods, such physical synthesis, electrochemical deposition, or sol-gel synthesis, can be used to prepare these materials (Hasany et al. 2013). Surface modification of these materials is usually done *in situ,* and the organic modifier is used as stabilizer. In the case of nanometal oxides synthesis, several physical and chemical methods have been investigated. At the laboratory scale, solution routes are the most widely used, affording a better control of size and shape, but at the industrial scale, gas phase reactions are the most commonly used. Both methods are widely used to produce different nanometal oxides such as TiO_2, CeO_2, ZrO_2, Al_2O_3, SiO_2 among others (Manimozhi et al. 2016; Liu et al. 2016; Yong et al. 2015; Asgharpour et al. 2015). Some solution routes are the precipitation method, microemulsion, electrochemical synthesis, the sonochemical method and the sol-gel method (Cherkezova-Zheleva et al. 2015).

Preparation of polymeric and ceramic nanofibres

Electrospinning methods are usually employed to produce nanofibres randomly oriented and connected by bonds or entanglements (Lavrynenko et al. 2018). In this method, a high voltage is used to produce the nanofibers. This methodology is also used during the synthesis of ceramic nanofibres.

In general, the production of nanofibres is usually carried on in closed reactors and occupational and environmental exposure have not been reported during this life cycle step (Jung et al. 2018; Gadzhimagomedov et al. 2018; Gao et al. 2018).

Industrial preparation of nanocomposites

A multiphase solid material with at least one of the phases in the nanoscale is defined as nanocomposite (Sapkota et al. 2018). The main difference between traditional composites and nanocomposites is the high surface of contact between the phases of the nanocomposite (Zhang et al. 2017). New materials with new properties are produced when nanomaterials are added to solid matrices, such as electroluminescent metals, conductive polymeric matrices, photo-luminescent textiles or semiconductor ceramics (Khalid et al. 2017). The properties of the nanocomposite so formed, depend on their constituents, and also on the degree of dispersion of the different phases (Wang et al. 2017; Bondarenko et al. 2017; Hamid et al. 2017; El Fewaty et al. 2017; Bari et al. 2016).

Moreover, new surface properties of materials can be obtained if a nanomaterial is added to the surface of the corresponding material. For instance, the addition could be carried out by *in situ* nanocomposite formation (surface treatment of ceramic tiles with a solution of nanosized titania); or by physical or chemical grafting of the nanomaterial on the surface, such as silver nanoparticles bonded to the fibres of a textile; or by thin layer direct deposition on the surface (preparation of solar cells of nanosized TiO_2 by direct CVD over the cell surface).

Other types of processes are the use of nano-additivated compositions such as dispersion of nanomaterial in emulsions for cosmetics and paints, or mixtures of solids for catalytic mixtures used in the treatment of gas emission (Arslan et al. 2017; Zhao et al. 2017; Gu et al. 2016).

Studies on occupational exposure to nanomaterials, in general, are centered in the preparation procedures (Fonseca et al. 2018). The preparation is usually carried out in closed reactors. Nevertheless, the nanomaterial weight and mix are usually done in open conditions. The effects of the exposition of workers during the nanomaterial cleaning and recovering, have been described in different reviews (Robichaud et al. 2015).

Nanocomposites based on polymeric-matrix

Due to the increasing volume of products commercially available, based on polymers such as artificial tissues, conductive materials, etc., the investigation on polymers has grown very rapidly (Antunes et al. 2016; Kim and Choi 2015).

Three different methods can be used to prepare polymeric matrix nanocomposites: solution casting, melt blending and *in situ* polymerization. During the solution casting, the polymer and the nanomaterial are mixed by sonication to generate the nanocomposite after solvent elimination. The melt blending consists in the vigorous combination of the polymer and the nanomaterial mixed at a temperature that allows the mobility of the polymer. During the *in situ* polymerization, the mixture of the nanomaterial and the monomers, under appropriate conditions, facilitate the polymerization. The selection of one of these methods and the reaction conditions is totally dependent on the type of nanomaterial and on the nature of the polymer. Additives are used to improve the compatibility and the surface modification of the nanomaterial to enhance its compatibility with the polymer (Tarasov et al. 2015; Prasanth et al. 2015; Trakakis et al. 2015).

Nanocomposites based on ceramic-matrix

Traditionally, the main disadvantage of ceramics is their fragility. The incorporation of nano-additives is used for the reinforcement of the ceramic materials allowing their use in new applications (Mazerat and Pailler 2018). Other properties can be also induced by nanomaterials, finding applications in electronic, optics or sensor (Liu et al. 2018). Conventionally, ceramic matrices were reinforced with metallic particles. Actually, they are reinforced with nanometal oxides, carbon based nanomaterials or quantum dots (Li et al. 2018; Jin et al. 2018). Principally, three methods can be used for the nanocomposite processing based on the ceramic matrix: powder process, polymer precursor process and sol-gel process. During the powder process, the different materials are mixed and milled together under wet conditions. It is a simple process that generates a heterogeneous material. The polymer precursor method is similar, but, in this case, the nanomaterial precursor is added to a polymer, and later, pyrolized. Finally, the sol-gel process consists in the hydrolysis and condensation of molecular precursors, which has been previously solved in an organic solvent to generate a sol-gel, and is later dried.

Nanocomposites based on metal-matrix

The traditional reinforcement of composites based on a metal-matrix is a well-known method, but the use of nanomaterials for this purpose has been developed recently (Popov et al. 2018). The metal matrices are thermally stable, resistant to abrasion and thermal and electrical conductors. They are difficult to prepare and the method is expensive. The materials so formed present ceramic and metal hybrid properties, which makes them ideal for multiple applications in light energy conversion, and structural materials employed in the aeronautic industry (Hu et al. 2018). Several methods are used for metal-matrix nanocomposites processing: spray pyrolysis, vapour phase processing, chemical and deformation processes or solidification. The cheapest and most used method is the solidification. During the process, the metal is melted to facilitate the nano-reinforcement and the rapid solidification of the melt by different processes. When liquid infiltration are used, similar results are obtained, but in this case, only the metal is melt surrounding the nanomaterial. Other methods employed include CVD, sol-gel synthesis, spray pyrolysis, etc. (Bundschuh et al. 2018).

Nanomaterials in textiles

Nanomaterials can be added to the textiles forming different types of new materials; nanotextiles, nanocomposite textiles and nanofibrous materials (Gibney 2015; Aized et al. 2018; Deravi et al. 2018). Addition of nanomaterials to textiles affords new and improved properties, with interesting applications. The most common applications of textile nanomaterials are due to their antimicrobial activity and capacity for UV-filtering. This new textiles are also employed as water repellent, photo-luminescent, electro-luminescent, self-cleaning, electrical conductor, etc. (Shahid-Ul-Islam et al. 2018; Harifi and Montazer 2015).

Although different methods have been employed for the modification of surfaces of fibre materials, the main problem is that the nanomaterial usually is released during the washing step because it is not well fixed. Binders particles are used to minimize this problem, to improve the nanomaterial's affinity for the textiles (Rivero et al. 2018; Voelker et al. 2015).

The concept of hormesis and applications of nanomaterials

The hormesis is considered as an adaptive response of biological systems to moderate environmental challenges (Calabrese 2008). The nano-toxicological interest derived from the production and application of new nanomaterials provides a general idea of the current knowledge on the dose–responses induced by the exposure to nanoparticles (Iavicoli et al. 2010). Dose-responses have a direct correlation with physical and chemical properties of nanoparticles. It is necessary further investigations to determine the nanomaterial risks, at low-dose range, for the human health (Iavicoli et al. 2014; Nascarella and Calabrese 2014).

The variety of applications of nanomaterials and their multiple uses in commercial products have grown very rapidly during the last decade. The industrial development of engineering nanomaterials has been very broad for the synthesis of different products: catalysts, optics, sensors, fine chemicals, food, agriculture, vehicles, plastic, rubbers, membranes, etc. (Kaphle et al. 2018). Nevertheless, the distribution of nanomaterials by different product categories is still not well known. The lack of information on the real use of nanomaterials in consumer products, is due to the continuous progress in the nanomaterial industry, and will change in the future (Iavicoli et al. 2018). Today, many countries and agencies are trying to establish general rules and lists of nanomaterials. In 2013, France was the first European country to require the identification of substances of nanoparticle nature. One year later, in June 2014, Denmark also requested the register to products which contain nanomaterials. From the list of data, it could be observed that the sectors with the highest number of applications were agriculture and forestry. Considering the uses in chemical products, the most frequently reported are rubber articles, paints, paint removers, coatings, cosmetics, electronics and plastic articles (Thiruvengadam et al. 2018; Tiwari and Dhoble 2018; Ou et al. 2018; Olajire 2018; Fan et al. 2018).

Some of these applications involve the release of nanomaterials, with subsequent human exposure; for example, the application of sunscreen on the skin or generation of a nanocoating on glass surfaces. The understanding of the fraction that reaches the target application point is still unknown. In other applications, nanomaterials do not become released during use. The amount of the nanomaterial released from the matrix during the use stage will depend on the length of the lifetime, the surface contact area and the frequency and duration of use. To investigate this effect, the increasing research on the release of nanomaterials during their applications has been developed.

Nevertheless, the studies on release of nanomaterials from nanocomposites are small and the values of amounts released differ among the different studies (Gottschalk et al. 2011).

By now, the existing investigation and literature show a partial view of the potential release of nanomaterials from consumer products. So, further studies are needed in the future to describe a model able to determine the release effects under different situations in different processes (Iavicoli et al. 2013).

Nano-risk assessment and end of life

Nanomaterials are present in wastes, but there is still limited information on the possible risks associated with their presence (Kuhlbusch et al. 2018; Shandilya et al. 2018). Wastes can be classified into hazardous or non-hazardous wastes. This classification does not include specific requirements for nanomaterials. Therefore, nanomaterials will be classified in the same categories as their bulk forms (Giese et al. 2018). Even when nanomaterials would be classified as hazardous, they may still be appropriate for use in some consumer products. Due to the use and environmental release of such chemicals, taking into account the importance of real environmental low-dose exposure, the evaluation of the biological dose–response relationships induced by low concentrations of ENMs is an important issue for the nanotoxicological studies. Considering the range of dose exposure, an important development has been produced focused on the concept of hormesis and also in the nano-risk assessment.

When the nanomaterial products reach the end of their life, they can be recycled or discarded. When technical products are synthetized by nanotechnological routes, waste products containing nanomaterials are formed, and the waste generated during the life cycle are potential sources of nanomaterials. The treatment of such wastes will determine the effects of their release on the environment. The development of appropriate management strategies for waste is critical. More investigation is needed to estimate the amount of nanomaterials in different recycling systems, and how the nanomaterials affect the recycled material.

Incineration is another possible alternative at the end of the life of the nanomaterial. The combustion of waste takes place in an oxidizing ambient at high temperatures. During the process, nanomaterials can be totally or partially transformed depending on the different conditions in the oven and the reactivity of the nanomaterial. Depending of the characteristics, the nanomaterials can also remain unaltered. The non altered part normally get released into the environment, but also can be retained in the filters. The treatment of the bottom ash formed is generally disposed in landfills, depending of the local regulations.

Municipal solid waste containing nanomaterials from domestic uses are also disposed off in landfills. Landfills are one of the most used treatments for solid wastes at the end of their life, but the disposal of nanomaterials is not yet clear, because the degree of mobility for different nanomaterials, depends on their pH and organic composition. It is important to estimate the influence of nanomaterials on the biological activity of the area.

Finally, there are relatively serious concerns related to the impact of nanomaterials on the biological systems within wastewater treatments. Domestic wastewater containing nanomaterials end up in sewage treatment plants. Several studies have investigated such processes. Different physico-chemical characteristics play a key role and can modify the dose–response relationship. Current regulations establish metal content limits without consideration of particle size. Further studies on the particle size are needed to analyze the impact of applying sewage sludge containing nanomaterials on soils.

Modelling of nanomaterial toxicity

There is no doubt that nanotechnology provides industries with many advantages, but there are serious concerns about the potential health risks associated with this progress. To estimate the number of worker exposed to the engineered nanomaterials, it is necessary the development of computational models. In this way, it will be possible to correlate physicochemical properties of such materials with their toxicity. In addition, models can also support public needs and industrial regulations for its future design.

Taking into account the high number of nanomaterials requiring toxicity evaluation, alternative approaches such as computational modelling methods can predict human health risks with low cost in a short time. Different computational techniques have been recently developed and used in toxicological studies. The (quantitative) structure–activity relationship ((Q)SAR) analysis is one of the most important approaches. It is employed to predict the toxicity of nanomaterials, and quantify the correlation between biological activity and the properties of different classes of nanomaterials. This method assumes that toxicity depends on physical and chemical properties such as surface, shape and size and crystal structure and hence can be predicted (Tran et al. 2017).

This section presents a brief description of the applicability of ((Q)SAR) methods to the modelling of toxicity of engineered nanomaterials. The models are the key components for a successful application of data-driven toxicity prediction techniques to engineered nanomaterials.

The (Q)SAR methods to model toxicity of nanomaterials have been rapidly developed due to the increasing affect of nanotoxicity observed on the environment and human health. Numerous studies confirm that most of the nanomaterials have side effects, and need to be properly managed.

The toxicity of compounds can be evaluated *in vivo*, *in vitro*, and *in silico* studies. In (Q)SAR analysis, it is the specific type of activity, such as cell viability or toxicity, that is going to be modelled and predicted.

Physical and chemical descriptors are experimental or theoretical parameters employed to correlate the structural parameters with the applications through a statistical method. In general, molecular descriptors representing physical and chemical properties of chemicals can be obtained by two ways: theoretical calculations and experimental analysis. Theoretical descriptors are usually obtained from semi-empirical techniques and different theories. The descriptors give information about the structure and composition in correlation which biological activity, and the molecular structures can be described and represented by more than 5000 descriptors.

The (Q)SAR models are mathematical systems that try to correlate the biological activities of a series of chemicals in a quantitative manner. In (Q)SAR models, it is assumed that the biological activity is correlated with the structure of compounds.

Independent of the method used, the validity models should be evaluated both internally and externally. Internal validation is based on the dataset used in the modelling process. The most used internal validation in (Q)SAR studies are least squares fit (R2), chisquared (χ2), root-mean squared error (RMSE), leave-one-out or leave-many-out cross-validation, bootstrapping and Y-randomization. External validation is always used in combination to internal validation methods and is recommended by researchers for the assessment of (Q)SAR model reliability. Normally, it is interesting to use more than one validation for quantitative measures. The last step is the definition of the applicability domain of the statistically validated model. There are several approaches, based on different algorithms to define the domain of statistical models.

Summary

Nanotechnology clearly provides interesting benefits to the industry and society, but also serious problems and risk to health.

Characterization of nanomaterials is a key concept in industries to predict and control the toxicological problems though all the steps of the production and application of different nanocompounds.

The target of researchers and companies is how to make nanomaterials safer. (Q)SAR models have been used to predict the hazardous properties of nanomaterials for more than 50 years and predictive models such as (Q)SAR provided interesting data on nanotoxicity. When the potential risks are identified, the following step is the adoption of risk reduction measures in the range of tolerable limits. The development of new descriptors able to express the specificity of nano-characteristics would be of interest. The generation of systematic datasets with specific types of nanomaterials will be useful to classify the different mechanisms of toxicity.

Different physico-chemical characteristics of nanomaterials play an important role in determining the dose–response relationship. The experimental conditions (exposure times, environmental conditions) should be investigated to clarify the influence of those factors determining hormetic responses in biological systems. These studies will provide a better understanding of the nanomaterial risk.

Acknowledgements

This work has been supported by Spanish Ministry (project CTM2014-56668-R). RMMA thanks Spanish Ministry for the Salvador de Madariaga Grant (PRX16/00390) at the Operando Molecular Spectroscopy and Catalysis Laboratory, Lehigh University (USA). PFR thanks International School of Doctorade of Universidad Nacional de Educación a Distacia (UNED) for the support of his PhD.

Reference cited

Aized, T., M.B. Khan, H. Raza and M. Ilyas. 2017. Production routes, electromechanical properties and potential application of layered nanomaterials and 2D nanopolymeric composites—A review. Int. J. Adv. Manuf. Technol. 93(9–12): 3449–3459.

Alsaleh, N.B. and J.M. Brown. 2018. Immune responses to engineered nanomaterials: Current understanding and challenges. Curr. Opin. Toxicol. 10: 8–14.

AlZoubi, T., H. Qutaish, E. Al-Shawwa and S. Hamzawy. 2018. Enhanced UV-light detection based on ZnO nanowires/graphene oxide hybrid using cost-effective low temperature hydrothermal process. Opt. Mater. 77: 226–232.

Antunes, M., G. Gedler, H. Abbasi and J.I. Velasco. 2016. Graphene nanoplatelets as a multifunctional filler for polymer foams. Materials Today: Proceedings 3: S233–S239.

Arshad, A., J. Iqbal, A. Alam, B. Khadija and R. Faryal. 2018. Synthesis, characterization, enhanced dielectric and antimicrobial properties of WxCu1−xO nanostructures. Ceram. Int. 44(6): 5894–5900.

Arslan, M. and M.A. Tasdelen. 2017. Polymer nanocomposites via click chemistry reactions. Polymers-Basel 9(10): art. no. 499.

Asgharpour, S., M.R. Vaezi and S.A. Tayebifard. 2015. Preparation of Al2O3/ZrO2 and Al2O3/ZrO2/ZrN nanocomposite by mechanical activated combustion synthesis: Effect of milling time and synthesis atmosphere. J. Optoelectron. Adv. M. 17(9–10): 1507–1514.

Bari, S.S., A. Chatterjee and S. Mishra. 2016. Biodegradable polymer nanocomposites: An overview. Polym. Rev. 56(2): 287–328.

Baweja, L. and A. Dhawan. 2018. Computational approaches for predicting nanotoxicity at the molecular level. Iss. Toxicol. 35: 304–327.

Blanco-Flores, A., N. Arteaga-Larios, V. Pérez-García, J. Martínez-Gutiérrez, M. Ojeda-Escamilla and I. Rodríguez-Torres. 2018. Efficient fluoride removal using Al-Cu oxide nanoparticles supported on steel slag industrial waste solid. Env. Sci. Pollut. R. 25(7): 6414–6428.

Bondarenko, M.V., T.A. Khalyavka, N.D. Shcherban and N.N. Tsyba. 2017. Mesoporous nanocomposites based on titanium dioxide and carbon as perspective photocatalysts for water purification. Nanosistemi, Nanomateriali, Nanotehnologii 15(1): 99–112.

Boraschi, D., B.J. Swartzwelter and P. Italiani. 2018. Interaction of engineered nanomaterials with the immune system: Health-related safety and possible benefits. Curr. Opin. Toxicol. 10: 74–83.

Bundschuh, M., J. Filser, S. Lüderwald, M.S. McKee, G. Metreveli, G.E. Schaumann, R. Schulz and S. Wagner. 2018. Nanoparticles in the environment: where do we come from, where do we go to? Environ. Sci. Eur. 30(1): art. no. 6.

Calabrese, E.J. 2008. Hormesis: Why it is important to toxicology and toxicologists. Environ. Toxicol. Chem. 27: 1451–1474.

Chatterjee, S., R. Mankamna Kumari and S. Nimesh. 2017. Nanotoxicology: Evaluation of toxicity potential of nanoparticles. Advances in Nanomedicine for the Delivery of Therapeutic Nucleic Acids, 188–201. Elsevier: WoodHead Publishing.

Chen, B., X. Fu, J. Tang, M. Lysevych, H.H. Tan, C. Jagadish and A.H. Zewail. 2017. Dynamics and control of gold-encapped gallium arsenide nanowires imaged by 4D electron microscopy. P. Natl. Acad. Sci. USA 114(49): 12876–12881.

Chen, G., M.G. Vijver, Y. Xiao and W.J.G.M. Peijnenburg. 2017. A review of recent advances towards the development of (quantitative) structure-activity relationships for metallic nanomaterials. Materials 10(9): art. no. 1013.

Cherkezova-Zheleva, Z.P., K.L. Zaharieva, M.P. Tsvetkov, V.S. Petkova, M.M. Milanova and I.G. Mitov. 2015. Impact of preparation method and chemical composition on physicochemical and photocatalytic properties of nano-dimensional magnetite-type materials. Am. Mineral. 100(5-6): 1257–1264.

De Marchi, L., C. Pretti, B. Gabriel, P.A.A.P. Marques, R. Freitas and V. Neto. 2018. An overview of graphene materials: Properties, applications and toxicity on aquatic environments. Sci. Total. Environ. 631-632: 1440–1456.

De Matteis, V. and R. Rinaldi. 2018. Toxicity assessment in the nanoparticle era. Advances in Experimental Medicine and Biology 1048: 1–19. Springer.

Deng, H., Y. Zhang and H. Yu. 2018. Nanoparticles considered as mixtures for toxicological research. J. Environ. Sci. Heal. C 36(1): 1–20.

Deravi, L.F., N.R. Sinatra, C.O. Chantre, A.P. Nesmith, H. Yuan, S.K. Deravi, J.A. Goss, L.A. MacQueen, M.R. Badrossamy, G.M. Gonzalez, M.D. Phillips and K.K. Parker. 2017. Design and fabrication of fibrous nanomaterials using pull spinning. Macromol. Mater. Eng. 302(3): art. no. 1600404.

Dilnawaz, F., S. Acharya and S.K. Sahoo. 2018. Recent trends of nanomedicinal approaches in clinics. Int. J. Pharm. 538(1-2): 263–278.

Ding, S., M. Xue, R. Wu, Y. Lai, Y. Men, X. Kong, L. Han, J. Han, W. Yang, Y. Yang, H. Du, C. Wang and J. Yang. 2018. Electron beam reduction induced self-assembly growth of Co/CoO nanocomposite materials. J. Alloy Compd. 744: 615–620.

Dong, R. and L. Liu. 2016. Preparation and properties of acrylic resin coating modified by functional graphene oxide. Appl. Surf. Sci. 368: 378–387.

El Fewaty, N.H., A.M. El Sayed and R.S. Hafez. 2016. Synthesis, structural and optical properties of tin oxide nanoparticles and its CMC/PEG–PVA nanocomposite films. Polym. Sci. Ser. A+ 58(6): 1004–1016.

Fàbrega, C., O. Casals, F. Hernández-Ramírez and J.D. Prades. 2018. A review on efficient self-heating in nanowire sensors: Prospects for very-low power devices. Sensor Actuat. B-Chem. 256: 797–811.

Fan, L., B. Zhu, P.-C. Su and C. He. 2018. Nanomaterials and technologies for low temperature solid oxide fuel cells: Recent advances, challenges and opportunities. Nano Energy 45: 148–176.

Fojtů, M., W.Z. Teo and M. Pumera. 2017. Environmental impact and potential health risks of 2D nanomaterials. Environ. Sci.-Nano. 4(8): 1617–1633.

Fonseca, A.S., E. Kuijpers, K.I. Kling, M. Levin, A.J. Koivisto, S.H. Nielsen, W. Fransman, Y. Fedutik, K.A. Jensen and I.K. Koponen. 2018. Particle release and control of worker exposure during laboratory-scale synthesis, handling and simulated spills of manufactured nanomaterials in fume hoods. J. Nanopart. Res. 20(2): art. no. 48.

Gadzhimagomedov, S.H., N.M.-R. Alikhanov, R.M. Emirov, D.K. Palchaev, Z.K. Murlieva, M.K. Rabadanov, S.A. Sadykov, M.M. Khamidov and A.D.H. Hashafa. 2017. Structure and properties of nanostructured YBa2Cu3O7-δ, BiFeO3, and Fe3O4. Semiconductors+ 51(13): 1686–1691.

Gajanan, K. and S.N. Tijare. 2018. Applications of nanomaterials. Mater Today-Proc 5(1): 1093–1096.

Gao, C., P. Feng, S. Peng and C. Shuai. 2017. Carbon nanotube, graphene and boron nitride nanotube reinforced bioactive ceramics for bone repair. Acta Biomater. 61: 1–20.

Gibney, E. 2015. Q and A: The nanomaterials designer. Nature 526(7574): 504.

Giese, B., Klaessig, B. Park, R. Kaegi, M. Steinfeldt, H. Wigger, A. Von Gleich and F. Gottschalk. 2018. Risks, release and concentrations of engineered nanomaterial in the environment. Sci. Rep.-UK 8(1): art. no. 1565.

Giovannetti, R., E. Rommozzi, M. Zannotti and C.A. D'Amato. 2017. Recent advances in graphene based TiO2 nanocomposites (GTiO2Ns) for photocatalytic degradation of synthetic dyes. Catalysts 7(10): art. no. 305.

Golchin, A., S. Hosseinzadeh and L. Roshangar. 2018. The role of nanomaterials in cell delivery systems. Med. Mol. Morphol. 51(1): 1–12.

Gottschalk, F. and B. Nowack. 2011. The release of engineered nanomaterials to the environment. J. Environ. Monit. 13: 1145–1155.

Gottschalk, F., T. Sun and B. Nowack. 2013. Environmental concentrations of engineered nanomaterials: Review of modeling and analytical studies. Environ. Pollut. 181: 287–300.

Gu, H., G. Ma, J. Gu, J. Guo, X. Yan, J. Huang, Q. Zhang and Z. Guo. 2016. An overview of multifunctional epoxy nanocomposites. J. Mater. Chem. C 4(25): 5890–5906.

Guggenheim, E.J., S. Milani, P.J.F. Röttgermann, M. Dusinska, C. Saout, A. Salvati, J.O. Rädler and I. Lynch. 2018. Refining *in vitro* models for nanomaterial exposure to cells and tissues. NanoImpact 10: 121–142.

Hamid, Z.A., A.Y. El-Etre and M. Fareed. 2017. Performance of Ni-Cu-ZrO2 nanocomposite coatings fabricated by electrodeposition technique. Anti-Corros. Methods M 64(3): 315–325.

Harifi, T. and M. Montazer. 2015. A review on textile sonoprocessing: A special focus on sonosynthesis by nano–bio–eco interactions. J. Environ. Sci. Heal. C 36(1): 21–42.

Hasany, S.F., N.H. Abdurahman, A.R. Sunarti and R. Jose. 2013. Magnetic iron oxide nanoparticles: Chemical synthesis and applications review. Curr. Nanosci. 9(5): 561–575.

He, X., P. Fu, W.G. Aker and H.-M. Hwang. 2018. Toxicity of engineered nanomaterials mediated by nano–bio–eco interactions. J. Environ. Sci. Heal. C 36(1): 21–42.

Hendriks, F.C., S. Mohammadian, Z. Ristanović, S. Kalirai, F. Meirer, E.T.C. Vogt, P.C.A. Bruijnincx, H.C. Gerritsen and B.M. Weckhuysen. 2018. Integrated transmission electron and single-molecule fluorescence microscopy correlates reactivity with ultrastructure in a single catalyst particle. Angew. Chem. Int. Edit. 57(1): 257–261.

Hu, Z., F. Chen, J. Xu, Q. Nian, D. Lin, C. Chen, X. Zhu, Y. Chen and M. Zhang. 2018. 3D printing graphene-aluminum nanocomposites. J. Alloys Compd. 746: 269–276.

Iavicoli, I., E.J. Calabrese and M.A. Nascarella. 2010. Exposure to nanoparticles and hormesis. Dose-Response 8: 501–517.

Iavicoli, I., L. Fontana, V. Leso and A. Bergamaschi. 2013. The effects of nanomaterials as endocrine disruptors. Int. J. Mol. Sci. 14: 16732–16801.

Iavicoli, I., L. Fontana, V. Leso and E.J. Calabrese. 2014. Hormetic dose-responses in nanotechnology studies. Sci. Total Environ. 487: 361–374.

Iavicoli, I., V. Leso, L. Fontana and E.J. Calabrese. 2018. Nanoparticle exposure and hormetic doseresponses: An update. Inter. J. Mol. Sci. 19(3): art. no. 805.

Ibrahim, E.M.M., L.H. Abdel-Rahman, A.M. Abu-Dief, A. Elshafaie, S.K. Hamdan and A.M. Ahmed. 2018. Electric, thermoelectric and magnetic characterization of γ-Fe2O3 and Co3O4 nanoparticles synthesized by facile thermal decomposition of metal-Schiff base complexes. Mater. Res. Bull. 99: 103–108.

Iwamura, S., K. Fujita, R. Iwashiro and S.R. Mukai. 2018. Efficient preparation of TiO2/C nanocomposite for electrode material through the liquid pulse injection technique. Mater. Today Comm. 14: 15–21.

Jin, H., Z. Li, L. Wang and Q. Zeng. 2018. Fabrication and properties of CNT/Ni/Y/ZrB2 nanocomposites reinforced in situ. J. Am. Ceram. Soc. 101(4): 1747–1753.

Jo, J., Y. Tchoe, G.-C. Yi and M. Kim. 2018. Real-time characterization using in situ rheed transmission mode and TEM for investigation of the growth behaviour of nanomaterials. Sci. Rep.-UK 8(1): art. no. 1694.

Jung, D.-H., A. Sharma and J.-P. Jung. 2018. Influence of dual ceramic nanomaterials on the solderability and interfacial reactions between lead-free Sn-Ag-Cu and a Cu conductor. J. Alloy Compd. 743: 300–313.

Kaphle, A., P.N. Navya, A. Umapathi and H.K. Daima. 2018. Nanomaterials for agriculture, food and environment: applications, toxicity and regulation. Environ. Chem. Lett. 16(1): 43–58.

Karimi, M., R. Sadeghi and J. Kokini. 2018. Human exposure to nanoparticles through trophic transfer and the biosafety concerns that nanoparticle-contaminated foods pose to consumers. Trends Food Sci. Tech. 75: 129–145.

Kaymakci, A., N. Ayrilmis, T. Gulec and M. Tufan. 2017. Preparation and characterization of highperformance wood polymer nanocomposites using multi-walled carbon nanotubes. J. Compos. Mater 51(9): 1187–1195.

Khalaj, M., M. Kamali, Z. Khodaparast and A. Jahanshahi. 2018. Copper-based nanomaterials for environmental decontamination—An overview on technical and toxicological aspects. Ecotox. Environ. Safe 148: 813–824.

Khalid, M., C.T. Ratnam, R. Walvekar, M.R. Ketabchi and M.E. Hoque. 2017. Reinforced natural rubber nanocomposites: Next generation advanced material. Green Energy and Technology, Green Biocomposites, 309–345. Springer.

Kim, H.-H., W. Han, K.-H. An and B.-J. Kim. 2016. Preparation of nickel coated-carbon nanotube/zinc oxide nanocomposites and their antimicrobial and mechanical properties. App. Chem. Eng. 27(5): 502–507.

Kim, J.W. and B.G. Choi. 2015. Preparation of three-dimensional graphene/metal oxide nanocomposites for application of supercapacitors. Appl. Chem. Eng. 26(5): 521–525.

Kuhlbusch, T.A.J., S.W.P. Wijnhoven and A. Haase. 2018. Nanomaterial exposures for worker, consumer and the general public. NanoImpact 10: 11–25.

Kumar, A., S. Singh, R. Shanker and A. Dhawan. 2018. Chapter 1: Nanotoxicology: Challenges for biologists. Iss. Toxicol. (35): 1–16.

Kumar, R., A. Umar, D.S. Rana, P. Sharma, M.S. Chauhan and S. Chauhan. 2018. Fe-doped ZnO nanoellipsoids for enhanced photocatalytic and highly sensitive and selective picric acid sensor. Mater. Res. Bull. 102: 282–288.

Laux, P., C. Riebeling, A.M. Booth, J.D. Brain, J. Brunner, C. Cerrillo, O. Creutzenberg, I. Estrela-Lopis, T. Gebel, G. Johanson, H. Jungnickel, H. Kock, J. Tentschert, A. Tlili, A. Schäffer, A.J.A.M. Sips, R.A. Yokel and A. Luch. 2018. Challenges in characterizing the environmental fate and effects of carbon nanotubes and inorganic nanomaterials in aquatic systems. Environ. Sci.-Nano. 5(1): 48–63.

Laux, P., J. Tentschert, C. Riebeling, A. Braeuning, O. Creutzenberg, A. Epp, V. Fessard, K.-H. Haas, A. Haase, K. Hund-Rinke, N. Jakubowski, P. Kearns, A. Lampen, H. Rauscher, R. Schoonjans, A. Störmer, A. Thielmann, U. Mühle and A. Luch. 2018. Nanomaterials: certain aspects of application, risk assessment and risk communication. Arch. Toxicol. 92(1): 121–141.

Lavrynenko, S., A.G. Mamalis and E. Gevorkyan. 2018. Features of consolidation of nanoceramics for aerospace industry. Mater. Sci. Forum 915: 179–184.

Leso, V., L. Fontana, M.C. Mauriello and I. Iavicoli. 2017. Occupational risk assessment of engineered nanomaterials: Limits, challenges and opportunities. Curr. Nanosc. 13: 55–78.

Levchenko, I., S. Xu, G. Teel, D. Mariotti, M.L.R. Walker and M. Keidar. 2018. Recent progress and perspectives of space electric propulsion systems based on smart nanomaterials. Nat. Commun. 9(1): art. no. 879.

Li, H., Y. Gou, S. Chen and H. Wang. 2018. Preparation and properties of a novel precursor-derived Zr-C-B-N composite ceramic via zirconocene and borazine. Ceram. Int. 44(4): 4097–4104.

Liu, D.G., L. Zheng, Y. Liang, H. Li, J.Q. Liu, L.M. Luo and Y.C. Wu. 2018. Preparation, biocompatibility, and biotribological properties of TiN-incorporated graphite-like amorphous carbon bio-ceramic composite films. Ceram. Int. 44(6): 6810–6816.

Liu, Z., Q. Li, H. Lin, L. Qin, Y. Li, M. Nie and G. Xie. 2016. Room temperature preparation of ZnO nanosheets by an environmental friendly method and their photocatalytic properties. Gongneng Cailiao/J. Funct. Mater. 47: 217–222.

Lv, M., Y. Liu, J. Geng, X. Kou, Z. Xin and D. Yang. 2018. Engineering nanomaterials-based biosensors for food safety detection. Biosens. Bioelectron. 106: 122–128.

Mackevica, A., M.E. Olsson and S.F. Hansen. 2018. Quantitative characterization of TiO2 nanoparticle release from textiles by conventional and single particle. ICP-MS J. Nanopart. Res. 20(1): art. no. 6.

Manimozhi, V., N. Partha, E.K.T. Sivakumar, N. Jeeva and V. Jaisankar. 2016. Preparation and characterization of ferrite nanoparticles for the treatment of industrial waste wáter. Dig. J. Nanomater. Bios. 11(3): 1017–1027.

Mazerat, S. and R. Pailler. 2018. Self-organized nano-scale multilayer coating on SiC fibers obtained by phosphating. J. Eur. Ceram. Soc. 38(6): 2486–2494.

Morère, J., E. Sánchez-Miguel, M.J. Tenorio, C. Pando and A. Cabañas. 2017. Supercritical fluid preparation of Pt, Ru and Ni/graphene nanocomposites and their application as selective catalysts in the partial hydrogenation of limonene. J. Supercrit. Fluid 120: 7–17.

Najafi, E., B. Liao, T. Scarborough and A. Zewail. 2018. Imaging surface acoustic wave dynamics in semiconducting polymers by scanning ultrafast electron microscopy. Ultramicroscopy 184: 46–50.

Nascarella, M.A. and E.J. Calabrese. 2012. A method to evaluate hormesis in nanoparticle dose–responses. Dose-Response 10: 344–354.

Naumenko, E.A. and R.F. Fakhrullin. 2017. Toxicological evaluation of clay nanomaterials and polymerclay nanocomposites. RSC Smart Materials 22: 399–419.

Oberd rster, G., V. Stone and K. Donaldson. 2007. Toxicology of nanoparticles: A historical perspective. Nanotoxicology 1(1): 2–25.

Olajire, A.A. 2018. Recent progress on the nanoparticles-assisted greenhouse carbon dioxide conversion processes. J. CO2 Util. 24: 522–547.

Ou, G., J. Zhao, P. Chen, C. Xiong, F. Dong, B. Li and X. Feng. 2018. Fabrication and application of noble metal nanoclusters as optical sensors for toxic metal ions. Anal. Bioanal. Chem. 410(10): 2485–2498.

Ovid'ko, I.A., R.Z. Valiev and Y.T. Zhu. 2018. Review on superior strength and enhanced ductility of metallic nanomaterials. Prog. Mater. Sci. 94: 462–540.

Popov, V.A., M. Burghammer, M. Rosenthal and A. Kotov. 2018. *In situ* synthesis of TiC nanoreinforcements in aluminum matrix composites during mechanical alloying. Compos. Part B- Eng. 145: 57–61.

Prasanth, R., P.S. Owuor, R.S. Ravi Shankar, J. Joyner, S. Kosolwattana, S.P. Jose, P. Dong, V.K. Thakur, J.H. Cho and M. Shelke. 2015. Eco-friendly polymer-layered silicate nanocomposite-preparation, chemistry, properties, and applications. Adv. Struct. Mat. 74: 1–42.

Pulido-Reyes, G., F. Leganes, F. Fernández-Piñas and R. Rosal. 2017. Bio-nano interface and environment: A critical review. Environ. Toxicol. Chem. 36(12): 3181–3193.

Ramos-Delgado, N.A., M.A. Gracia-Pinilla, R.V. Mangalaraja, K. O'Shea and D.D. Dionysiou. 2016. Industrial synthesis and characterization of nanophotocatalysts materials: titania. Nanotechnol. Rev. 5(5): 467–479.

Ravi, S. and S. Vadukumpully. 2016. Sustainable carbon nanomaterials: Recent advances and its applications in energy and environmental remediation. J. Environ. Chem. Eng. 4(1): 835–856.

Rivero, P.J., A. Urrutia, J. Goicoechea and F.J. Arregui. 2015. Nanomaterials for functional textiles and fibers. Nanoscale Res. Lett. 10(1): art. no. 501.

Robichaud, C.O., D. Tanzie, U. Weilenmann and M.R. Wiesner. 2005. Relative risk analysis of several manufactured nanomaterials: An insurance industry context. Environ. Sci. Technol. 39(22): 8985–8994.

Saba, F., S.A. Sajjadi, M. Haddad-Sabzevar and F. Zhang. 2018. TiC-modifie carbon nanotubes, TiC nanotubes and TiC nanorods: Synthesis and characterization. Ceram. Int. 44(7): 7949–7954.

Saha, S., P. Samanta, N.C. Murmu and T. Kuila. 2018. A review on the heterostructure nanomaterials for supercapacitor application. J. Energy Storage 17: 181–202.

Salieri, B., D.A. Turner, B. Nowack and R. Hischier. 2018. Life cycle assessment of manufactured nanomaterials. Where are we? NanoImpact 10: 108–120.

Sánchez-Calvo, A., E. Núñez-Bajo, M.T. Fernández-Abedul, M.C. Blanco-López and A. Costa García. 2018. Optimization and characterization of nanostructured paper-based electrodes. Electrochim. Acta 265: 717–725.

Sanchez, C., C. Boissiere, S. Cassaignon, C. Chaneac, O. Durupthy, M. Faustini, D. Grosso, C. Laberty-Robert, L. Nicole, D. Portehault, F. Ribot, L. Rozes and C. Sassoye. 2014. Molecular engineering of functional inorganic and hybrid materials. Chem. Mater. 26(1): 221–238.

Sapkota, J., J.C. Natterodt, A. Shirole, E.J. Foster and C. Weder. 2017. Fabrication and properties of polyethylene/cellulose nanocrystal composites. Macromol. Mater. Eng. 302(1): art. no. 160.

Sardoiwala, M.N., B. Kaundal and S.R. Choudhury. 2018. Toxic impact of nanomaterials on microbes plants and animals. Environ. Chem. Lett. 16(1): 147–160.

Shahid-Ul-Islam, B.S. Butola and F. Mohammad. 2016. Silver nanomaterials as future colorants and potential antimicrobial agents for natural and synthetic textile materials. RSC Adv. 6(50): 44232–44247.

Shandilya, N., T. Ligthart, I. van Voorde, B. Stahlmecke, S. Clavaguera, C. Philippot, Y. Ding and H. Goede. 2018. A nanomaterial release model for waste shredding using a Bayesian belief network. J. Nanopart. Res. 20(2): art. no. 33.

Sharma, N., H. Ojha, A. Bharadwaj, D.P. Pathak and R.K. Sharma. 2015. Preparation and catalytic applications of nanomaterials: a review. RSC Adv. 5(66): 53381–53403.

Sharma, S.K., G.S. Ghodake, D.Y. Kim, D.-Y. Kim and O.P. Thakur. 2018. Synthesis and characterization of hybrid Ag-ZnO nanocomposite for the application of sensor selectivity. Curr. Appl. Phys. 18(4): 377–383.

Shin, Y.-J., W.-M. Lee, J.I. Kwak and Y.-J. An. 2018. Dissolution of zinc oxide nanoparticles in exposure media of algae, daphnia, and fish embryos for nanotoxicological testing. Chem. Ecol. 34(3): 229–240.

Singh, S. 2018. Synthesis of nanoparticles for biomedical applications. Issues in Toxicology 35: 39–93.

Song, Y., H. Jiang, B. Wang, Y. Kong and J. Chen. 2018. Silver-incorporated mussel-inspired polydopamine coatings on mesoporous silica as an efficient nanocatalyst and antimicrobial agent. ACS Appl. Mater. Inter. 10(2): 1792–1801.

Sukhanova, A., S. Bozrova, P. Sokolov, M. Berestovoy, A. Karaulov and I. Nabiev. 2018. Dependence of nanoparticle toxicity on their physical and chemical properties. Nanoscale Res. Lett. 13: art. no. 44.

Swierczewska, M., R.M. Crist and S.E. McNeil. 2018. Evaluating nanomedicines: Obstacles and advancements. pp. 3–16. *In*: McNeil, S. (ed.). Characterization of Nanoparticles Intended for Drug Delivery. Methods in Molecular Biology, vol. 1682. Humana Press, New York, NY.

Tarasov, V.A., M.A. Komkov, N.A. Stepanishchev, V.A. Romanenkov and R.V. Boyarskaya. 2015. Modification of polyester resin binder by carbon nanotubes using ultrasonic dispersión. Polym. Sci. Ser. D+ 8(1): 9–16.

Thambiraj, S. and D. Ravi Shankaran. 2017. Preparation and physicochemical characterization of cellulose nanocrystals from industrial waste cotton. Appl. Surf. Sci. 412: 405–416.

Thiruvengadam, M., G. Rajakumar and I.-M. Chung. 2018. Nanotechnology: current uses and future applications in the food industry. 3 Biotech. 8(1): art. no. 74.

Tiwari, A. and S.J. Dhoble. 2018. Recent advances and developments on integrating nanotechnology with chemiluminescence assays. Talanta 180: 1–11.

Trakakis, G., G. Anagnostopoulos, L. Sygellou, A. Bakolas, J. Parthenios, D. Tasis, C. Galiotis and K. Papagelis. 2015. Epoxidized multi-walled carbon nanotube buckypapers: A scaffold for polymer nanocomposites with enhanced mechanical properties. Chem. Eng. J. 281: 793–803.

Tran, L., M.A. Bañares and R. Rallo (eds.). 2017. Modelling the toxicity of nanoparticles. Advances in Experimental Medicine and Biology 947, Springer.

Villaseñor, MJ. and A. Ríos. 2018. Nanomaterials for water cleaning and desalination, energy production, disinfection, agriculture and green chemistry. Environ. Chem. Lett. 16(1): 11–34.

Vlastou, E., M. Gazouli, A. Ploussi, K. Platoni and E.P. Efstathopoulos. 2017. Nanoparticles: Nanotoxicity aspects. J. Phys.: Conf. Ser. 931(1): art. no. 012020.

Voelker, D., K. Schlich, L. Hohndorf, W. Koch, U. Kuehnen, C. Polleichtner, C. Kussatz and K. Hund-Rinke. 2015. Approach on environmental risk assessment of nanosilver released from textiles. Environ. Res. 140: 661–672.

Vogli, E.D., O. Turkarslan, S.M. Iconomopoulou, D. Korkmaz, A. Soto Beobide and G.A. Voyiatzis. 2018. From lab-scale film preparation to up-scale spinning fibre manufacturing of multiwall carbon nanotube/poly ethylene terephthalate composites. J. Ind. Text. 47(6): 1241–1260.

Wang, C. and D. Astruc. 2018. Recent developments of metallic nanoparticle-graphene nanocatalysts. Prog. Mater. Sci. 94: 306–383.

Wang, Q., Y. Wang, Q. Meng, T. Wang, W. Guo, G. Wu and L. You. 2017. Preparation of high antistatic HDPE/polyaniline encapsulated graphene nanoplatelet composites by solution blending. RSC Adv. 7(5): 2796–2803.

Wu, C., X. Tong, Y. Ai, D.-S. Liu, P. Yu, J. Wu and Z.M. Wang. 2018. A review: Enhanced anodes of Li/Na-Ion batteries based on yolk–shell structured nanomaterials. Nano-Micro. Lett. 10(3): art. no. 40.

Xu, F., J. Hu, J. Zhang, X. Hou and X. Jiang. 2018. Nanomaterials in speciation analysis of mercury, arsenic, selenium, and chromium by analytical atomic/molecular spectrometry. Appl. Spectrosc. Rev. 53(2-4): 333–348.

Xu, J., Z. Cao, Y. Zhang, Z. Yuan, Z. Lou, X. Xu and X. Wang. 2018. A review of functionalized carbon nanotubes and graphene for heavy metal adsorption from water: Preparation, application, and mechanism. Chemosphere 195: 351–364.

Yan, X.-Y. 2018. Nanozyme: A new type of artificial enzyme. Progress in Biochemistry and Biophysics 45(2): 101–104. DOI:10.16476/j.pibb.2018.0041.

Yang, Y., Z. Qin, W. Zeng, T. Yang, Y. Cao, C. Mei and Y. Kuang. 2017. Toxicity assessment of nanoparticles in various systems and organs. Nanotechnol. Rev. 6(3): 279–289.

Yata, V.K., B.C. Tiwari and I. Ahmad. 2018. Nanoscience in food and agriculture: research, industries and patents. Environ. Chem. Lett. 16(1): 79–84.

Yin, Y. and D. Talapin. 2013. The chemistry of functional nanomaterials. Chem. Soc. Rev. 42(7): 2484–2487.

Yong, H.E., K. Krishnamoorthy, K.T. Hyun and S.J. Kim. 2015. Preparation of ZnO nanopaint for marine antifouling applications. J. Ind. Eng. Chem. 29: 39–42.

Zaharescu, M., L. Predoana and J. Pandele. 2018. Relevance of thermal analysis for sol-gel-derived nanomaterials. J. Sol-Gel Sci. Techn. 86(1): 7–23.

Zhang, L., M. Chen, Y. Jiang, M. Chen, Y. Ding and Q. Liu. 2017. A facile preparation of montmorillonitesupported copper sulfide nanocomposites and their application in the detection of H2O2. Sens Actuators B: Chem. 239: 28–35.

Zhao, J., Z. Lu, X. He, X. Zhang, Q. Li, T. Xia, W. Zhang and C. Lu. 2017. Fabrication and characterization of highly porous Fe(OH)3@cellulose hybrid fibers for effective removal of congo red from contaminated water. ACS Sustain. Chem. Eng. 5(9): 7723–7732.

Zhao, J.Z.-B., D.-M. Zhang, L. Tai, P.-F. Jiang and Y. Jiang. 2017. A facile and low-cost preparation of durable amphiphobic coatings with fluoride–silica@poly(methacrylic acid) hybrid nanocomposites. J. Coat. Technol. Res. 14(6): 1369–1380.

Part II
Value Products

Chapter 2

Bifunctional Porous Catalysts in the Synthesis of Valuable Products
Challenges and Prospects

Elena Pérez-Mayoral,* Marina Godino-Ojer and
Daniel González-Rodal

INTRODUCTION

The development of new chemical processes characterized by reduced environmental and economic impacts is one of the challenges of the current century. The green chemistry is a valuable tool focused not only on the search and discovery of new catalytic, efficient and selective chemical reactions but also the development of new methodologies able to minimize the waste production or even the recycling or valourization of these reactions (i.e., biomass residues). In this sense, at the end of the 1990s, nanocatalysis emerged as a discipline which could be considered as the interface between homogeneous and heterogeneous catalysis. J.M. Thomas recently revises the advantages of using heterogeneous catalysts which offer the possibility of multifunctionalization. Furthermore, the author highlights that certain chemical processes only takes efficiently place in the presence of heterogeneous systems (Thomas 2010). Due to the severe regulations for environmental protection, the development of new ecofriendly processes for the production of fine chemicals which can be commercialized are desirable. Heterogeneous catalysis can then highly contribute to the sustainability and economy of the industrial chemical process (Holderic 2000). Major efforts at this respect have been made towards the design of new heterogeneous nanocatalysts by immobilizing the homogeneous ones onto the surfaces or even intercalating them within, so that they exhibit improved catalytic

Departamento de Química Inorgánica y Química Técnica, Facultad de Ciencias, Universidad Nacional de Educación a Distancia, UNED, Paseo Senda del Rey 9, E-28040-Madrid, Spain.
Emails: godino.marina@gmail.com; dgonzalezrodal@gmail.com
* Corresponding author: eperez@ccia.uned.es

performances but also improved the understanding of the chemical processes (Collins and Horváth 2011).

The use of both metal nanoparticles and nanomaterials as supports of active phases has also displayed spectacular progress. Much work concerning the impact of the intrinsic properties of the nanomaterials on catalysis—shape, size, interparticle distance, composition, among others—influencing the efficiency, selectivity, stability and even the recycling of the materials has been reported. The nanomaterials are then considered excellent nanocatalysts extremely useful in organic synthesis. In recent years, a great variety of nanomaterials showing different structures, morphologies and compositions with improved catalytic properties has been reported (Chng et al. 2013; Barbaro et al. 2012). Climent et al. (2011) recently revised a multitude of heterogeneous catalysts showing single acid or basic active sites or even single-site metal catalysts involved in multistep sequential processes useful in the synthesis of fine chemicals. In this context, a large variety of crystalline or amorphous inorganic solids exhibiting acid properties, especially zeolites, aluminosilicates and aluminophosphates and structured mesoporous materials containing metals within the framework structure or even in extra-framework positions have been described. Basic inorganic solids including magnesium oxides, layered magnesium aluminates, and also supported alkaline oxides have been also successfully developed as catalysts for one-pot multistep processes although in a lesser degree.

In many cases the design of new catalysts is based on a bioinspired approach. It is well documented that enzymes denoted as biocatalysts are the most potent catalysts known. Enzymes are multifunctional catalysts which operate under mild reaction conditions at the physiologist pH catalysing the chemical processes in living organisms (Copeland 2000). Enzymes are able to accelerate the chemical transformations activating the corresponding substrates by interaction with the appropriated positioned functional groups which act in cooperation. A great variety of functional groups such as metal centres, nucleophiles, acids and bases, hydrogen bond donors and acceptors located at the fixed distances in enzymes are able to interact by electrostatic, hydrogen and even covalent bonding strongly affecting to the reactivity and selectivity of the chemical processes.

An important type of biocatalysts is the metalloenzymes often constituted by huge protein containing a small and well-defined transition-metal complex in the active site in which the catalytic activity relays onto the metals. These biocatalysts or functional mimic analogues able to activate molecular oxygen can serve as catalytic models for the development of new nanocatalysts for specific oxidation reactions (Xuereb et al. 2012). In this regard, many efforts aimed towards the heterogenization of a functional mimic of the single catalytic active site in metalloenzymes have been made. An example recently published is that comprising manganese(II) complexes grafted on the surface of a mesostructured porous silica bioinspired on the model of the manganese-dependent dioxygenase and useful in the selective oxidation of the solid multifunctionalization requires that incompatible active catalytic sites (i.e., acid or basic sites) are be spatially separated avoiding mutual interactions but at the adequate distance for acting in cooperation. It is important to stress that cooperation in catalysis is a term referred to a system where at least two different catalytic functions can act together increasing the reaction rate beyond the sum of the rates from the individual catalytic site alone.

The concept of cooperativity has been investigated in homogeneous systems by using small molecules; however, this is not trivial when working with heterogeneous catalysts. In this context, slight changes on the distances between both catalytic sites can have a significant impact on the catalytic performance. This circumstance was already studied by Corma et al. (2011) while investigating a new class of recyclable bifunctional acid–base organocatalysts containing ionic liquid (IL) functional groups. The authors explored the effect of the separation and spatial configuration of the active catalytic sites on different condensation reactions such as Knoevenagel and Claisen–Schmidt condensations. The investigated molecules combined the presence of both basic and acid sites spatially separated by different number of carbon atoms: one, two or three (Fig. 1). DFT calculations concerning the reaction mechanism for the synthesis of *trans*-chalcones (important compounds for the pharmaceutical industry), demonstrated that this appropriated acid–base bifunctional catalyst followed a mechanism inspired by natural aldolases.

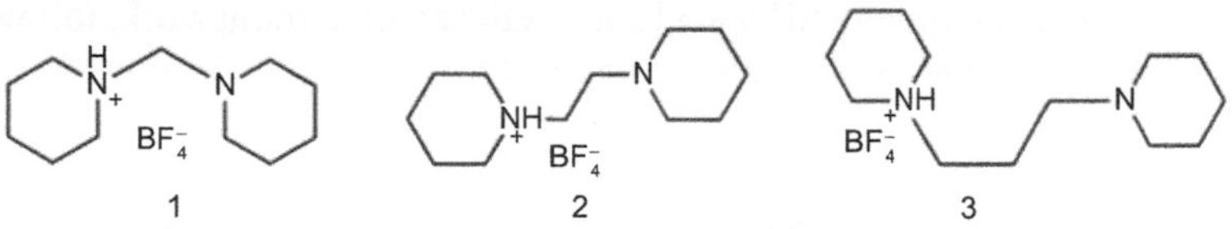

Fig. 1. ILs investigated on the effect of the separation of the active site on their catalytic performance.

Bifunctional porous nanocatalysts

A great variety of functional porous catalysts (in this case bifunctional heterogeneous catalysts), and their catalytic application in the synthesis of valuable compounds has been reported. Obviously, the development of this type of nanocatalysts will have industrial and social repercussions related to the environmental concerns (End and Schöning 2004). Interestingly, some examples of basic-metal bifunctional and bi- and multimetallic catalysts for organic synthesis have been revised by Climent et al. (2011).

An important issue in the development of bifunctional porous catalysts is the choice of the support of the active phases. In general, multisite catalysts comprise well-structured microporous and mesoporous organic–inorganic hybrids, silicas being extensively investigated for heterogeneous catalytic applications, highlighting the periodic mesoporous silicas. Other types of interesting porous materials suitable to be functionalized are the metal organic frameworks (MOFs) and carbon-based materials including purely porous organic polymers and, particularly, the porous aromatic frameworks (Ben et al. 2009; Lee et al. 2009; Thomas 2010; Kaur et al. 2011; Zhang and Riduan 2012). In this context, one of the most commonly studied bifunctional nanocatalysts is the one composed by different acid or basic active sites or when both are simultaneously tethered to the porous surface. The interaction of both components often leads to inactive salts in homogenous systems whereas these functions immobilized over solid surfaces are able to coexist (Shylesh and Thiel 2011). The synergic effect of both acidic and basic sites located on the same catalytic structure is an attractive concept demonstrated in the bond-forming processes for the synthesis of complex organic molecules (Motokura et al. 2008).

When modifying solid supports, the functions often are randomly distributed at the surface. In this respect, a significant number of papers are focused on the control of the structural properties of the multifunctional nanocatalysts. Several different approaches have been used to prepare bifunctional heterogeneous catalysts by using various supports where distinct functional groups are responsible for improved catalytic activity. Although zeolites are the preferred materials widely used in the petrochemical industry, these solids show some limitations because of their small pore sizes which restrict the scope of catalytic reactions. One example of layered zeolitic hybrid organic–inorganic material comprising acid and base centres located in the zeolitic counterpart and the organic component, respectively, is that reported by Corma et al. (2010). While the acid sites are located within micropores in the zeolitic material, the amine functions were incorporated on the bridged benzene groups in the interlayer space. This zeolitic hybrid was proved in the cascade reactions involving the acetal deprotection promoted by acid sites in the zeolitic framework, followed by consecutive Knoevenagel condensation catalysed by basic sites constituted by amine functions.

Based on that, this chapter aims to illustrate some interesting bifunctional nanocatalysts able to catalyze interesting organic transformations and the synthetic strategies used for the improvement of their catalytic performances. The subsequent sections will summarize some of those as functions of the solid support, emphasizing the silica-based materials but also showing some examples comprising MOFs.

Bifunctional silica-based nanocatalysts

Silica-based materials show a huge number of applications in a great variety of research fields: for instance the catalysis, which are being extensively used as catalytic supports (Pagliaro 2009).

Among the most relevant properties of amorphous silica which make it an appropriate support are its high surface area and relative low cost. However, the non-uniform surface and pore structure strongly affect their applications. In this sense, ordered mesoporous silicas (OMSs) such as SBA-15 and MCM-41 and others are also easily to functionalize by using two important synthetic approaches: (i) direct synthesis or (ii) post-synthetic grafting procedures, producing OMSs with notable differences. OMSs functionalized by direct synthesis exhibit a higher content of functional groups, well distributed over the silica support, resulting frequently as a less well-ordered material. However, an interesting and determinant factor during the post-synthetic functionalization is the reactivity of the modifier silane. OMSs modified in the presence of highly reactive silanes cause the inclusion of higher organic loadings but are less well-distributed on the surface. Interestingly, the presence of mesopores in OMSs allows the reduction of the reactions times in some chemical processes in comparison with other supporting structures (Stein et al. 2000). Different silane coupling agents—hydrocarbons or even containing functional groups such as amines, halides, nitriles, thiols and epoxides—have been extensively investigated for the functionalization of the OMSs surface through the available Si-OH groups.

In the last years, the use of mesoporous silica for catalytic applications has exponentially grown. OMSs have been extensively explored in the synthesis of complex molecules, through cascade reactions, particularly heterocyclic systems with pharmacological relevant properties (Pérez-Mayoral et al. 2014; 2015). The use of OMSs for catalytic purposes offers the possibility of the surface multi-functionalization allowing the synthesis of great molecular diversity (Pérez-Mayoral et al. 2017). Nevertheless, these nanomaterials used as supports of active phases provide high catalytic performance and also selectivity mainly due to the spectacular increasing of the contact surface contributing to the overall yields of the desired products (Thomas 2009; Martínez and Corma 2011).

In this section, we emphasize some relevant tasks concerning the cooperative catalysis when using OMSs, paying attention to the different interactions between organic functions and support but also including the cooperative catalysis by silica-supported organic functional groups (Margelefsky et al. 2008).

Organic-inorganic hybrids containing amine functions

There are a lot of studies focused on the catalytic nucleophilic addition reactions. The nucleophile activation often is produced by the abstraction of acidic protons onto carbonylic systems and related compounds containing α-hydrogen atoms, whereas the activation of the electrophile occurs by the lowering of the LUMO levels by interaction with acidic catalysts exhibiting Brønsted or Lewis characters. In this context, organic amines immobilized onto the inorganic solid–acid surfaces provide highly active acid–base bifunctional catalysts active in various organic transformations.

Based on the acid-base cooperative effects commonly involved in enzymatic catalysis, OMSs modified with amine groups, prepared by using both the direct synthesis or grafting procedures, have been proposed as the simplest bifunctional porous catalysts (Brunelli et al. 2012). Silanol groups showing weak Brönsted acidity in combination with amine functions over the silica surface are recognized as bifunctional catalysts in several interesting chemical processes (Motokura et al. 2008). Both amine and silanols groups acting in cooperation show higher catalytic activity than the silica functionalized with amine and stronger carboxylic acids. These cooperative effects between silanols and amine groups have been reported in the Michael addition and Henry reaction (Scheme 1) (Bass et al. 2006). Figure 2 depicts a schematic representation for the amine groups anchored over the silica surface offering a broad range of distances between catalytic sites for cooperativity. In this sense, the authors have used a synthetic approach for the preparation of organic-inorganic hybrid catalysts in order to elucidate the independent dielectric and acid-base bifunctional cooperative effects, demonstrating that the outer-sphere environment is a key factor in the reaction mechanism and selectivity.

Interestingly, Xie et al. (2009) reported a study of optimization of the amino functions grafted to the mesoporous silica MCM-41. In order to explore the

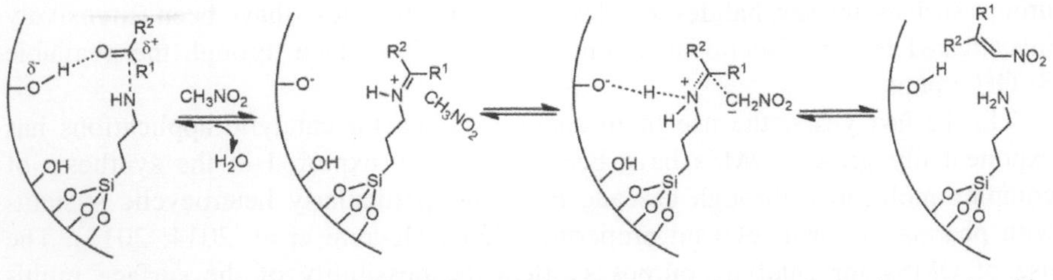

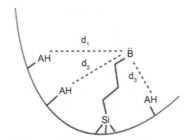

Scheme 1. Reaction mechanism of the Henry reaction between acetone and nitromethane catalyzed by amines on silica.

Fig. 2. Range of acid-base distances for cooperation between amine and silanol groups.

importance of silanol groups in the cooperative and catalytic aldol reaction, some control experiments were carried out using homogenous amine catalyst and methyl-capped MCM-41 silica. Among the investigated silicas modified with primary, secondary and tertiary amines, the most active basic catalyst was the MCM-41 functionalysed with the secondary amine groups (MCM41-MAPI). A plausible reaction mechanism could reveal cooperative effects between the isolated amine groups and residual silanols, Si-OH groups involved not only in the enamine formation but also favouring the nucleophilic addition. Mesoporous silicas modified with primary amines and tertiary amines barely showed any catalytic activity.

The functionalization on OMSs can sometimes change the operative reaction pathway influencing both reactivity and selectivity. Such is the case of Friedländer reaction between 2-amino aryl ketones and 1,3-dicarbonyl compounds catalysed by amino-grafted MCM-41 mesoporous silicas (Scheme 2) (Domínguez-Fernández et al. 2009). Interesting selectivity differences were found as a function of the used catalyst and its acid-base character. While the reaction catalysed by mesoporous silica with acid character yielded a mixture of quinolines/quinolones, dramatic selectivity changes were observed by basic catalysts, quinolin-2(1*H*)-ones being regioselectively isolated as unique reaction products, probably because the reaction occurs through another reactive reaction route. In this case, the optimum catalyst was also the MCM-41 incorporating secondary amine functions, particularly methyl aminopropyl groups. The initial step of the reaction could consist of the amidation reaction between the amino group over the *o*-aminoaryl ketone and ethyl acetoacetate, followed by aldolization reaction and subsequent dehydration.

In the same context, Aider et al. (2014) reported the combined experimental and theoretical study focused on the synthesis of coumarins (a class of heterocyclic

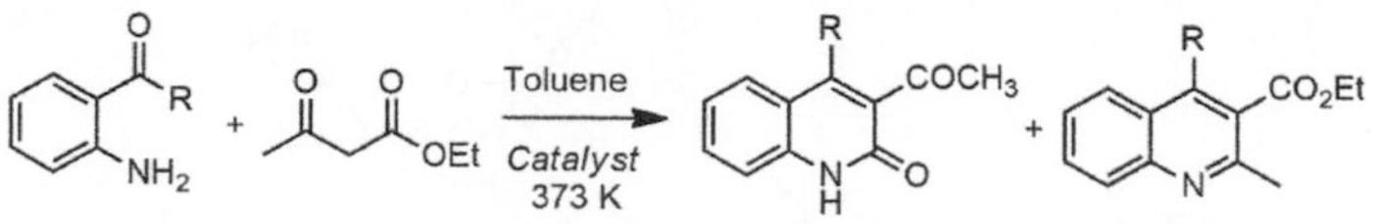

R = Ph, Me

Scheme 2. Condensation reaction of *o*-aminoaryl ketones with ethyl acetoacetate, catalyzed by amino-grafted MCM-41.

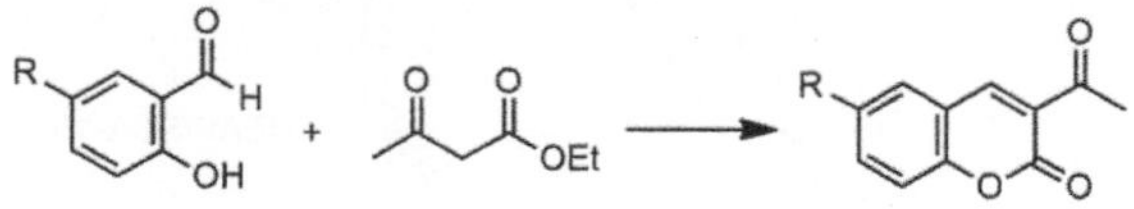

R= H, Cl, OMe and NO$_2$

Scheme 3. Synthesis of coumarins from substituted 2-hydroxybenzaldehydes and ethyl acetoacetate catalyzed by amino-grafted SBA-15 mesoporous silicas, under solvent-free conditions.

systems with pharmacological interest). Coumarins were then successfully prepared from 5-substituted-2-hydroxybenzaldehydes and ethyl acetoacetate in the presence of amino-grafted SBA-15 materials (Scheme 3).

The experimental observations suggested that the reaction starts with the Knoevenagel condensation between reagents, in the absence of any solvent and under mild conditions, followed by a non-catalytic lactonization step. Amino-grafted SBA-15 incorporating the secondary amine groups, MAP/SBA-15, was found to be the most efficient and totally recyclable catalyst as compared with the mesoporous silica including tertiary amine groups, DEAP/SBA-15 (where MAP and DEA are methyl aminopropyl and diethyl aminopropyl groups, respectively) which results as an almost inactive catalyst. At this regard, the theoretical study using the most reduced models of the catalysts, including exclusively the basic active sites supported on the small silica cluster, confirmed that the steric congestion and the absence of NH protons in DEAP/SBA-15 are both key factors for the increasing of the activation barrier (Fig. 3). Although, the selected models did not represent the involvement of the silanol groups in the catalytic performance, they provided valuable information concerning the role of the amine functional groups and its implications in the catalytic process. The computational study demonstrated that MAP/SBA-15 can be considered as a dual acid–base catalyst for the coumarins synthesis, through Knoevenagel condensation, assisting the approaching of the reagents to each other and strongly stabilizing the transition structure. Furthermore, the substitution in 2-hydroxybenzaldehydes at 5-position notably affects the electrophilic character of –CHO group but also the acidity of –OH groups.

The diverse synthetic procedures for grafting of mesoporous silicas with organosilanes and particularly the influence of the used solvent have been also analysed. A great number of post-grafting procedures, including some mentioned above, use excess of the corresponding organosilane in the presence of non-polar solvents, particularly toluene at reflux. These materials often show a high concentration of covalently bound organic functions densely distributed over the silica surface. In these cases, the surface areas and pore volumes of the materials are notably reduced and the silanol groups are residual, thus contributing to the

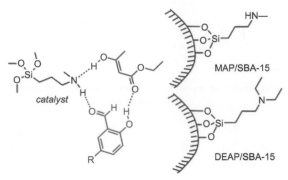

Fig. 3. Reactant complex involving salicylaldehyde, ethyl acetoacetate and the reduced models of the two amino functionalized mesoporous silicas based on the SBA-15, MAP/SBA-15 and DEAP/SBA-15.

worse catalytic behavior. Sharma and Asefa 2007; Sharma et al. 2008a reported the synthesis of bifunctional mesoporous catalysts which incorporate spatially distributed organoamine and silanol groups also tested as efficient catalysts in the Henry reaction. The post-grafting experimental protocols involved an excess amount of 3-aminopropyl silane using a polar solvent, ethanol, at a lower temperature and, on the contrary, using a smaller amount of the aminoorganosilane in the presence of toluene, a non-polar solvent, during a shorter reaction time. The authors concluded that the grafted materials synthesized in the presence of ethanol resulted in increased catalytic efficiency for the Henry reaction (99% of yield in only 15 min) in comparison to those prepared in toluene.

In the same context, Sharma et al. (2008b) systematically studied the influence of the site isolation, concentration and separation distance of the functional groups by synthetizing different series of organoamine functionalized MCM-41 mesoporous silicas and analyzing their catalytic activity in the Henry reaction between *p*-hydroxybenzaldehyde and nitromethane. The investigated materials incorporate grafted linear, flexible monoamines, diamines and triamines and also rigid *meta*- and *para*-substituted aromatic amines in different degrees of site isolation. The authors observed cooperative effects between amine and silanol groups over the silica surface when the materials were prepared in the presence of ethanol. The reaction was completed within 15 min of reaction time, an optimum concentration of grafted amine functions between 0.8–1.5 mmol being recognized.

Molecular imprinting techniques also allow the rational design of porous solid materials modified with chemical functionalities adequately positioned by covalent (Whitcombe et al. 1995) or non-covalent interactions (Vlatakis et al. 1993) using imprint molecules during the synthetic process. The created functions after the removal of the imprint molecules offer the possibility of small molecules recognition, making them as promising useful materials for separations, chemical sensing and catalysis applications. An interesting synthetic strategy for the anchoring of two or even three spatially organized identical amine functions was illustrated by Katz and Davis (2003). The molecular template includes the corresponding carbamate protecting amine functions which were imprinted into the amorphous bulk silica by using sol–gel polymerization method (Fig. 4). In this case, the removal of the template molecule was by reaction with trimethylsilyl iodide.

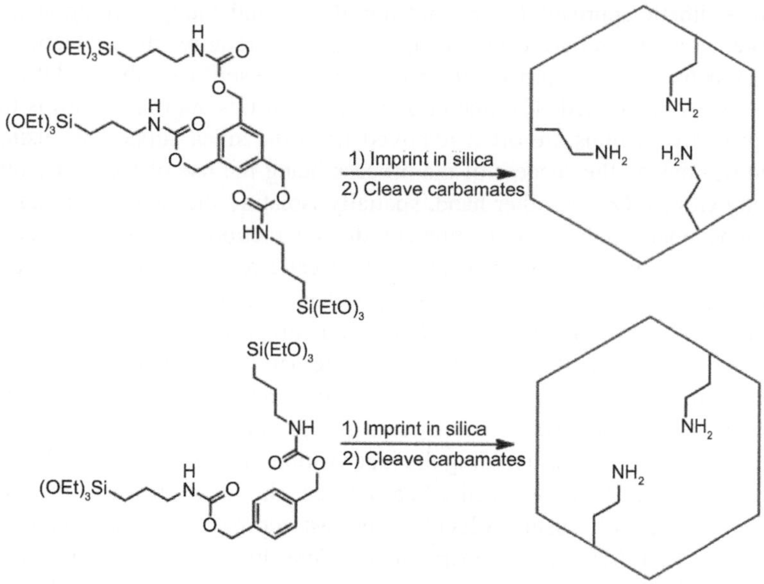

Fig. 4. Molecular imprinting of two or three carbamate groups as precursors of positioned primary amine functions.

Cooperative effects between different amine functions have also been reported. As above illustrated, primary amines immobilized over silica surfaces efficiently catalyse the aldol-based condensation reactions in which the carbonyl compound could be activated through the formation of imine intermediates. On the other hand, it is well-known that tertiary amines are recognized as Lewis and Brønsted bases for the activation of nucleophiles. Combining both concepts, Motokura et al. (2008) immobilized both amine functions, tertiary and primary amines (SA-NH$_2$-NEt$_2$) over silica-alumina for the efficient one-pot synthesis of 1,3-dinitroalkanes between benzaldehyde and nitromethane. Interestingly, the initial formation of 1,3-dinitro-2-phenylpropane occurred with notable enhancement of conversion rate in the presence of the bifunctional SA-NH$_2$-NEt$_2$ catalyst as compared when using a physical mixture of SA-NH$_2$ and SA-NEt$_2$ or even operating under homogeneous reaction conditions and using equivalent amine loadings.

Organic-inorganic hybrids containing two different organic functions

Positioning two different functional groups over a solid matrix is an aspect notably more difficult since it frequently involves different chemical steps including protecting and cleavage of the functions.

Acid/Base organic-inorganic hybrids

Some interesting fine chemical reactions have been reported in the presence of bifunctional OMSs containing at least two different catalytic sites. The functions are often distributed randomly over the silica surface, some of them acting in cooperation but others as individual catalytic sites (Kuschel et al. 2010).

One synthetic approach for the grafting of a second catalytic component when incorporating acid and base catalytic sites frequently requires the protection of the reactive groups during the process in order to avoid the self-assembly and the mutual inactivation as mentioned. The main disadvantage of this approximation is that the major protective groups are often removed from the silica support by using acid or basic treatments, the success of the process being the use of thermally unstable protecting groups. On the other hand, spatially isolating the two functional groups in a bifunctional catalyst would prevent the interaction of both. However, this constitutes a proper strategy for catalysing cascade reactions but not cooperative reactions. In this context, the control and optimization of the relative positions of the catalytic centres constitute the major challenge in preparing multifunctional OMSs.

Co-condensation method represents an alternative synthetic approach for the incorporation of two functional groups. The viscosity and acid character of the formed gel allow the co-habitation of both functional groups introduced in the polymerization mixture at the initial stages avoiding the inactivation of those. Remarkably, in order to do not alter the functionalized silica surface, the structure directing agent is commonly extracted by using solvents. This last approximation has been used by Nasab and Kiasat (2016) in the preparation of ordered bifunctional mesoporous silica, SBA-Im/SO$_3$Hactiveinthemulticomponentsynthesisof4*H*-chromen-4-ylphosphonate derivatives, under ultrasound irradiation, at room temperature. The organic linkers were 3-chloropropyltrimethoxysilane and 3-mercaptopropyltrimetoxysilane in such a manner that the oxidant treatment by using H$_2$O$_2$ provided the formation of –SO$_3$H from –SH groups and the –Cl group was able to react with imidazole (Scheme 4).

Novel ionic liquid (IL) and sulfonic acid based bifunctional periodic mesoporous organosilica (BPMO–IL–SO$_3$H), which is highly recyclable, has been recently reported in the one-pot Biginelli reaction of a variety of different aldehydes with alkylacetoacetates and urea, under solvent-free conditions, for the synthesis of dihydropyridines (Elhamifar et al. 2014a). The authors followed a similar synthetic strategyforthepreparationofthenanocatalystbutusing1,3-bis(trimethoxysilylpropyl)-imidazolium iodide and 3-mercaptopropyltrimethoxysilane. A similar bifunctional sample showing differences concerning the preparation method was reported by some of these authors as a nanocatalyst for the esterification of carboxylic acids (Elhamifar et al. 2014b).

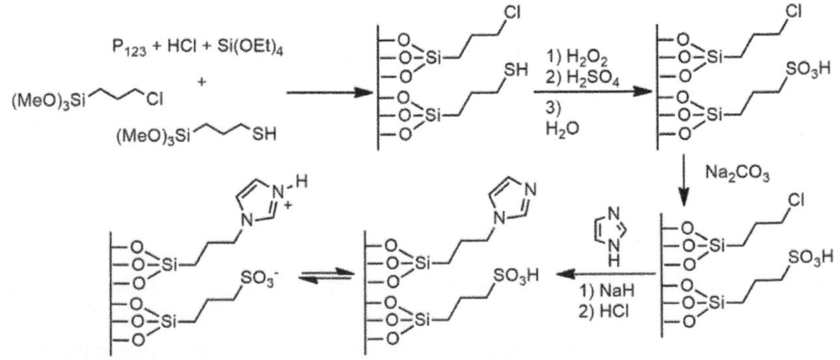

Scheme 4. Synthetic approach for the preparation of SBA-IM/SO$_3$H.

Elhamifar et al. (2016) have used a similar synthetic approach for preparing amine functionalized IL-based mesoporous organosilica showing good structural regularity and is highly efficient in the Knoevenagel condensation as compared with the individual counterparts, IL-free SBA-15-Pr-NH$_2$ and SBA-15-IL, both catalysts exhibiting inferior catalytic activity (Fig. 5). The authors attributed the enhanced activity, on one hand, to the lipophilicity conferred by the presence of IL skeleton offering better diffusion of substrates into mesochannels during the reaction, but also to the acid character of the hydrogen on the imidazolium ring activating the carbonylic group of the benzaldehydes during the catalytic process.

Interesting studies concerning the development of mesoporous silicas bearing separate acidic functions and amine groups have been recently reported. As an example of periodic mesoporous organosilicas containing sulfonic acid and amines has been reported by Alazun et al. (2006). The synthetic approach for the preparation of the bifunctional catalyst denoted as MSO$_3$H-NH$_2$ is shown in Scheme 5. This strategy is used in order to maintain both acid and basic groups isolated from one another. Both organic groups co-exist independently separated and maintaining their properties in the presence of aprotic solvents, whereas the protonation of the NH$_2$ groups was observed when using protic solvents.

Many efforts are currently made focused onto the development of the immobilized acid/base pairs in such a manner that both functions can act in cooperation. Zeidan et al. (2006; 2007b) prepared three different mesoporous silicas using SBA-15, functionalizing the surface with aminopropyl groups and acids of varied strengths, –SO$_3$H (pK$_a$ – 2), –PO$_3$H$_2$ (pK$_a$ ~ 3) and –CO$_2$H (pK$_a$ ~ 5), by using the one-pot synthesis procedure. The sample functionalized with –CO$_2$H and amine groups was found to be the most effective catalyst, proved in the aldol reaction between acetone and *p*-nitrobenzaldehyde, affording mixtures of the corresponding aldol and the

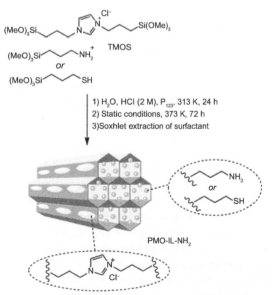

Fig. 5. IL based bifunctional periodic mesoporous organosilicas.

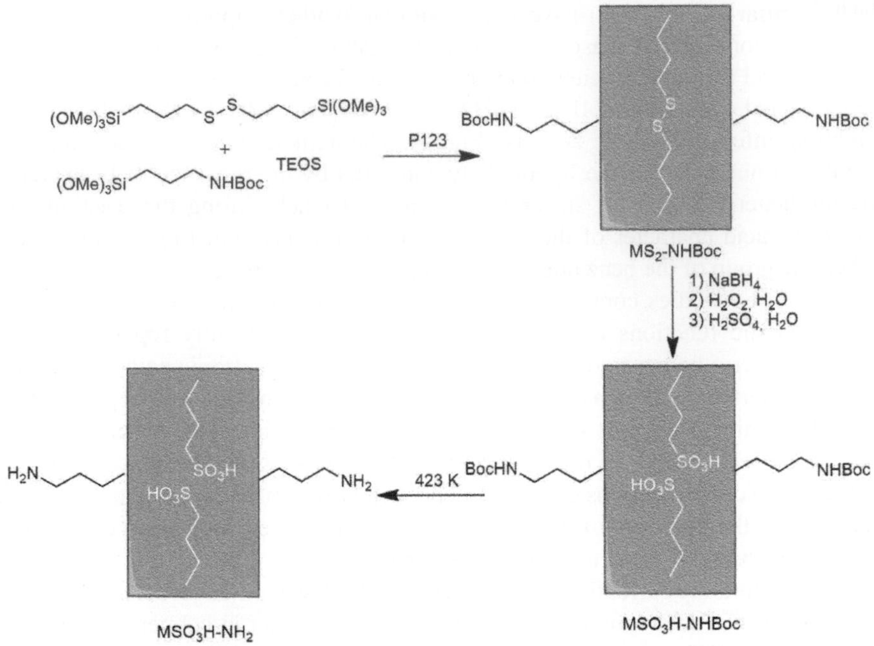

Scheme 5. Synthesis of bifunctional catalyst MSO$_3$H-NH$_2$.

dehydrated product (Table 1). The dramatically enhanced reactivity was attributed to the –CO$_2$H groups cooperatively acting with the amine functions while the stronger acidic functions operate neutralizing the amino groups.

Another reported synthetic approach comprises the grafting of an organic linker, containing both functions, over the mesoporous silica surface. In this sense, López-Sanz et al. (2012) have developed, for the first time, a mesoporous catalyst based on SBA-15 mesoporous silica, SBA-15/APS, combining at the same time basic (secondary amino groups) and acid functions (–SO$_3$H groups) and explored in the Friedländer condensation of 2-aminoaryl ketones with ethyl acetoacetate to produce quinolines with good yields (Schemes 2 and 6). SBA-15/APS resulted in the best selective catalyst in the formation of the corresponding quinoline whereas SBA-15/S, showing only –SO$_3$H functions, was found to be the most active catalyst. Remarkably, while the use of acid mesoporous silicas in this condensation reaction led to mixtures of corresponding quinolones and quinolines, the amino grafted SBA-15 materials afforded quinolones with excellent yields and total selectivity (Domínguez-Fernández et al. 2009). These observations suggested the dual contribution of sulfonic acids and the amine functions in the SBA-15/APS catalyst.

An innovative synthetic approach for the co-existence of amine and sulfonic functional groups in a periodic mesoporous organosilica with strict control for positioning both functions was developed by Shylesh et al. (2010). The synthetic strategy implies (i) the preparation of mesoporous phenylene-bridged silica material, PMO-NH$_2$, from 1,4-bis(triethoxysilyl)benzene and 3-aminopropyltrimethoxysilane, both hydrolysed in the presence of cetyltrimethylammonium bromide (CTAB),

Table 1. Henry reaction from *p*-nitrobenzaldehyde and acetone catalyzed by SBA-15 modified with amine and acid functions.

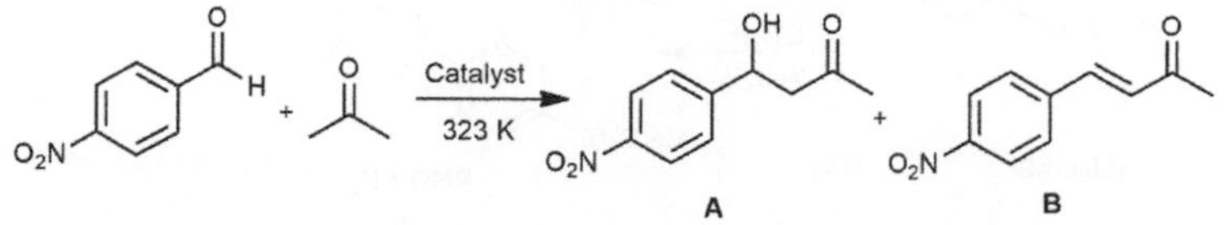

Catalyst	A (%)	B (%)	Total conversión (%)
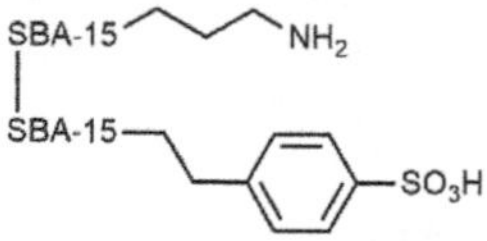 SBA-15⌒⌒NH₂ / SBA-15—⟨⟩—SO₃H	45	17	62
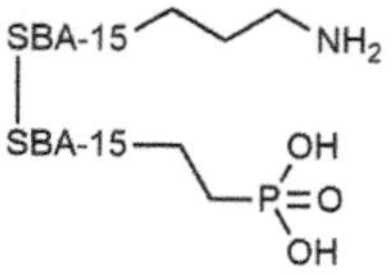 SBA-15⌒⌒NH₂ / SBA-15—P(=O)(OH)OH	62	16	78
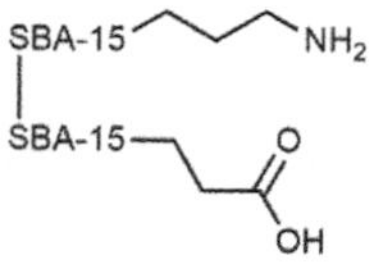 SBA-15⌒⌒NH₂ / SBA-15—CH₂C(=O)OH	75	24	99
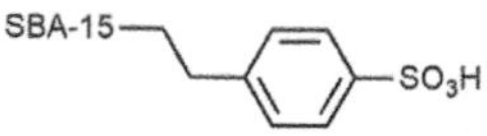 SBA-15—⟨⟩—SO₃H	8	8	16
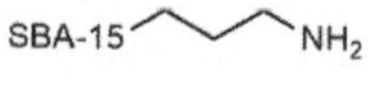 SBA-15⌒⌒NH₂	25	8	33

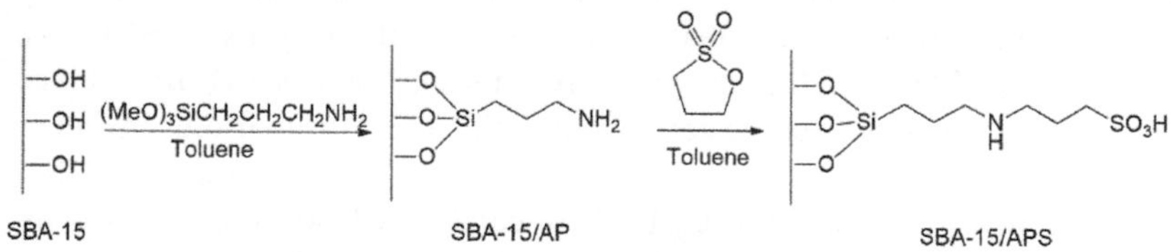

Scheme 6. Synthetic approach for the preparation of SBA-15/APS.

(ii) subsequent protection of amine functions by using di-*tert*-butyl-dicarbonate, (iii) sulfonation of bridging phenylene units with chlorosulfonic acid and finally (iv) the thermal cleavage of the amino groups (Scheme 7). The resulting bifunctional mesoporous catalyst showed the acidic groups located in the framework walls, whereas the basic functions are in the channel pores. The mesoporous phenylene silica showing a precise concentration and location of acid and basic functions was

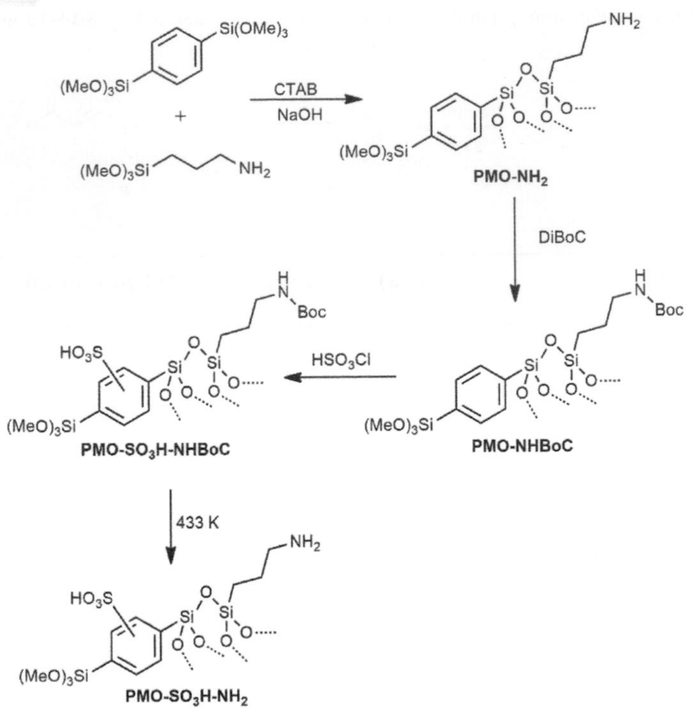

Scheme 7. Synthesis of the bifunctional mesoporous catalyst in which the acidic groups and basic functions are precisely located.

successfully checked in the cooperative catalysis through consecutive reactions from dimethyl acetal of benzaldehyde and nitromethane.

Much more recently, ordered and non-ordered bifunctional hybrids simultaneously containing strong basic and acid groups, particularly 1,8-bis(dimethylamino) naphthalene (DMAN) (Fig. 6) and sulfonic acids, synthetized by using both one pot co-condensation and post-synthetic grafting procedures, useful in consecutive catalytic transformations have been reported (Gianotti et al. 2013). Remarkably, the precursor thiol groups in these materials were quantitatively oxidized to sulfonic acids, by using H_2O_2, avoiding the protective steps of the basic DMAN molecules. These hybrids materials were tested as bifunctional catalysts useful in interesting chemical processes involving two catalytic steps, firstly comprising the acid promoted deacetalization, followed by base-catalyzed Knoevenagel condensation or nitroaldol reaction. Interestingly, the non-ordered material showed more superior catalytic activity than the ordered analogues probably due to the higher structural flexibility of the non-ordered silica network but also containing a higher amount of the silanol groups, which could act in cooperation for the electrophilic activation during the second catalytic step. The authors then reported a facile and more sustainable synthetic route for the preparation of non-ordered bifunctional hybrid, through fluoride-catalyzed sol–gel process, under mild reaction conditions, neutral pH and low temperatures, avoiding the use of structure directing agents, offering new changes to the rational design of advanced multifunctional hybrid materials.

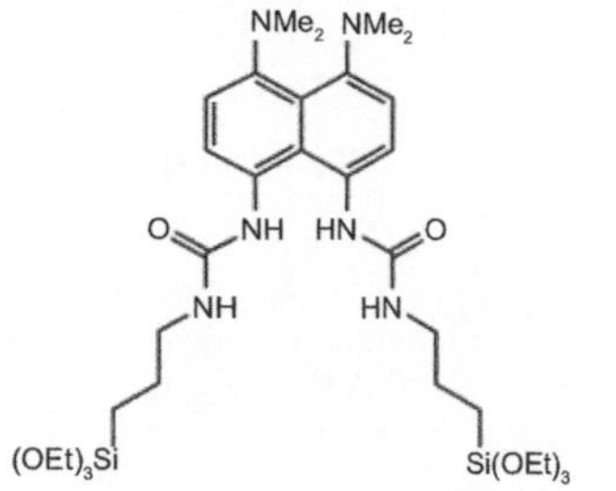

Fig. 6. Silylated base precursor containing 1,8-bis(dimethylamino)naphthalene (DMAN).

Acid or basic bifunctional organic-inorganic hybrids

Enhanced reactivity and regioselectivity in the transformation of phenol and acetone to produce *p,p'*-bisphenol in the presence of multiple sulfonic acid functionalized SBA-15 silicas was reported by Dufaud and Davis (2003), the acidic mesoporous silica prepared from the cleavage and oxidation of a dipropyl disulfide modified SBA-15 (Scheme 8). Attending to the difficulty on the quantitative oxidation of thiol to sulfonic groups when using H_2O_2, it is reasonable to think that sulfonic acid and thiol groups can act in cooperation enhancing the catalytic behaviour. In order to explore if this hypothesis could be assumable, Zeidan et al. (2006) demonstrated that the immobilization of thiol and sulfonic acid groups over SBA-15, by using the one pot synthetic method, causes an increase of the reaction rate and regioselectivity as compared when using mixtures of these two functional groups under homogeneous and heterogeneous conditions or combining those. The results strongly suggest that both functions must be in proximity to one another enhancing the catalytic activity. The authors proposed a reaction mechanism involving the sequential activation of acetone by protonation followed by a thiol attack leading to a highly electrophilic sulfonium intermediate (Scheme 9).

In the same context, Margelefsky et al. (2007) explored the effect of the distance between both functionalities, sulfonic acid and thiol groups, on the catalytic activity and selectivity in the condensation reaction of acetone and phenol to bisphenol A, by developing a new family of bifunctional nanocatalysts in which both functions are contained as part of the same organic linker (Scheme 10). Sultone ring as the precursor of alkylsulfonic acid sites was initially anchored to the SBA-15 silica surface. Subsequent ring opening with nucleophiles such as hydrosulfide ion or the monoanion of a dithiol originated the formation of sulfonic acids and thiol groups at fixed distances as a function of the used nucleophile. The activity and selectivity

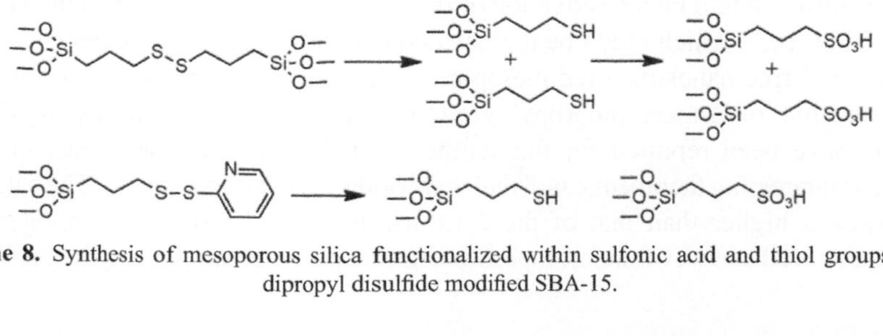

Scheme 8. Synthesis of mesoporous silica functionalized within sulfonic acid and thiol groups from dipropyl disulfide modified SBA-15.

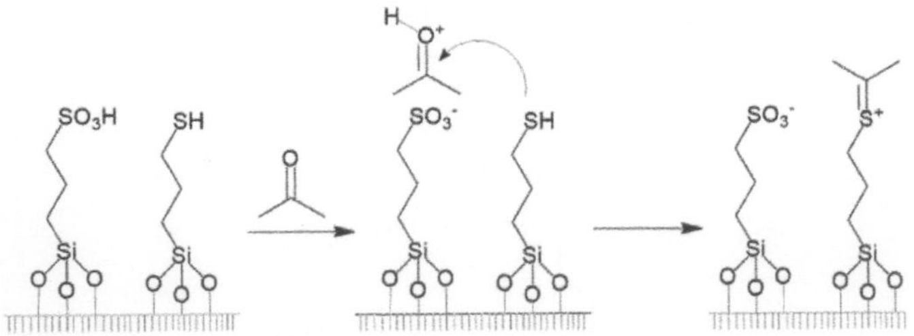

Scheme 9. Tentative reaction mechanism for acid/thiol cooperatively activation of acetone, by forming a highly electrophilic sulfonium intermediate, in the synthesis of bisphenol A.

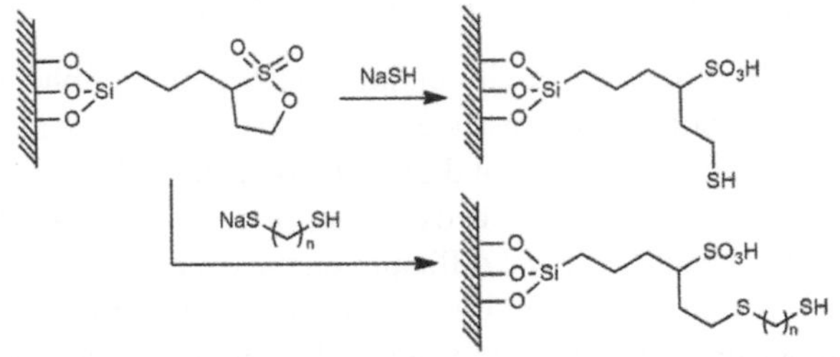

Scheme 10. Synthetic approach for the preparation of acid-thiol paired silica materials varying the spacer lengths.

of the acid–thiol paired materials to bisphenol A intimately depend on the distance implying both functions as close to each other as possible; the catalyst in which the acid–thiol paired are separated by three carbon atoms being the most active sample than that incorporating randomly distributed functions.

Huh et al. (2005) investigated a new cooperative catalytic system by the functionalization of mesoporous silica nanospheres with ureidopropyl groups and also basic centres constituted by 3-[2-(2-aminoethylamino)ethylamino]-propyl linkers efficiently promoting the aldol based reaction, and also, cyanosilylation. Based on the high capacity of the urea functions and analogues as activating agents for the carbonyl compounds, by double hydrogen bonding interactions (Petri 2004), the work was focused onto the synthesis of modified solids, through the co-condensation method, using different molar ratios of the corresponding organosilanes. The study comprises a case of dual cooperative catalysis in which the ureido species and amine groups contribute to the electrophile and nucleophile activations, respectively (Fig. 7). The results strongly suggested that the concentration of surface ureidopropyl functions is a key factor controlling the observed reactivity. The main role of these functionalities should be the activation of the 4-nitrobenzaldehyde for each explored chemical process, which might be the rate determining step on the model catalysts.

A metal-free nanostructured mesoporous organocatalyst prepared by thiol–ene click reaction of 3-mercaptopropyl grafted SBA-15 and 2,4,6-triallyloxy-1,3,5-triazine have been reported for the synthesis of 2-amino-4H-chromenes in good yields, as interesting therapeutic molecules (Mondal 2012) (Scheme 11). The catalytic activity was higher than that of the 2,4,6-triallyloxy-1,3,5-triazine homogeneous catalyst, as well as that of the mesoporous organic polymer supporting the backbone

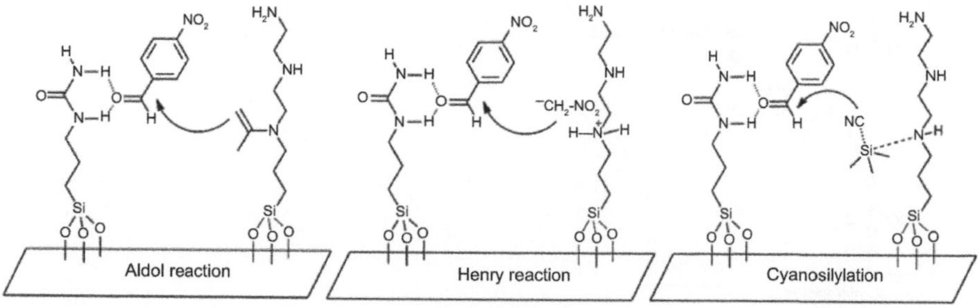

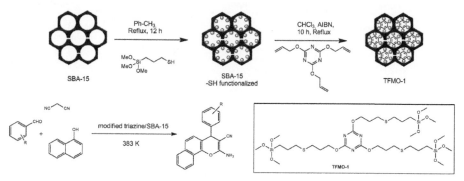

Fig. 7. Proposed cooperative activation modes of nucleophile and electrophile by mesoporous silica nanospheres functionalized with ureidopropyl and amine functions efficiently catalyzing the aldol-based reaction, and also, cyanosilylation.

Scheme 11. Synthesis of 2-amino-chromenes over triazine functionalised mesoporous organocatalyst.

of the 2,4,6-triallyloxy-1,3,5-triazine system, MPTAT-1, an excellent support for immobilizing Pd(II) showing very good catalytic activity in several C–C cross-coupling reactions in a water medium (Modak et al. 2011). The authors attributed the observed catalytic behaviour to the existence of sulphur atoms inside the pores of the catalyst, where they apparently enhanced the electrophilicity of the aldehyde and cyanide groups, promoting the Michael addition and the subsequent intramolecular cyclization reaction.

L-serine functionaling mesoporous silica prepared by using post-synthetic modification of benzene bridged organosilica has been recently reported (Huybrechts et al. 2017). The synthetic route of this organosilica which allow the introduction of the organic functions over the benzene ring is as follow (Scheme 12): (i) nitration of benzene rings, (ii) reduction of nitro to amine groups, (iii) reaction with amino protected and carboxy-activated L-serine and, finally, (iv) Bz-NH cleavage under acid treatment using HCl. The authors proposed that as a more robust nanocatalyst for the aldol condensation between acetone and 4-nitrobenzaldehyde when operating in water media, suggesting a cooperative mechanism involving the alcohol and amine functions leading to a very high activity and selectivity towards the aldol product.

Metal bifunctional silica-based catalysts

A great variety of organic-inorganic hybrids comprising metallic centres have been developed. This chapter, in subsequent sections, illustrates some interesting

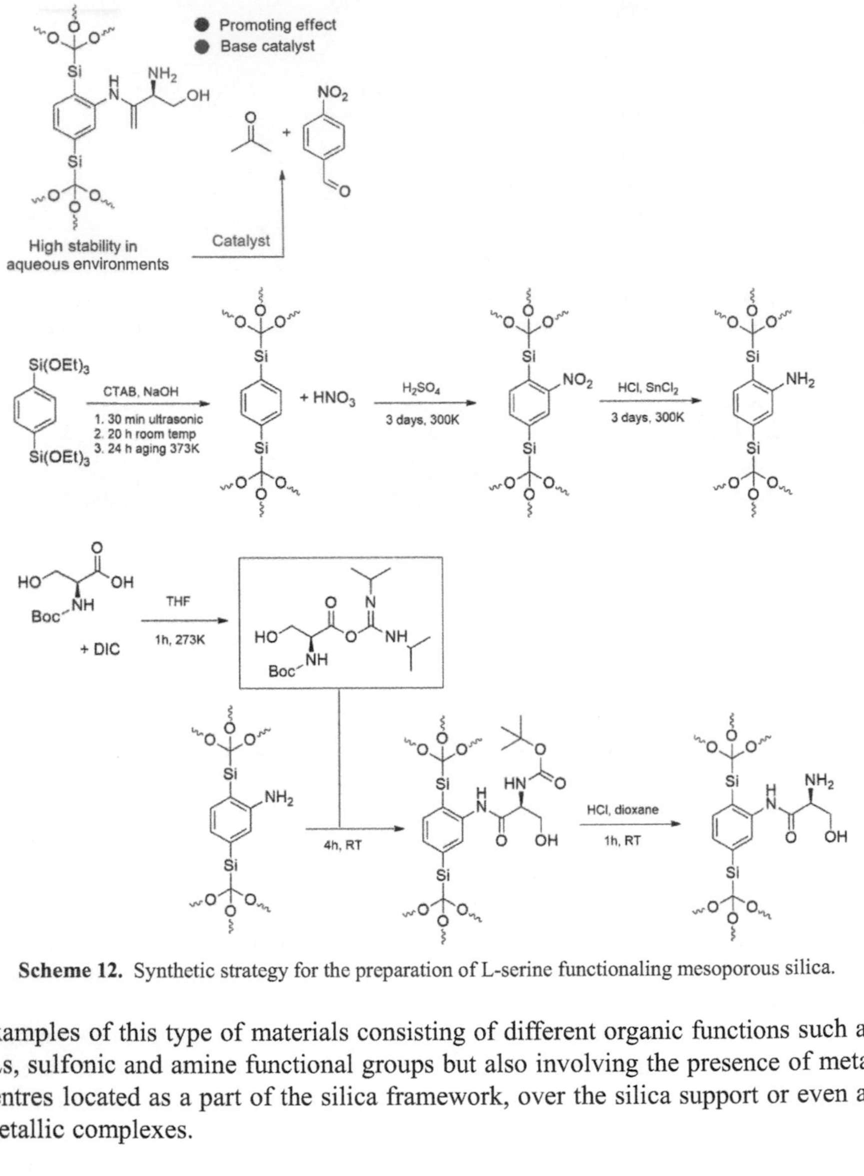

Scheme 12. Synthetic strategy for the preparation of L-serine functionaling mesoporous silica.

examples of this type of materials consisting of different organic functions such as ILs, sulfonic and amine functional groups but also involving the presence of metal centres located as a part of the silica framework, over the silica support or even as metallic complexes.

IL-based/Metal organic-inorganic hybrids

ILs are known as designer solvents being an environmentally friendly alternative to the volatile organic solvents. ILs are valuable compounds exhibiting unique chemical and physical properties (Chiappe et al. 2005; Greaves et al. 2008; Lei et al. 2017), which make them promising candidates for multiple applications, the catalysis among them (Welton 2004; Vekariya 2017; Dai et al. 2017). However, ILs used as reaction media under homogeneous reaction conditions show some negative effects from economic and environmental points of views; some of them are (i) expensive compounds, (ii) difficult to separate and extract the reaction products and

(iii) potential toxicity and low biodegradability, etc. Furthermore, the high viscosity of those causes slow substrate diffusion, the reaction often proceeding in the interphase of the IL-based catalyst and, hence, limiting their industrial catalytic application.

Another important issue when using ILs as reaction media for catalytic purposes is that all potential catalytic active species do not participate in the reaction because they are used as super-stoichiometric amounts. In this context, ILs-based heterogeneous catalysts emerge as ideal and an interesting alternative to the ILs used as bulk solvents (Gua and Li 2009).

ILs have also gained attention over the last two decades as vehicles for the immobilization of transition metal-based catalysts. In this context, these systems have also been widely used for the functionalization of OMSs maintaining the advantages of ILs and the silica support. Then, IL-based mesoporous silicas are also recognized as interesting materials for the immobilization of different metal catalytic species (Elhamifar et al. 2016b). Keeping this in mind several IL functionalyzed OMS materials immobilizing transition metal-based catalysts have been investigated in numerous interesting catalytic chemical transformations. For instance, palladium-supported SBA-15 based on alkylimidazolium IL has been reported as nanocatalyst exhibiting an extremely high catalytic activity for the Suzuki–Miyaura coupling reaction for a great variety of aryl halides and arylboronic acids in a water medium (Karimi et al. 2010) and also in the Heck cross-coupling reaction between aryl halides and olefins (Elhamifar 2013). In these studies, the parent organosilica counterpart was prepared *via* co-condensation method using 1,3-bis(3-trimethoxysilylpropyl) imidazolium chloride resulting in highly ordered organosilica, the Pd catalyst being subsequently introduced by treatment of IL-based mesoporous silica with palladium acetate. The authors suggested that the presence of IL units in this material plays an important role in releasing and recapturing the Pd nanoparticles into the silica mesochannels avoiding the extensive Pd agglomeration. Some of these authors also demonstrated the efficiency of this catalyst in the aerobic oxidation of several alcohols in the production of corresponding carbonyl compounds (Karimi et al. 2011).

A perruthenate anion was also immobilized by using an ion-exchange method on several alkyl imidazolium based periodic mesoporous organosilicas (Karimi et al. 2012). The authors found that the structural and morphologic properties of these materials are strongly dependent on the IL precursor initial concentration (10–35 mol% of IL), all of the samples maintaining the typically hexagonal ordered structure. Interestingly, the Ru exchanged mesoporous silica containing 10% of imidazolium salt was found to be an effective and recyclable catalyst for the aerobic oxidation of a huge number of alcohols. This type of IL modifying mesoporous silicas has been used by other authors for the immobilization of Lewis acidic chloroindate(III) involved in the Friedel–Crafts reaction between benzene and benzyl chloride (Zhao et al. 2009).

Very recently, new and recyclable phenyl IL-based bifunctional organosilica modified within copper acetate (Cu@BPMO-Ph-IL) has been reported in the one-pot Hantzsch condensation between aldehydes, ammonium acetate, alkylacetoacetates and dimedone, under solvent-free conditions, for the clean synthesis of polyhydroquinoline derivatives (Elhamifar and Ardeshirfard 2017). The synthesis of the Cu@PMO-Ph-IL nanocatalyst includes the following steps:

(i) hydrolysis and condensation of tetramethoxysilane and bis(triethoxysilyl) benzene, in the presence of pluronic P123, under acidic conditions, (ii) grafting of 1-methyl-3-(3-trimethoxysilylpropyl)-imidazolium chloride over the silica surface and (iii) treatment with copper acetate.

Acid/Metal organic-inorganic hybrids

Tang et al. (2010) demonstrated that the combination of mesoporous organic–inorganic hybrid silica functionalized with organosulfonic acids and Pt supported nanoparticles (Pt/SBA15-PrSO₃H) can efficiently catalyse the sequential hydrogenation and esterification reaction from actetaldehyde and acetic acid in the presence of hydrogen using 1,4-dioxane as the solvent. Pt/SBA15-PrSO₃H showed strong acid character, even much more than the SBA15-PrSO₃H precursor, that the authors attributed to the sulfonic acid and Pt particles interactions forming a possible [Pt$_\pi$–H]$^+$ adduct as also reported for sulfonated resin-supported Pd catalysts (Seki et al. 2008) (Fig. 8). The Pt loading was also explored indicating that although high Pt contents make the particle dispersion difficult, an improvement was observed in the catalytic performance.

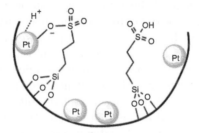

Fig. 8. Representation of the possible bifunctional interactions between sulfonic groups and Pt particles.

Basic/Metal organic-inorganic hybrids

Aminopropyl organosilicas functionalized with transition metals centres, especially niobium, have been widely investigated by Ziolek and Sobczak (2017). The niobium interactions with aminopropyl functions depend on the silica structure and the used method for functionalisation, as demonstrated in the incorporation of aminopropyltrimethoxysilanes into MCM-41 (Blasco-Jimenez et al. 2009; Blasco-Jimenez et al. 2010) and SBA-15 (Olejniczak and Ziolek 2014), and when using the post-synthetic method, and also in the one-pot synthesis of MCF materials (Stawicka et al. 2012). The thermal stability of organosilicas is also depending on the used silane precursor—3-aminopropyl-trimethoxysilane (APMS), [3-(2-aminoethylamino) propyl]trimethoxysilane (2APMS) and 3-[2-(2-aminoethylamino)ethylamino] propyl-trimethoxysilane (3APMS). In this way, amino-grafted materials modified with 2 APMS or 3APMS organic linkers resulted in superior thermal stability than the APMS ones attributed to the formation of hydrogen bonding between NH₂ functions of the modifier and silanol groups in the support already reported by Das and Sayari (2007) (Fig. 9). These interactions have two interesting effects consisting of an increase of the amine stability but also a decreasing of the basicity of the

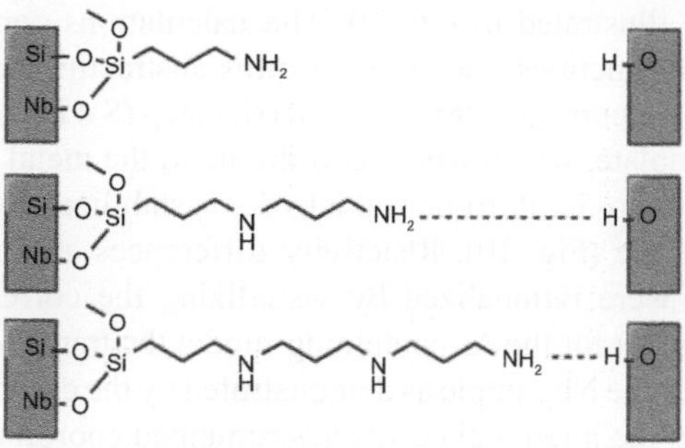

Fig. 9. Interaction models between amine functions and hydroxyl groups.

amine functions. Furthermore, the presence of niobium into the silica framework in NbMCM-41 strongly influences the basicity of the amine groups, amino-grafted NbMCM-41 being more active than APMS-MCM-41 catalyst, as demonstrated in the test reactions such as isopropanol decomposition and the Knoevenagel condensation.

The higher basic character of APMS on NbMCM-41 silica could be explained by the interaction of niobium species with free methoxy groups from APMS forming Nb-O-Si bonds and enhancing the basicity of the amine functions. In the case of APMS-NbSBA-15 silica in which the pore size is greater than in 2APMS-NbSBA-15, the main contribution is in the stabilization of aminosilane being dominated by metal species.

These types of mesoporous silicas have been successfully tested in carbon-carbon double bond isomerization reaction for the synthesis of interesting compounds useful in perfume manufacture such as isomerization of safrole (Sobczak et al. 2012). Different TMCM-41 mesoporous silicas (where T is Nb, Si or Al) functionalized within organotrialkoxysilanes by using the post-synthesis grafting method were studied, the isomerization of safrole being strongly dependent on the nature of the silica and the amine functions. Among the silicas modified with APMS organosilane, the APMS/AlMCM-41 results in the best catalytic performance, whereas the presence of longer ethylene diamine chains on AlMCM-41 produced a decreasing of that. Amino groups tethered over these mesoporous materials are probably the basic active sites which promote the isomerization reaction, the acid sites on the support probably enhancing the basicity of the amine functions in APMS/TMCM-41 (T = Al or Nb).

This strategy has been also reported for the preparation of new families of amino functionalized NbMCF mesoporous silica, by using both the post-synthesis and one-pot synthesis methods, efficiently catalysing the synthesis of relevant nitrogen heterocyclic systems, such as substituted quinolines and naphtiridines, through the Friedländer reaction (Smuszkiewicz et al. 2013a; Smuszkiewicz et al. 2013b). Similarly, the presence of Al or Nb in TMCF silica samples produced an enhancement of the yield to the corresponding quinoline as compared with that when using MCF. In order to explain the observed reactivity, the authors proposed a cooperative catalysis mechanism involving the amine groups and metallic centres as activating agents of the reagents. In this sense, a computational study concerning the first step of the reaction comprising the aldol condensation between reagents was reported. The selected models are illustrated in Fig. 10. The calculations confirmed cooperative effects between both functions, the amine groups abstracting the proton from enol form of the keto ester occurring within a low barrier step (8.4 kcal/mol) and providing the corresponding enolate, which would coordinate to the metal centre, whereas the electrophile, the 2-amino-5-chloro-benzaldehyde, is stabilized by hydrogen bonding to the protonated amine (Fig. 10). Reactivity differences as function of the metal on TMCF matrices were rationalized by visualizing the corresponding transition structures indicating that for the Al-containing model the transition structure is more advanced than that for the Nb sample as demonstrated by the computed C-C distances. The alcohol group in the intermediate species remained coordinated to the Nb atom through the carbonyl group, this feature being not observed for the Al-containing model. If assuming that the first step of the reaction could be the Knoevenagel condensation between reagents, the heterocyclization of both resulting in *cis*- or *trans*-isomers would lead to the corresponding quinolone or quinoline, respectively (Scheme 2). However, both intermediate compounds showed similar energy values (*trans*-isomer 0.6 kcal/mol more stable than *cis*-isomer), these data do not explain the total reaction regioselectivity to the corresponding quinoline. In this regard, it was especially interesting to observe the effect of the water on the heterocyclization step as suggested by the combination of experimental and theoretical observations. It seems that the involvement of a water molecule notably decreases the barrier for the

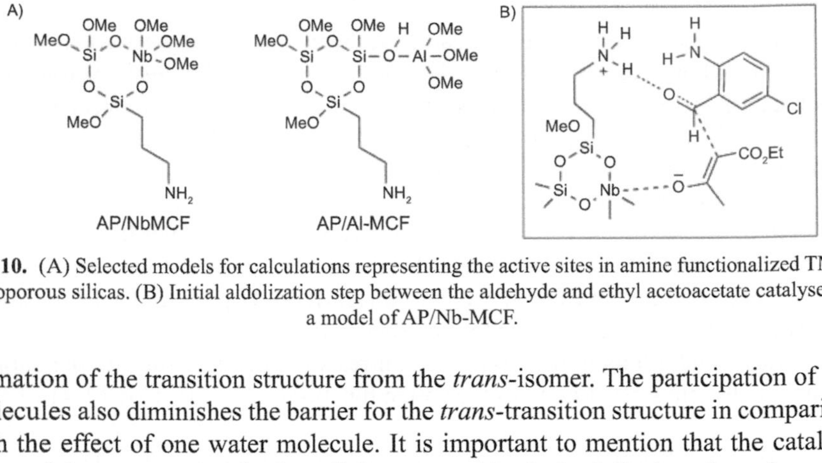

Fig. 10. (A) Selected models for calculations representing the active sites in amine functionalized TMCF mesoporous silicas. (B) Initial aldolization step between the aldehyde and ethyl acetoacetate catalysed by a model of AP/Nb-MCF.

formation of the transition structure from the *trans*-isomer. The participation of two molecules also diminishes the barrier for the *trans*-transition structure in comparison with the effect of one water molecule. It is important to mention that the catalytic effect of the water molecules is well documented in the hydrolysis and condensation processes (Tantillo and Houk 1999; Zhan et al. 2000; Zhan and Landry 2001; Fu and Lin 2011), the water molecule being recognized as a bifunctional acid-base catalyst.

SBA-15 mesoporous silica doped with Zr, Nb and Mo species and also functionalized within aminopropyl groups have been also investigated as acid-base bifunctional heterogeneous catalysts in the Knoevenagel condensation between benzaldehyde and active methylene compounds, the nature of metal species strongly influencing the catalytic performance (Calvino-Casilda et al. 2016).

Metal complexes immobilized on organic-inorganic hybrids

Metal complexes immobilized on the silica surface have been widely investigated. Some reported examples of different synthetic strategies immobilizing metals on the silica matrix are summarized in this section. At this regard, Ti-grafted polyamidoamine dendritic hybrid silica has been recently reported by Sinija and Sreekumar (2015). This bifunctional catalyst exhibits Lewis acid and Lewis base characters and has been used for the synthesis of 3,4-dihydropyrimidin-2(1*H*)-one derivatives, through Biginelli reaction, and also the pyranopyrazole synthesis (Scheme 13).

The dendritic structure was prepared from the corresponding amino-grafted mesoporous silica (BS-NH$_2$) in which the amine group can undergo a Michael addition reaction to methyl methacrylate, followed by amidation with ethylenediamine, resulting in the formation of the first generation of mesoporous silica. Thus, the appropriated repetition of Michael addition + amidation reaction allowed the obtention of the corresponding generation of anchored dendrimer able to coordinate titanium(IV) isopropoxide. The functionalisation provoked a diminished meso-scale periodicity and also strongly decreasing the surface area, pore volume and diameter. The authors suggested that the dual nature of the catalyst played an important role in the reagent activation for both processes.

Pd(II) chelating more complex organic functions over mesoporous silica has been recently reported by Surmiak et al. (2017) as a novel bifunctional catalyst explored in the alkynes or alkenes hydrogenation reactions. The catalyst was synthetized by the co-condensation method using tetraethyl orthosilicate and organosilanes containing azides or alkoxyamines functions, which were selectively modified by using a click chemical method and radical nitroxide exchange, generating bidentate triazole-sulfoxide ligands at the silica surface suitable for palladation with PdCl$_2$ (Scheme 14). The authors confirmed the formation of Pd^0 nanoparticles adsorbed to the silica surface, the chelated Pd^{2+} species being reduced to Pd^0. Remarkably,

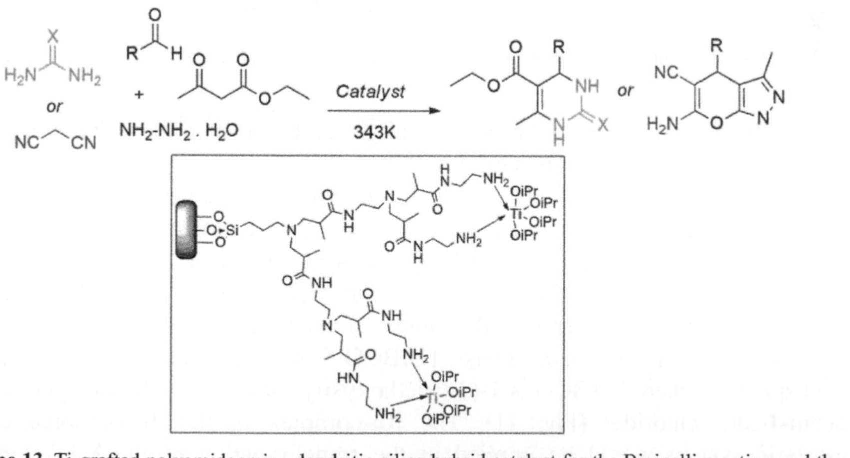

Scheme 13. Ti-grafted polyamidoamine dendritic silica hybrid catalyst for the Biginelli reaction and the pyranopyrazoles synthesis.

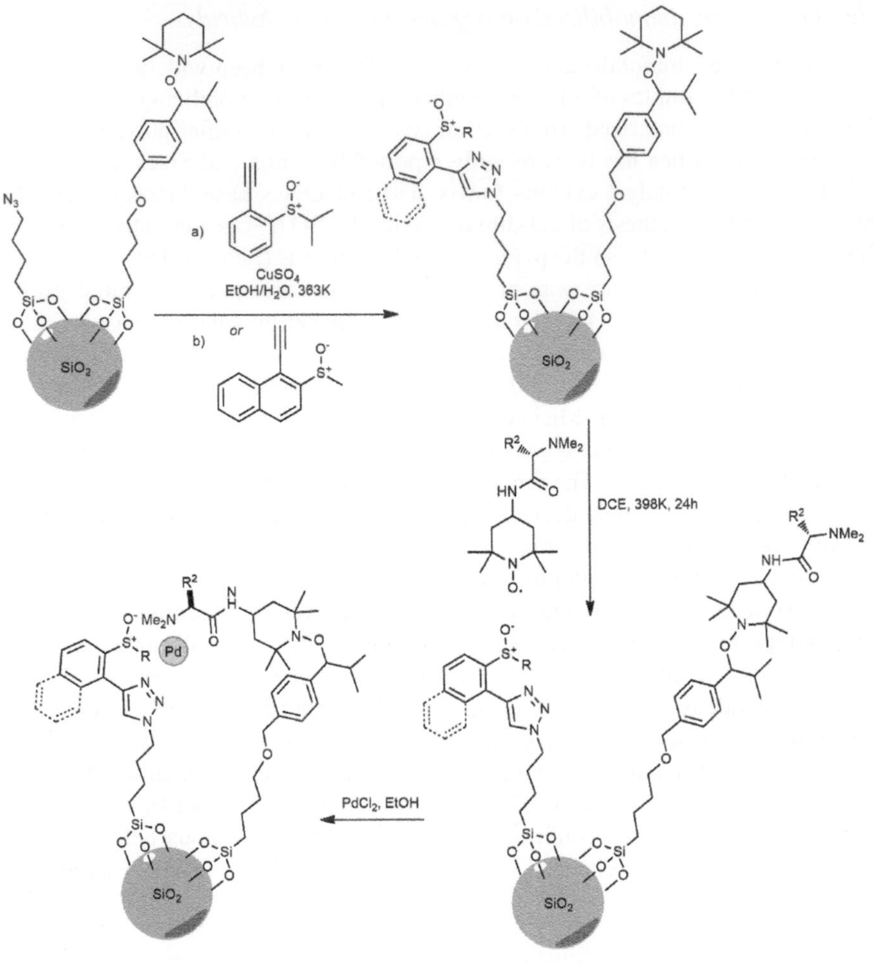

Scheme 14. Synthetic approach for the preparation of Pd^{2+} complexes over mesoporous silica.

ligands functionalising the silica surface plays an important role on the growth and stabilization of the metal nanoparticles.

A novel base–rhodium/diamine-bifunctionalized heterogeneous catalyst promoting the tandem asymmetric transfer hydrogenation of α-haloketones/ epoxidation, particularly 2-bromophenylethanone, in water, have been recently published (Liao et al. 2017). In this case, the cationic rhodium/diamine complex was assembled by hydrogen bounding interactions through BF$_4^-$ anion within silanol groups over the silica surface of basic DABCO functionalized mesostructured silica nanospheres, where DABCO is 1-(3-(triethoxysilyl)propyl)-1,4-diazabicyclo[2.2.2] octan-1-ium chloride) (Fig. 11). The Ru-complex in this bifunctional chiral catalytic system would be implied in the transformation of the haloketone to haloethanol whereas the DABCO moiety promotes the epoxidation reaction of the formed haloethanol. Because these tandem transformations normally occur in

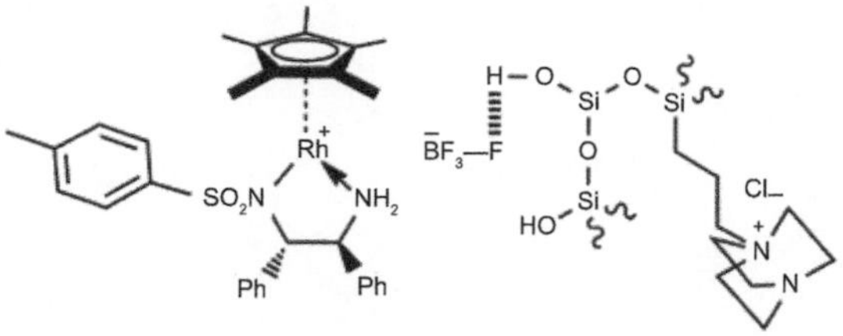

Fig. 11. Synthetic approach for the preparation of the rhodium/diamine–bifunctionalized heterogeneous catalyst.

Pd@mestyleneRuArDPEN@PMO

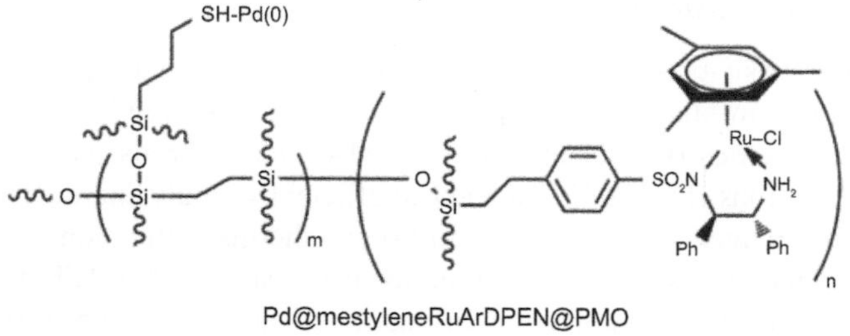

Fig. 12. Synthetic approach for the preparation of the bifunctional catalyst Pd@mestyleneRuArDPEN@PMO.

water, under homogeneous conditions, the presence of a phase-transfer catalyst as cetyltrimethylammonium bromide is required. In this context, since the chiral heterogeneous catalyst was prepared by using the templating method employing that, the enhancement of the catalytic performance could be attributed to the residual template acting as a phase-transfer agent.

Another structurally different bifunctional catalyst recently reported comprises two metal centres chelated by different organic functions attached on the silica surface. Such is the case of Pd(0)–Ru(III)/diamine-bifunctionalized periodic mesoporous organosilica, Pd@mestyleneRuArDPEN@PMO (Fig. 12), published by Zhao et al. (2017). Some relevant mesoporous silica-based bifunctional catalysts have been successfully developed for tandem reactions. However, these reports are commonly focused onto the design of new bifunctional silicas suitable for the synthesis of non-chiral molecules. In this context, the authors described a new and reusable bifunctional heterogeneous catalyst within ethylene bridged by immobilization of palladium nanoparticles and well defined chiral single site ruthenium species able to efficiently catalyse the asymmetric transfer hydrogenation–Sonogashira coupling one pot tandem reaction between 4-iodoacetophenone and ethynylbenzene (Scheme 15). While the Ru-complex catalysed, the asymmetric transfer hydrogenation, the Pd species are involved in the Sonogashira coupling leading to various chiral conjugated alkynols in high yields and enantioselectivity (97%). Some features such as the high ethylene bridged hydrophobicity, uniform distribution of small size Pd nanoparticles and single-site chiral ruthenium catalytic active sites are responsible for the catalytic performance.

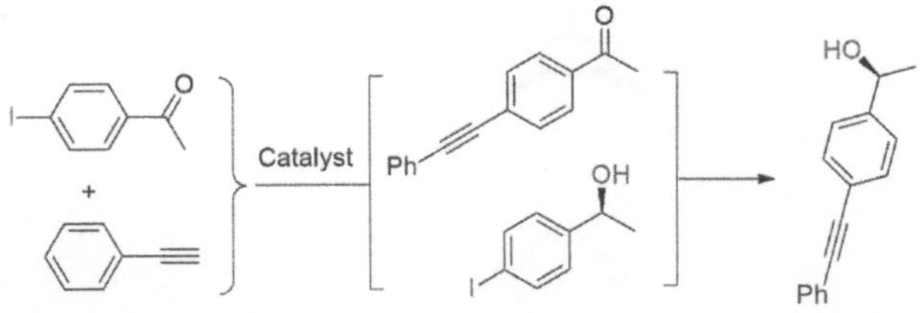

Scheme 15. Asymmetric transfer hydrogenation–Sonogashira coupling one pot tandem reaction between 4-iodoacetophenone and ethynylbenzene involving Pd nanoparticles and Ru-complex.

Other bimetallic basic silica-based catalysts

In one sense, basicity can be introduced on Nb doped mesoporous silicas by modification within alkaline cations. In this respect, Calvino-Casilda et al. (2010) reported a new series of Nb-containing MCM-41 mesoporous materials modified with alkaline cations (Li, Na, K, Rb and Cs) efficiently catalysing the isomerization of eugenol to isoeugenol under sonochemical and thermal activation.

Interestingly, others mesoporous niobiosilicates such as NbMCF modified with alkali metals have been described as bifunctional catalysts active in the 2-amino-4*H*-chromene synthesis, as mixtures of diastereomers *erythro/threo*, with good to excellent yields, from substituted salycilaldehydes and ethyl cyanoacetate, at room temperature (Scheme 16) (Smuszkiewicz et al. 2016). The authors prepared two kinds of niobium containing materials by using different metal sources, ammonium niobate(V) oxalate hydrate or niobium etoxide, the NbMCF samples subsequently impregnated with aqueous solutions of MCH_3CO_2 (M = Li, Na, K, Rb, Cs). The NbMCF samples showed different acid-base properties depending on the Nb source and alkaline metals type and loading. The results indicated that the metal loading on the catalysts, the acid-base character and the texture parameters are probably determining factors in the reactivity and reaction diastereoselectivity. The involvement of Nb species and the role of the alkalines oxides on NbMCF samples were determined by computational methods, based on DTF calculations, concerning to the first step of the reaction consisting of the aldol condensation between reagents. Thus, the presence of alkaline oxides on M/NbMCF catalysts, showing strong basicity, activates the nucleophile and initiates the reaction, whereas the Nb centre promotes the approaching of the reactants also stabilizing the aldolization transition structure. It is important to note that the combined experimental and theoretical results suggested an optimum basicity of the samples, particularly for Na and K/NbMCF samples. Therefore, the reaction requires a compromise between basicity

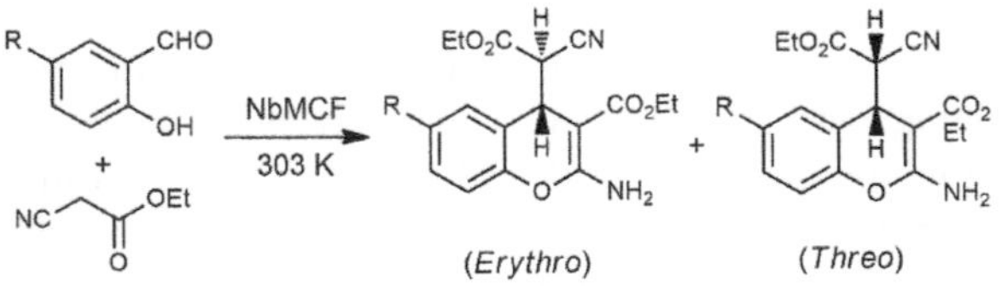

Scheme 16. Synthesis of 2-amino-4H-chromenes over NbMCF silicas modified with alkaline metals.

and the alkaline metal size on M/NbMCF materials because of the bulkiness of the alkaline centres (Rb, Cs) can limit the formation and the stability of the reactant complex, restricting the effective aldolization.

Metal organic frameworks (MOFs)

MOFs are considered as a new class of crystalline porous materials consisting of inorganic building units often constituted by metal ions or clusters connected within organic linkers by coordination bonds of moderate strength. The synthesis and characterization of MOFs is an attractive area that is rapidly growing: firstly by designing new structures due to the multitude of inorganic builders from metal centres and also a large number of organic linkers; followed by the understanding their characteristic and properties on developing novel materials with potential applications—adsorption and separation, storage, health, the food industry and electronic, among others (Gascon et al. 2014).

Great attention is shown to MOFs acting as heterogeneous catalysts because of their high surface area and pore volume and also to the high metallic centres' content, bridging the zeolites to enzymes gap (Farrusseng et al. 2009; Lee et al. 2009). In this respect, interesting a lot of examples of MOFs within metal active sites, reactive functional groups and as host matrices or nanometric reaction cavities within demonstrated catalytic efficiency have been reported (Corma et al. 2010b). More recently, Dhakshinamoorthy et al. (2013a) revised some examples of MOFs catalysing the synthesis of fine chemicals or even condensation reactions involving carbonyl components (Dhakshinamoorthy et al. 2013).

Regarding the catalytic properties, some of the characteristic of MOFs such as large internal surface areas and uniform pore and cavity sizes are shared with zeolites. However, it has been established that there are some relevant distinctions: (i) greater variety of MOFs, (ii) lower thermal stability and (iii) although many MOFs exhibit permanent microporousity like zeolites, often others collapse after solvent removal, these properties limiting their specific catalytic applications. Interestingly, García-García et al. (2014) have reported a comparative study concerning the catalytic activity of MOFs and their homogeneous counterparts such as metal benzoates, acetates, metal halides or even exploring others commonly investigated organic-inorganic hybrid materials and inorganic solid catalysts, concluding about advantages and limitations of using MOFs. The main conclusion of the work is that although MOFs were catalytically active in the transformations under study, these catalysts cannot compete with simple metal, acetates or benzoates barely showing activity differences or ever reaching lower values.

Bifunctional metal organic framework

Some MOFs able to act as bifunctional catalysts in a great variety of chemical processes have also been designed. This section summarizes some interesting examples of different bifunctional MOFs including acid–base, acid–metal and metal–metal systems.

Acid–Base bifunctional MOFs

Vermoortele et al. (2011) described as an acid–base bifunctional MOF, zirconium aminoterephthalate UiO-66-NH$_2$, highly effective and selective in the aldol condensation reaction from benzaldehyde and heptanal. The authors demonstrated that the appropriate thermal treatment produced the reversible dehydroxylation of the $[Zr_6O_4(OH)_4]^{12+}$ clusters to $[Zr_6O_6]^{12+}$, generating coordinatively unsaturated sites on Zr which could be operated in cooperation with the amine groups. Similarly, NH$_2$-MIL-101(Al), a MOF containing Lewis acid and Brönsted basic sites, constituted by coordinatively unsaturated Al^{3+} sites and amino groups, respectively, was reported as an bifunctional acid-base catalyst for tandem Meinwald rearrangement of epoxides to aldehydes and Knoevenagel condensation, using highly activated methylene compounds, more specifically malononitrile (Srirambalaji et al. 2012).

A multi-functional MOF, PCN-124, was reported for the selective adsorption of CO$_2$ over CH$_4$ but also as acid-base bifunctional MOF showing excellent catalytic activity in a tandem one-pot deacetalization–Knoevenagel condensation reaction (Park et al. 2012). 5,5'-((Pyridine-3,5-dicarbonyl)bis-(azanediyl))diisophthalate containing two isophthalate and one pyridine groups connected through amide bonds was prepared to build the copper MOF as depicted in Fig. 13. This MOF is composed by two essential catalytic functions acting in cooperation, the weakly acidic coordinatively unsaturated Cu^{2+}, catalysing the deacetalization of benzaldehyde dimethyl acetal to benzaldehyde, and basic amide groups implicated in the second step of the reaction consisting of the Knoevenagel condensation between the formed benzaldehyde and malononitrile.

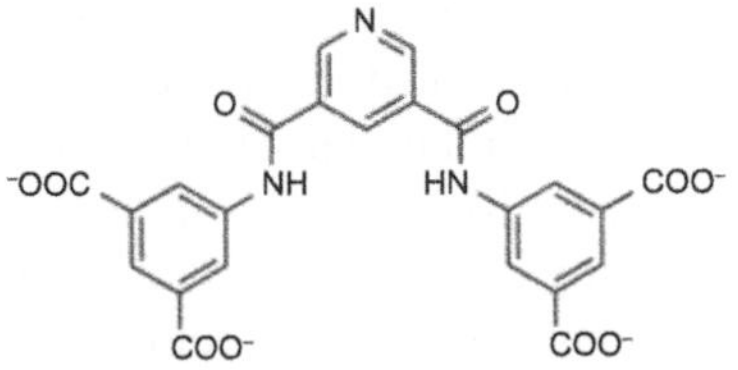

Fig. 13. 5,5'-((Pyridine-3,5-dicarbonyl)bis-(azanediyl))diisophthalate.

Acid–Metal bifunctional MOFs

Pan et al. (2010) reported a Cr(III)-based MOF, MIL-101(Cr^{3+}) containing Pd nanoparticles, as a bifunctional Lewis acid/metal catalyst for the one pot synthesis of methyl isobutyl ketone from acetone, the catalytic activity being higher on metal oxides or zeolites than on palladium (Scheme 17). The reaction sequence consisted of the condensation reaction of acetone involving the Lewis acid sites and subsequent hydrogenation of the corresponding α,β-unsaturated compound catalysed by Pd nanoparticles. The authors demonstrated that Lewis acids centres and Pd nanoparticles must be closely located. Thus, the increased concentrations of Pd provoked the deposition of Pd nanoparticles at the external surface of the MOF, whereas the Cr^{3+} centres are located inside the pore, notably influencing the selectivity to methyl isobutyl ketone. Similar MOF has been explored for the cyclization of citronellal to isopulegol and for the one-pot tandem isomerization/hydrogenation of citronellal to menthol (Cirujano et al. 2012).

Interestingly, a Ir-based bifunctional MOF consisting of amino containing UiO-66-NH$_2$ or IRMOF-3 with additional Ir^I coordinately bound (Scheme 18) has been described for the catalytic synthesis of *N*-alkyl amines from nitro arenes and alkyl or aryl aldehydes through reductive amination (Pintado-Sierra et al. 2013). The process takes place through cascade reactions involving the reduction of nitro arenes in the presence of H$_2$, condensation reaction between the formed aniline and the corresponding aldehyde and, finally, the hydrogenation of resulting imine. Both reduction steps occur over the Ir centers whereas the imination reaction is promoted by the MOF acid sites (Scheme 19).

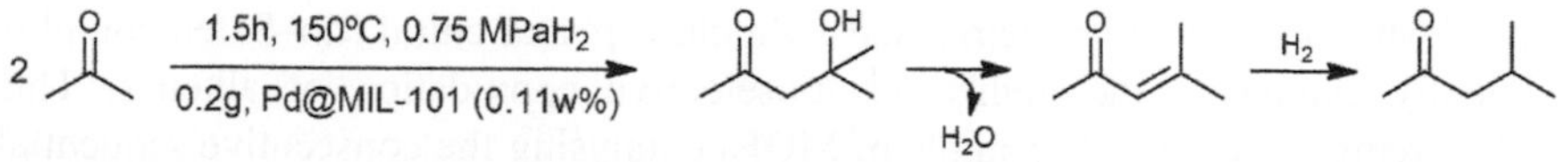

Scheme 17. Synthesis of methyl isobutyl ketone from acetone in the presence of MIL-101(Cr^{3+}) containing Pd nanoparticles catalyst.

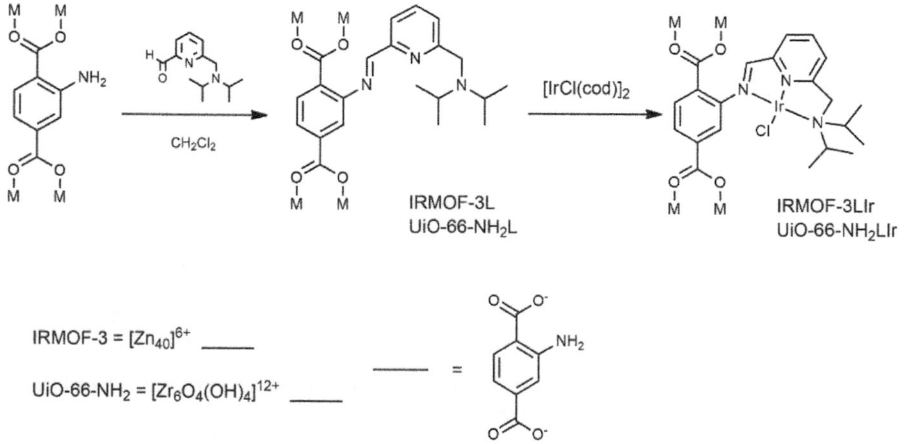

Scheme 18. IRMOF-3 and UiO-66-NH$_2$ post-modification.

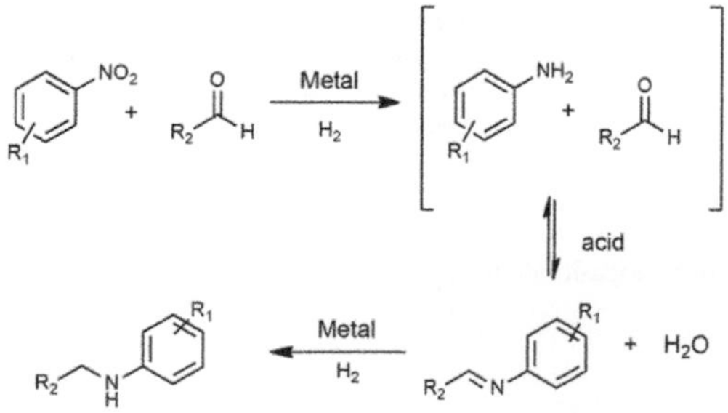

Scheme 19. Catalytic synthesis of *N*-alkyl amines from nitro arenes and alkyl or aryl aldehydes.

Metal−Metal bifunctional MOFs

Luz et al. (2010) recently reported several copper containing MOFs demonstrating that it is possible the functionalisation of coordinatively unsaturated copper sites by immobilizing small organic molecules such as pyridine species grafted on the surface and useful chelating agents for the introduction of Pd centers (Scheme 20). This bifunctional MOFs was tested in the synthesis of 1,4-disubstituted 1,2,3-triazoles by Click reaction consisting of the 1,3-dipolar cycloaddition reaction of alkynes to azides and also in the preparation of the starting azide. Later, Arnanz et al. (2012) used the Pd-grafted Cu-BTC catalysts for the one-pot Sonogashira/click reactions of 2-iodobenzyl bromide, NaN_3 and alkynes showing good activities (Scheme 21).

Song et al. (2011) proposed a metal-metal bifunctional MOF as porous catalytic system able to mediate regio- and stereoselective sequential reactions for the synthesis of complex valuable molecules. In this context, the authors synthetized a chiral MOF constituted by tetranuclear Zn clusters and chiral Mn−Salen complex efficiently catalysing the highly enantioselective epoxidation of alkenes. This catalyst represents the first example of MOFs catalysing the consecutive sequential asymmetric alkene epoxidation/epoxide ring-opening reactions (Scheme 22).

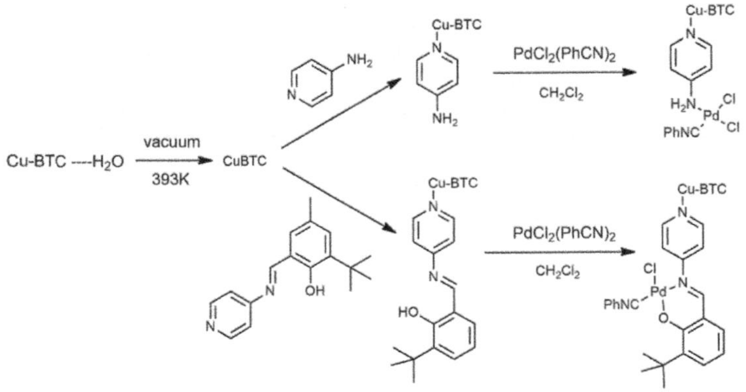

Scheme 20. Synthetic approach for the functionalization of coordinatively unsaturated copper sites by immobilizing pyridine species.

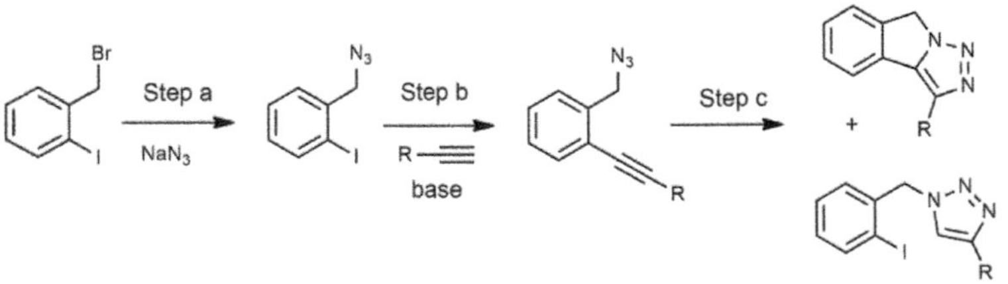

Scheme 21. One-pot Sonogashira/click reactions of 2-iodobenzyl bromide, NaN_3 and alkynes.

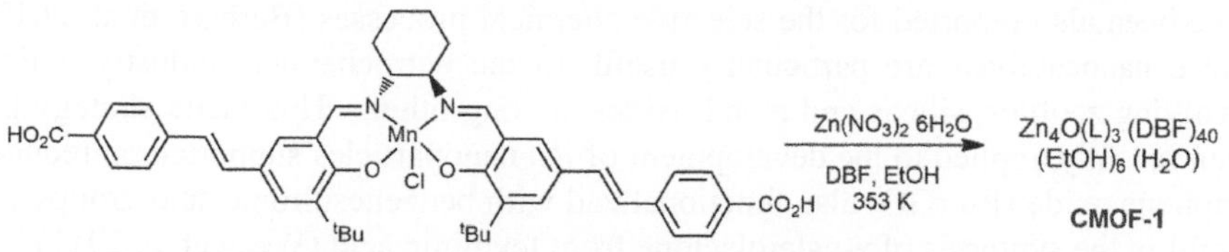

Scheme 22. Synthesis of the bifunctional chiral catalyst, CMOF-1.

Summary

In the last decades, based on the accumulated knowledge concerning the enzymes actuation modes, the development of bioinspired acid-base bifunctional heterogeneous catalysts acting as activating agents of both electrophiles and nucleophiles reactants have paying great attention. It is reasonably easy to prepare nanostructured catalysts functionalized with different catalytic sites to be of use in sequential organic reactions in an efficient and selective manner revealing the advantages of using heterogeneous catalysis. Thus, the use of inorganic solids allows the coexistence of incompatible functions tethered to the solid surface which can cooperatively act, being at the appropriate distances and producing the notable enhancement of the catalytic performance in interesting chemical processes.

A great impulse for designing synthetic strategies to develop new and more effective and selective bifunctional nanocatalysts involved on the process optimization have been made. In this context, many efforts are aimed to the co-location of acidic and basic sites, particularly onto mesoporous silica supports. In this regard, cooperative interactions between amino functional groups and silanols from SBA-15 mesoporous silica but also with acidic functionalities by varying their acid strengths have been demonstrated. As examined to a limited extent in this chapter, other functions such as thiols or ureido groups, among others, can act in cooperation with acids or amine functions in important organic transformations.

Furthermore, the co-habitation of acid and basic functions is also possible using others porous supports. This chapter summarizes some types concerning the bifunctional catalysts related to mesoporous silicas and MOFs because both have been widely investigated, but this topic is not limited to those. Another interesting example is the modification of graphene oxide with aminoproyl groups, by using a silylation approach, applied in one-pot cascade reactions comprising sequential acetal hydrolysis and Knoevenagel condensation (Zhang et al. 2014).

Although the development of bifunctional solid catalysts has impressed a spectacular progress, many more efforts are needed to investigate the synergetic effects between multiple functions emphasizing on the control and optimization of the relative positions of the catalytic sites. This issue is also being studied by introducing specifically positioned acid and basic sites on porous polymeric aromatic frameworks and applied in catalytic cascade-type reactions (Merino et al. 2013); a novel class of supports showing high porosity and surface area, and high physicochemical stability, all of them relevant characteristics in heterogeneous catalysis.

Bifunctional catalysts including both acidic and metal nanoparticles, in which the proximity of both functions is a determinant factor for the optimal efficiency, have been also reported for the selective chemical processes (Barbaro et al. 2012). These nanocatalysts are particularly useful in the petrochemical industry mainly involving zeolites, silicas and metal oxides, among others. This facile strategy has been recently applied to the development of Ru nanoparticles supported on reduced graphene oxide (Ru/rGO) also functionalized with benzenesulfonic acid groups and useful in the synthesis of γ-valerolactone from levulinic acid (Wang et al. 2017).

An important aspect to be considered for the application, in general, of inorganic solids in catalysis is that both chemical functions and the porous architecture strongly influence the catalytic performance. The introduction of functionalities onto the porous system often produces a decreasing of both the surface area and pore volume restricting the reactant interactions. However, this circumstance is changed when functionalizing the surface with metals, where the porosity is frequently maintained after impregnation with metal salts and subsequent thermal treatments.

The presence of cooperative catalytic sites spatially separated can also contribute to the control of the reaction sequence in cascade reactions by coupling the separate catalytic activities (Corma 2016). In this context, Parlett et al. (2016) have recently reported the preparation of spatially orthogonal bifunctional silica-based catalysts in which different metal nanoparticles are spatially compartmentalized, by selective functionalization of silica mesopores and macropores, and applied to the synthesis of cinnamic acid by Pd/Pt-catalyzed oxidation of cinnamyl alcohol.

On the other hand, new advanced heterogeneous catalysts differing from the conventional ones are the integrated nanocatalysts (INCs), showing a sophisticated nanoarchitecture and unique catalytic properties (Gawande et al. 2015). INCs are often built using "bottom-up" methodologies consisting of self-assembly of preformed building blocks as individual catalytic components, allowing the synthesis of multifunctional catalytic systems (Zeng 2013).

Based onto the great knowledge concerning bifunctional or even multifunctional catalysis through the different developed methodologies, one of the current challenges for the chemists could consist on the development of chiral heterogeneous catalysts but designed no by immobilizing the chiral homogeneous counterparts.

Acknowledgments

This work has been supported by the Spanish Ministry (CTM 2014-56668-R project).

Reference cited

Aider, N., A. Smuszkiewicz, E. Pérez-Mayoral, E. Soriano, R.M. Martín-Aranda, D. Halliche and S. Menad. 2014. Amino-grafted SBA-15 material as dual acid-base catalyst for the synthesis of coumarin derivatives. Catal. Today 227: 215–222.

Alauzun, J., A. Mehdi, C. Reyé and R.J.P. Corriu. 2006. Mesoporous materials with an acidic framework and basic pores. A successful cohabitation. J. Am. Chem. Soc. 128: 8718–8719.

Arnanz, A., M. Pintado-Sierra, A. Corma, M. Iglesias and F. Sánchez. 2012. Bifunctional metal organic framework catalysts for multistep reactions: MOF-Cu (BTC)-[Pd] Catalyst for one-pot heteroannulation of acetylenic compounds. Adv. Synth. Catal. 254: 1347–1355.

Barbaro, P., F. Liguori, N. Linares and C. Moreno Marrodan. 2012. Heterogeneous bifunctional metal/acid catalysts for selective chemical processes. Eur. J. Inorg. Chem. 24: 3807–3823.

Bass, J.D., A. Solovyov, A.J. Pascall and A. Katz. 2006. Acid-base bifunctional and dielectric outer-sphere effects in heterogeneous catalysis: A comparative investigation of model primary amine catalysts. J. Am. Chem. Soc. 128: 3737–3747.

Ben, T., H. Ren, S. Ma, D. Cao, J. Lan, X. Jing, W. Wang, J. Xu, F. Deng, J. M. Simmons, S. Qiu and G. Zhu. 2009. Targeted synthesis of a porous aromatic framework with hight stability and exceptionally high surface area. Angew. Chem. Int. Ed. 48: 9457–9460.

Blasco-Jimenez, D., A.J. Lopez-Peinado, R.M. Martin-Aranda, M. Ziolek and I. Sobczak. 2009. Sonocatalysis in solvent-free conditions: An efficient eco-friendly methodology to prepare N-alkyl imidazoles using amino-grafted NbMCM-41. Catal. Today 142: 283–287.

Blasco-Jimenez, D., I. Sobczak, M. Ziolek, A.J. Lopez-Peinado and R.M. Martin-Aranda. 2010. Amino-grafted metallosilicate MCM-41 materials as basic catalysts for eco-friendly processes. Catal. Today 152: 119–125.

Brunelli, N.A., K. Venkatasubbaiah and C.W. Jones. 2012. Cooperative catalysis with acid-base bifunctional mesoporous silica: Impact of grafting and co-condensation synthesis methods on material structure and catalytic properties. Chem. Mater. 24: 2433–2422.

Calvino-Casilda, V., E. Pérez-Mayoral, R.M. Martín-Aranda, Z. Zienkiewicz, I. Sobczak and M. Ziolek. 2010. Isomerization of eugenol under ultrasound activation catalyzed by alkali modified mesoporous NbMCM-41. Top Catal. 53: 179–186.

Calvino-Casilda, V., M. Olejniczak, R.M. Martin-Aranda and M. Ziolek. 2016. The role of metallic modifiers of SBA-15 supports for propyl-amines on activity and selectivity in the Knoevenagel reactions. Microporous Mesoporous Mater. 224: 201–207.

Chaignon, J., M. Gourgues, L. Khrouz, N. Moliner, L. Bonneviot, F. Fache, I. Castro and B. Albela. 2017. A bioinspired heterogeneous catalyst based on the model of the manganese-dependent dioxygenase for selective oxidation using dioxygen. RSC Adv. 7: 17336–17345.

Chiappe, C. and D. Pieraccini. 2005. Ionic liquids: solvent properties and organic reactivity. J. Phys. Org. Chem. 18: 275–297.

Chng, L.L., N. Erathodiyil and J.Y. Ying. 2013. Nanostructured catalysis for organic transformations. Acc. Chem. Res. 46: 1825–1837.

Cirujano, F.G., F.X. Llabrés i Xamena and A. Corma. 2012. MOFs as multifunctional catalysts: One-pot synthesis of menthol from citronellal over a bifunctional MIL-101 catalyst. Dalton Trans. 41: 4249–4254.

Climent, M.J., A. Corma and S. Iborra. 2011. Heterogeneous catalysis for the one-pot synthesis of chemicals and fine chemicals. Chem. Rev. 111: 1072–1133.

Collins, A.E.C. and I. Horváth. 2011. Heterogenization of homogeneous catalytic systems. Catal. Sci. Technol. 1: 912–919.

Copeland, R.A. 2000. Enzymes. A practical introduction to structure, mechanism and data structure. 2nd Edition. Wiley-VCH.

Corma, A., U. Díaz, T. García, G. Sastre and A. Velty. 2010. Multifunctional hybrid organic-inorganic catalytic materials with a hierarchical system of well-defined micro- and mesopores. J. Am. Chem. Soc. 132: 15011–15021.

Corma, A., H. García and F.X. Llabre's i Xamena. 2010. Engineering metal organic frameworks for heterogeneous catalysis. Chem. Rev. 110: 4606–4655.

Corma, A., M. Boronat, M. Climent, S. Iborra, R. Montón and M.J. Sabater. 2011. A recyclable bifunctional acid-base organocatalyst with ionic liquid character. The role of site separation and spatial configuration of site separation and spatial configuration on different condensation reactions. Phys. Chem. 13: 17255–17261.

Corma, A. 2016. Separate to accumulate. Nat. Mater. 15: 134–136.

Dai, C., J. Zhang, C. Huang and Z. Lei. 2017. Ionic liquids in selective oxidation: catalysts and solvents. Chem. Rev. 117: 6929–6983.

Das, D.D. and A. Sayari. 2007. Amine grafted pore-expanded MCM-41 as base catalysts. Stud. Surf. Sci. Catal. 170: 1197–1204.

Dhakshinamoorthy, A., M. Opanasenko, J. Cejka and H. Garcia. 2013. Metal organic frameworks as solid catalysts in condensation reactions of carbonyl groups. Adv. Synt. Catal. 355: 247–268.

Dhakshinamoorthy, A., M. Opanasenko, J. Čejka and H. Garcia. 2013. Metal organic frameworks as heterogeneous catalysts for the production of fine chemicals. Catal. Sci. Technol. 3: 2509–2540.

Domínguez-Fernández, F., J. López-Sanz, E. Pérez-Mayoral, D. Bek, R.M. Martín-Aranda, A.J. López-Peinado and J. Cejka. 2009. Novel basic mesoporous catalysts for the Friedländer reaction from 2-aminoaryl ketones: Quinolin-2(1H)-ones versus quinolines. ChemCatChem. 1: 241–243.

Dufaud, V. and M.E. Davis. 2003. Design of heterogeneous catalysts via multiple active site positioning in organic-inorganic hybrid materials. J. Am. Chem. Soc. 125: 9403–9413.

Elhamifar, D., B. Karimi, J. Rastegar and M.H. Banakar. 2013. Palladium-containing ionic liquid-based ordered mesoporous organosilica: an efficient and reusable catalyst for the Heck reaction. ChemCatChem. 5: 2418–2424.

Elhamifar, D., M. Nasr-Esfahani, B. Karimi, R. Moshkelgosha and A. Shábani. 2014a. Ionic liquid and sulfonic acid based bifunctional periodic mesoporous organosilica (BPMO-IL-SO3H) as a highly efficient and reusable nanocatalyst for the Biginelli reaction. ChemCatChem. 6: 2593–2599.

Elhamifar, D., B. Karimi, A. Moradi and J. Rastegar. 2014b. Synthesis of a novel sulfonic acid containing ionic-liquid-based periodic mesoporous organosilica and study of its catalytic performance in the esterification of carboxylic. ChemPlusChem. 79: 1147–1152.

Elhamifar, D., S. Kazempoora and B. Karimib. 2016. Amine-functionalized ionic liquid-based mesoporous organosilica as a highly efficient nanocatalyst for the Knoevenagel condensation. Catal. Sci. Technol. 6: 4318–4326.

Elhamifar, D., F. Shojaeipoor and O. Yari. 2016. Thiopropyl containing ionic liquid based periodic mesoporous organosilica as a novel and efficient adsorbent for the removal of Hg(II) and Pb(II) ions from aqueous solutions. RSC Adv. 6: 58658–58666.

Elhamifar, D. and H. Ardeshirfard. 2017. Phenyl and ionic liquid based bifunctional periodic mesoporous organosilica supported copper: An efficient nanocatalyst for clean production of polyhydroquinolines. J. Colloid Interface Sci. 505: 1177–1184.

End, N. and K. Schöning. 2004. Immobilized catalysis in industrial research and application. Top. Curr. Chem. 242: 241–271.

Farrusseng, D., S. Aguado and C. Pinel. 2009. Metal–organic frameworks: opportunities for catalysis. Angew. Chem. Int. Ed. 48: 7502–7513.

Fu, C.W. and T.H. Lin. 2011. The effects of water on β-d-xylose condensation reactions. J. Phys. Chem. A 115: 13523–13533.

García-García, P., M. Müller and A. Corma. 2014. MOF catalysis in relation to their homogeneous counterparts and conventional solid catalysts. Chem. Sci. 5: 2979–3007.

Gascon, J., A. Corma, F. Kapteijn and F.X. Llabrés. 2014. Metal organic framework catalysis: Quo vadis. ACS Catal. 4: 361–378.

Gawande, M.B., R. Zboril, V. Malgrasb and Y. Yamauchi. 2015. Integrated nanocatalysts: a unique class of heterogeneous catalysts. J. Mater. Chem. A 3: 8241–8245.

Gianotti, E., U. Díaz, A. Veltya and A. Corma. 2013. Designing bifunctional acid-base mesoporous hybrid catalysts for cascade reactions. Catal. Sci. Technol. 3: 2677–2688.

Greaves, T.L. and C.J. Drummond. 2008. Protic ionic liquids: properties and applications. Chem. Rev. 108: 206–237.

Gua, Y. and G. Li. 2009. Ionic liquids-based catalysis with solids: state of the art. Adv. Synth. Catal. 351: 817–847.

Hoelderic, W.F. 2000. Environmentally begin manufacturing of fine and intermediate chemicals. Catal. Today 62: 115–130.

Huh, S., H.T. Chen, J.W. Wiench, M. Pruski and V.S.Y. Lin. 2005. Cooperative catalysis by general acid and base bifunctionalized besoporous silica nanospheres. Angew. Chem. Int. Ed. 44: 1826–1830.

Huybrechts, W., J. Lauwaert, A. De Vylder, M. Mertens, G. Mali, J.W. Thybaut, P. Van Der Voort and P. Cool. 2017. Synthesis of L-serine modified benzene bridged periodic mesoporous organosilica and its catalytic performance towards aldol condensations. Microporous and Mesoporous Mater. 251: 1–8.

Katz, A. and M.E. Davis. 2000. Molecular imprinting of bulk, microporous silica. Nature 403: 286–289.

Kaur, P., J.T. Hupp and S.T. Nguyen. 2011. Porous organic polymers in catalysis: opportunities and challenges. ACS Catal. 1: 819–835.

Karimi, B., D. Elhamifar, J.H. Clark and A.J. Hunt. 2010. Ordered mesoporous organosilica with ionic-liquid framework: An efficient and reusable support for the palladium-catalyzed Suzuki–Miyaura coupling reaction. Chem. Eur. J. 16: 8047–8053.

Karimi, B., D. Elhamifar, J.H. Clark and A.J. Hunt. 2011. Palladium containing periodic mesoporous organosilica with imidazolium framework (Pd@PMO-IL): an efficient and recyclable catalyst for the aerobic oxidation of alcohols. Org. Biomol. Chem. 9: 7420–7426.

Karimi, B., D. Elhamifar, O. Yari, M. Khorasani, H. Vali, J.H. Clark and A.J. Hunt. 2012. Synthesis and characterization of alkyl imidazolium based periodic mesoporous organosilicas: A versatile host for the immobilization of perruthenate (RuO_4^-) in the aerobic oxidation of alcohols. Chem. Eur. J. 18: 13520–13530.

Kuschel, A., M. Drescher, T. Kuschel and S. Polarz. 2010. Bifunctional mesoporous organosilica materials and their application in catalysis: cooperative effects or not? Chem. Mater. 22: 1472–1482.

Lee, J., O.K. Farha, J. Roberts, K.A. Scheidt, S.T. Nguyen and J.T. Hupp. 2009. Metal-organic framework materials as catalysts. Chem. Soc. Rev. 38: 1450–1459.

Lei, Z., Y.M. Koo, B. Chen and D.R. MacFarlane. 2017. Introduction: ionic liquids. Chem. Rev. 117: 6633–6635.

Lewis, M. and R. Glaser. 2003. Synergism of catalysis and reaction center rehybridization. A novel mode of catalysis in the hydrolysis of carbon dioxide. J. Phys. Chem. A 107: 6814–6818.

Liao, H., Y. Chou, Y. Wang, H. Zhang, T. Cheng and G. Liu. 2017. Multi–step organic transformations over base–rhodium/diamine–bifunctionalized mesostructured silica nanoparticles. ChemCatChem. 9: 3197–3202.

López-Sanz, J., E. Pérez-Mayoral, E. Soriano, M. Sturm, R.M. Martín-Aranda, A.J. López-Peinado and J. Cejka. 2012. New inorganic-organic hybrid materials based on SBA-15 molecular sieves involved in the quinolines synthesis. Catal. Today 187: 97–103.

Luz, I., F.X. Llabres i Xamena and A. Corma. 2010. Bridging homogeneous and heterogeneous catalysis with MOFs: "Click" reactions with Cu-MOF catalysts. A. J. Catal. 276: 134–140.

Martínez, C. and A. Corma. 2011. Inorganic molecular sieves: preparation, modification and industrial application in catalytic processes. Coord. Chem. Rev. 255: 1558–1580.

Margelefsky, E.L., V. Dufaud, R.K. Zeidan and M.E. Davis. 2007. Organized surface functional groups: Cooperative catalysis via thiol/sulfonic acid pairing. J. Am. Chem. Soc. 129: 13691–13697.

Margelefsky, E.L., R.K. Zeidanb and M.E. Davis. 2008. Cooperative catalysis by silica-supported organic functional groups. Chem. Soc. Rev. 37: 1118–1126.

Merino, E., E. Verde-Sesto, E.M. Maya, M. Iglesias, F. Sánchez and A. Corma. 2013. Synthesis of structured porous polymers with acid and basic sites and their catalytic application in cascade-type reactions. Chem. Mater. 25: 981–988.

Modak, A., J. Mondal, M. Sasidharanb and A. Bhaumik. 2011. Triazine functionalized ordered mesoporous polymer: a novel solid support for Pd-mediated C–C cross-coupling reactions in water. Green Chem. 13: 1317–1331.

Mondal, J., A. Modak, M. Nandi, H. Uyama and A. Bhaumik. 2012. Triazine functionalized ordered mesoporous organosilica as a novel organocatalyst for the facile one-pot synthesis of 2-amino-4H-chromenes under solvent-free conditions. RSC Adv. 2: 11306–11317.

Motokura, K., M. Tada and Y. Iwasawa. 2008. Acid-base bifunctional catalytic surfaces for nucleophilic addition reactions. Chem. Asian J. 3: 1230–1236.

Nasab, M.J. and A.R. Kiasat. 2016. Designing bifunctional acid-base mesoporous organosilica nanocomposite and its application in green synthesis of 4H-chromen-4-yl phosphonate derivatives under ultrasonic irradiation. Microporous Mesoporous Mater. 223: 10–17.

Olejniczak, M. and M. Ziolek. 2014. Comparative study of Zr, Nb, Mo containing SBA-15 grafted with amino-organosilanes. Microporous Mesoporous Mater. 196: 243–253.

Pagliaro, M. 2009. Silica-based materials for advance applications. Royal Society of Chemistry, Cambridge.

Pan, Y., B. Yuan, Y. Li and D. He. 2010. Multifunctional catalysis by Pd@MIL-101: one-step synthesis of methyl isobutyl ketone over palladium nanoparticles deposited on a metal–organic framework. Chem. Commun. 46: 2280–2282.

Park, J., J.R. Li, Y.P. Chen, J. Yu, A.A. Yakovenko, Z.U. Wang, L.B. Sun, P.B. Balbuena and H.C. Zhou. 2012. A versatile metal–organic framework for carbon dioxide capture and cooperative catalysis. Chem. Commun. 48: 9995–9997.

Parlett, C.M.A., M.A. Isaacs, S.K. Beaumont, L.M. Bingham, N.S. Hondow, K. Wilson and A.F. Lee. 2016. Spatially orthogonal chemical functionalization of a hierarchical pore network for catalytic cascade reactions. Nat. Mater. 15: 178–182.

Pérez-Mayoral, E., V. Calvino-Casilda, M. Godino and R.M. Martín-Aranda. 2014. Chapter 15: Porous catalytic systems in the synthesis of bioactive heterocycles and related compounds. pp. 377–408. *In*: Brahmachari, G. (ed.). Green Synthetic Approaches for Biologically Relevant Heterocycles. Elsevier, Amsterdam, Netherlands.

Pérez-Mayoral, E., E. Soriano, R.M. Martín-Aranda and F.J. Maldonado-Hódar. 2015. Mesoporous catalytic materials and fine chemistry. *In*: Aliofkazraei, M. (ed.). Comprehensive Guide for Mesoporous Materials, Volume 1: Synthesis and Characterization. Nova Science Publishers, Inc., New York, USA.

Pérez-Mayoral, E., E. Soriano, V. Calvino-Casilda, M.L. Rojas-Cervantes and R.M. Martín-Aranda. 2017. Silica-based nanocatalysts in the C-C and C-heteroatom bond forming cascade reactions for the synthesis of biologically active heterocyclic scaffolds. Catal. Today 285: 65–88.

Petri, M.P. 2004. Activation of carbonyl compounds by double hydrogen bonding: An emerging tool in asymmetric catalysis. Angew. Chem. Int. Ed. 43: 2062–2064.

Pintado-Sierra, M., A. Rasero-Almansa, A. Corma, M. Iglesias and F. Sanchez. 2013. Bifunctional iridium-(2-aminoterephthalate)–Zr-MOF chemoselective catalyst for the synthesis of secondary amines by one-pot three-step cascade reaction. J. Catal. 299: 137–145.

Sankaranaravanapillai, S. and W.R. Thiel. 2011. Bifunctional acid-base cooperativity in heterogeneous catalytic reactions: Advances in Silica supported organic functional groups. ChemCatChem. 3: 278–287.

Seki, T., J.D. Grundwaldt, N. van Vegten and A. Baiker. 2008. Palladium supported on an acidic resin: A unique bifunctional catalyst for the continuous catalytic hydrogenation of organic compounds in supercritical carbon dioxide. Adv. Synth. Catal. 350: 691–705.

Sharma, K.K. and T. Asefa. 2007. Efficient bifunctional nanocatalysts by simple postgrafting of spatially isolated catalytic groups on mesoporous materials. Angew. Chem. Int. Ed. 46: 2879–2882.

Sharma, K.K., A. Anan, R.P. Buckley, W. Ouellette and T. Asefa. 2008a. Toward efficient nanoporous catalysts: controlling site-isolation and concentration of grafted catalytic sites on nanoporous materials with solvents and colorimetric elucidation of their site-isolation. J. Am. Chem. Soc. 130: 218–228.

Sharma, K.K., R.P. Buckley and T. Asefa. 2008b. Optimizing acid-base bifunctional mesoporous catalysts for the Henry reaction: Effects of the surface density and site isolation of functional groups. Langmuir 24: 14306–14320.

Shylesh, S., A. Wagener, A. Seifert, S. Ernst and W.R. Thiel. 2010. Mesoporous organosilicas with acidic frameworks and basic sites in the pores: an approach to cooperative catalytic reactions. Angew. Chem. Int. Ed. 49: 184–187.

Sinija, P.S. and K. Sreekumar. 2015. Facile synthesis of pyranopyrazoles and 3,4-dihydropyrimidin-2-(1*H*)-ones by a Ti-grafted polyamidoamine dendritic silica hybrid catalyst *via* a dual activation route. RSC Adv. 5: 101776–101788.

Smuszkiewicz, A., E. Pérez-Mayoral, E. Soriano, I. Sobczak, M. Ziolek, R.M. Martín-Aranda and A.J. López Peinado. 2013. Bifunctional mesoporous MCF materials as catalysts in the Friedländer condensation. Catal. Today 218: 70–75.

Smuszkiewicz, A., J. López-Sanz, E. Pérez-Mayoral, E. Soriano, I. Sobczak, M. Ziolek and A.J. López Peinado. 2013. Amino-grafted mesoporous materials based on MCF structure involved in the quinoline synthesis. Mechanistic insights. J. Mol. Catal. A: Chem. 378: 38–46.

Smuszkiewicz, A., J. Lopez-Sanz, I. Sobczak, M. Ziolek, R.M. Martín-Aranda, E. Soriano and E. Pérez-Mayoral. 2016. Mesoporous niobiosilicate NbMCF modified with alkali metals in the synthesis of chromene derivatives. Catal. Today 277: 133–142.

Sobczak, I., M. Ziolek, E. Pérez-Mayoral, D. Blasco-Jiménez, A.J. López-Peinado and R.M. Martín-Aranda. 2012. Efficient isomerization of safrole by amino-grafted MCM-41 materials as basic catalysts. Catal. Today 179: 159–163.

Srirambalaji, R., S. Hong, R. Natarajan, M. Yoon, R. Hota, Y. Kim, Y.H. Ko and K. Kim. 2012. Tandem catalysis with a bifunctional site-isolated Lewis acid–Brønsted base metal–organic framework, NH₂-MIL-101(Al). Chem. Commun. 48: 11650–11652.

Stawicka, K., I. Sobczak, M. Trejda, B. Sulikowski and M. Ziolek. 2012. Organosilanes affecting the structure and formation of mesoporous cellular foams. Microporous Mesoporous Mater. 155: 143–152.

Stein, A., B.J. Melde and R.C. Schroden. 2000. Hybrid inorganic-organic mesoporous silicates— Nanoscopic reactors coming of age. Adv. Mater. 12: 1403–1419.

Song, F., C. Wang and W. Lin. 2011. A chiral metal–organic framework for sequential asymmetric catalysis. Chem. Commun. 47: 8256–8258.

Surmiak, S.K., C. Doerenkamp, P. Selter, M. Peterlechner, A.H. Schfer, H. Eckert and A. Studer. 2017. Palladium nanoparticle loaded bifunctional silica hybrid material: preparation and applications as catalyst in hydrogenation reactions. Chem. Eur. J. 23: 6019–6028.

Tang, Y., S. Miao, B.H. Shanks and X. Zheng. 2010. Bifunctional mesoporous organic–inorganic hybrid silica for combined one-step hydrogenation/esterification. App. Catal. A: Gen. 375: 310–317.

Tantillo, D.J. and K.N. Houk. 1999. Fidelity in hapten design: how analogous are phosphonate haptens to the transition states for alkaline hydrolyses of aryl esters. J. Org. Chem. 64: 3066–3076.

Thomas, A. 2010. Functional materials: from hard to soft porous frameworks. Angew. Chem. Int. Ed. 9: 8328–8344.

Thomas, J.M. 2009. Handbook of heterogeneous catalysis. Angew. Chem. Int. Ed. 48: 3390–3391.

Thomas, J.M. 2010. The advantages of exploring the interface between heterogeneous and homogeneous catalysis. ChemCatChem. 2: 127–132.

Vekariya, R.L. 2017. A review of ionic liquids: applications towards catalytic organic transformations. J. Mol. Liq. 227: 44–60.

Vermoortele, F., R. Ameloot, A. Vimont, C. Serre and D.E. De Vos. 2011. An amino-modified Zr-terephthalate metal–organic framework as an acid–base catalyst for cross-aldol condensation. Chem. Commun. 47: 1521–1523.

Vlatakis, G., L.I. Anderson, R. Muèller and K. Mosbach. 1993. Drug assay using antibody mimics made by molecular imprinting. Nature 361: 645–647.

Wang, Y., Z. Rong, Y. Wang, T. Wang, Q. Du, Y. Wang and J. Qu. 2017. Graphene-based metal/acid bifunctional catalyst for the conversion of levulinic acid to γ-valerolactone. ACS Sustain. Chem. Eng. 5: 1538–1548.

Welton, T. 2004. Ionic liquids in catalysis. Coord. Chem. Rev. 248: 2459–2477.

Whitcombe, M.J., M.E. Rodríguez, P. Villar and E. Vulfson. 1995. A new method for the introduction of recognition site functionality into polymers prepared by molecular imprinting. J. Am. Chem. Soc. 117: 7105–7111.

Xie, Y., K.K. Sharma, A. Anan, G. Wang, A.V. Biradar and T. Asefa. 2009. Efficient solid-base catalysts for aldol reaction by optimizing the density and type of organoamine groups on nanoporous silica. J. Catal. 265: 131–140.

Xuereb, D.J, J. Dzierzak and R. Raja. 2012. From zeozymes to bio-inspired heterogeneous solids: Evolution of design strategies for sustainable catalysis. Catal. Today 198: 19–34.

Yong-Lee, J., O.K. Farha, J. Roberts, K.A. Scheidt, S. Binh, T. Nguyen and j.T. Hupp. 2009. Metal–organic framework materials as catalysts. Chem. Soc. Rev. 38: 1450–1459.

Zeng, H.C. 2013. Integrated nanocatalysts. Acc. Chem. Res. 46: 226–235.

Zhan, C.G., D.W. Landry and R.L. Ornstein. 2000. Reaction pathways and energy barriers for alkaline hydrolysis of carboxylic acid esters in water studied by a hybrid supermolecule-polarizable continuum approach. J. Am. Chem. Soc. 122: 2621–2627.

Zhan, C.G. and D.W. Landry. 2001. Theoretical studies of competing reaction pathways and energy barriers for alkaline ester hydrolysis of cocaine. J. Phys. Chem. A 105: 1296–1301.

Zhang, F., J. Huangyong, X. Li, X. Wu and H. Li. 2014. Amine-functionalized GO as an active and reusable acid–base bifunctional catalyst for one-pot cascade reactions. ACS Catal. 4: 394–401.

Zhang, Y. and S.N. Riduan. 2012. Functional porous organic polymers for heterogeneous catalysis. Chem. Soc. Rev. 41: 2083–2094.

Zhao, H., N. Yu, J. Wang, D. Zhuang, Y. Ding, R. Tan and D. Yin. 2009. Preparation and catalytic activity of periodic mesoporous organosilica incorporating Lewis acidic chloroindate(III) ionic liquid moieties. Microporous Mesoporous Mater. 122: 240–246.

Zhao, Y., R. Jin, Y. Chou, Y. Li, J. Lin and G. Liu. 2017. Asymmetric transfer hydrogenation– Sonogashira coupling one-pot enantioselective tandem reaction catalysed by Pd(0)–Ru(III)/ diaminebifunctionalized periodic mesoporous organosilica. RSC Adv. 7: 22592–22598.

Ziolek, M. and I. Sobczak. 2017. The role of niobium component in heterogeneous catalysts. Catal. Today 285: 211–225.

Chapter 3

State-of-the-Art in Nanocatalysts for the Transformation of Glycerol into High Added Value Products

Vanesa Calvino Casilda* and Eugenio Muñoz Camacho

INTRODUCTION

Different industries (petrochemical, pharmaceuticals, refinery, etc.) are progressively using metal-nanoparticle-catalysed reactions, for instance hydrogenations, oxidations, cycloadditions and so on (Bert and Van de Voorde 2017). The use of heterogeneous catalysts using nanoparticles of different compounds has been used in industries progressively due to the great reactivity of the nanospecies involved (Philippot and Sherp 2013; Ricciardi et al. 2015). Atom clusters at a scale of nanometers (1–50 nm) form metallic nanoparticles that possess properties between molecular structures and bulk metals. This quality provides them unique physical and chemical properties that are very valuable in the field of catalysis. The benefits of using heterogeneous catalysis are very well known, especially in the field of Green Chemistry (Sharma and Mudhoo 2011). The nanocatalysts are striking candidates to be used as catalysts due to their great surface/volume ratio, high activity and long life-time (Astruc 2008). The nanocatalysts' structure is very complex; their activity (conversion and selectivity) is influenced not only by the type of nanometal and support but also by their composition and the shape and size of their components. Different parameters size, shape and dispersion of their components or their electronic configuration, among others can be adjusted in order to obtain a very active and selective nanocatalyst. Nanocatalysis can be considered as the link between heterogeneous and homogeneous catalysis since it combines a great reactivity and selectivity of the homogeneous catalysis and the easy separation and reutilization of the solid catalyst, also complying with the goals of the green catalysis (Sharma and Mudhoo 2011).

Departamento de Ingeniería Eléctrica, Electrónica, Control, Telemática y Química Aplicada a la Ingeniería, ETS de Ingenieros Industriales, c/Juan del Rosal, 12, 28040-Madrid, Spain.
* Corresponding author: vcalvino@ieec.uned.es

The current chapter describes the valorization of glycerol to high added value products using nanocatalysts; selective examples of glycerol reactions are reported herein. The transformation of glycerol into chemical products of industrial interest has aroused great attention in the last decades and particularly in the field of nanocatalysis in recent years (Polshettiwar and Asefa 2013). Glycerol is a by-product obtained in the manufacture of biodiesel (10 wt %) by fats transesterification and vegetable oils. Taking into account the economic viability of the biodiesel industries, glycerol needs to be valourized, in both liquid and gas phases, into other products of high added value (Scheme 1). The recent works published in literature seems to indicate that the combination of nanocatalysts and valorization of glycerol is a promising route of research with a great future.

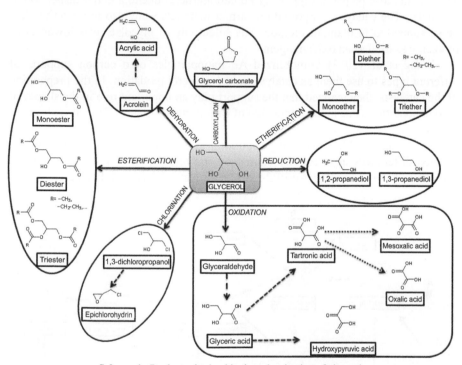

Scheme 1. Products obtained in the valourization of glycerol.

Glycerol oxidation reactions

One of the most important procedures to obtain high added value products derived from glycerol is through its oxidation (Mota et al. 2017). The main problem in glycerol oxidation reactions is the different products formed during the process, therefore getting a selective process that allows researchers to obtain the desired product is still a challenge of great interest.

Carrettin et al. (2002) promoted for the first time the use of Au nanoparticles to catalyse the glycerol oxidation. Since then, this research topic has been deeply studied using Au nanoparticles supported on carbon materials or inorganic supports

(Corma and García 2008). The glycerol oxidation follows a tangled reaction system that yields a variety of products (see Scheme 2) (Katryniok et al. 2011; Kapkowski et al. 2014). The glycerol oxidation is usually carried out in the liquid phase due to the high boiling point of glycerol.

Gold nanoparticles have proved to be really effective in different reactions of organic synthesis and in particular in reactions of transformation of glycerol (Heiz and Landman 2008). Sankar et al. (2009) carried out the oxidation of glycerol to obtain glycolate using metal nanoparticles supported on different materials, carbon materials and titanium dioxide, in an autoclave reactor in presence of hydrogen peroxide. The yields achieved were up to 60%, increasing the yields with time and temperature of reaction. Moreover, the catalysts prepared supporting Au nanoparticles were more active than those prepared supporting Pd nanoparticles due to a better dispersion of Au metal. The catalysts supported on carbon materials were also more active than those supported in titanium dioxide. The selectivity to glycolate was lower when oxygen was used as an oxidant agent.

Wang et al. (2013) immobilized Au nanoparticles onto carbon supports of different types to use them as catalysts in the glycerol oxidation. In the preparation of the catalysts, they realized that the support play an important role in the shapes of gold nanoparticles due to the interaction of nanoparticle-support. They also observed that this fact influenced its catalytic activity in the glycerol oxidation due to the existing structural difference. They exhibited that catalysts with similar activities showed differences in selectivity caused by the orientation of gold nanoparticles and their shape. Thus, the Au/PR24-LHT catalyst showed a better selectivity to

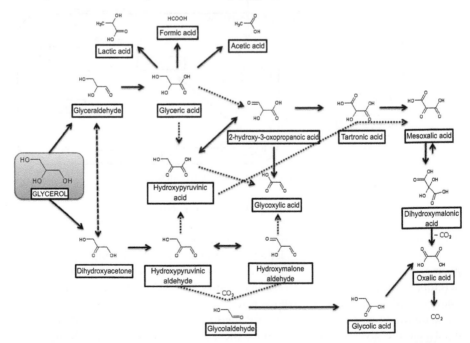

Scheme 2. The complicated reaction system of glycerol oxidation (Kapkowski et al. 2014).

the cleavage of C-C bonds in which gold nanoparticles exhibited a greater reactive surface.

Kapkowski et al. (2014) carried out the liquid phase oxidation of glycerol using SiO_2-, Ni-, Cu-supported Au nanoparticles as a catalyst in a non-concentrated H_2O_2/H_2O phase. Until then, nobody had studied the influence of the polarity of the glycerol-H_2O_2/H_2O-SiO_2 system. The catalytic system proposed in this work was highly selective to acetic acid (C_2 oxygenated glycerol) in diluted liquid phase (100% conversion and 90% yield). When the concentration of glycerol increased in the system, the conversion decreased to 40% and 20% acetic acid yield. However, the addition of acetonitrile enhanced the acetic acid yield up to 40%. The bimetallic catalysts obtained by nano-Au transfer, these are Au/Cu and Au/Ni, were very active in the glycerol oxidation but they were deactivated by the leaching of the active phase Cu or Ni.

Zhao et al. (2014) immobilized Au nanoparticles decorated with different amounts of Pd on carbon to obtain active and selective catalysts in the glycerol oxidation. These catalysts were also strongly resistant to deactivation depending on the coverage of Pd. The activity showed by the bimetallic Pd-on-Au/C catalysts was higher than that corresponding to the monometallic Au/C and Pd/C catalysts, reaching the total glycerol conversion at 80 surface coverages percentage. Above these values of surface coverages, the catalyst can show deactivation maybe due to the active sites that were poisoned by CO or fouled by adsorbates. Pd-on-Au catalysts showed the best yields to glyceric acid for cumulative surface coverages, being a dozen times more active than monometallic ones.

In the case of Hong et al. (2014), they dispersed Cu_2O rhombic and cubic nanocrystals onto graphene oxide supports obtaining catalysts highly active in the oxidation of glycerol into lactic acid. The original process of synthesis in aqueous media consisted of controlling the shape of the Cu_2O nanocrystals, employing as template and surfactant graphene oxide. It seems that thanks to the graphene oxide, Cu_2O nanocrystals are more stable and the catalytic activity enhances considerably. The hybrid system Cu_2O-graphene oxide shows that the catalytic activity depends on the shape of the nanocrystals. At the same time, Mimura et al. (2014) stabilized effectively Au-Pd nanoparticles in an anion-exchange resin employing the ion-exchange method (Mimura et al. 2014). The preparation of the catalytic structure was controlled in order to obtain a very good dispersion of the nanoparticles. The bimetallic catalyst presented higher catalytic activity in the liquid phase glycerol oxidation with O_2 than the monometallic ones. In addition, the bimetallic catalyst showed very good stability using a fixed bed flow reactor, achieving conversions of about 50% and a steady-state selectivity to tartronic acid (30%) and glyceric acid (60%).

Later, Rogers et al. (2015) prepared Au nanoparticles supported on titanium dioxide catalysts to carry out the oxidation of glycerol in the aqueous phase. The obtaining of the Au nanoparticles was performed applying colloidal methods varying the temperature and the solvent ratio of the mixed water/ethanol. It seems that the temperature, together with solvent system effects, influences the particle diameter of the Au nanoparticles. Thus, Au nanoparticles prepared in water at 1°C led to a higher activity than other Au/TiO_2 catalysts reported in the literature at comparable reaction conditions. Schünemann et al. (2015) reported for the first time the plasmonic photocatalysis of glycerol oxidation catalysed by Au and AuCu nanoparticles supported on mesoporous silica. Au and AuCu nanoparticles were supported on different mesoporous silica and its catalytic activity was studied in the presence and absence of visible light. It seemed that the Au nanoparticles dispersed onto monodispersed mesoporous silica spheres were more active than Au nanoparticles supported on ordered mesoporous silica. This fact was due to the Au particle size and the improved interaction of the light in the case of the small silica spheres. The addition of Cu nanoparticles to Au catalysts resulted in an improvement of its catalytic activity due to a synergistic effect of Au-Cu that considerably increased the glycerol conversion and the selectivity to dihydroxyacetone, in comparison with the catalyst without copper.

Yuan et al. (2015) studied the effect of the support on catalysts prepared supporting Au nanoparticles in the base-free water glycerol oxidation. They proposed a sequence of $MgO-Al_2O_3$ supported catalysts with a different molar Mg/Al ratio. The catalysts with the most acidity and the least basicity showed the better activity in the glycerol oxidation, and the better selectivity to dihydroxyacetone. When the surface basicity of the support increased or its acidity decreased, the selectivity to glyceric acid increased, decreasing also the selectivity to dihydroxyacetone. Thus, the authors established that the acid-base properties of the support are key to control the selectivity of Au nanoparticles on these supported catalysts for the glycerol oxidation.

The liquid phase oxidation of glycerol using AuPd nanoparticles supported on various carbon materials and also on titanium dioxide was investigated by Chan-Thaw et al. 2015. Activated carbon and carbon nanofibers functionalized with nitrogen were the supports with the best catalytic activity. The existence of AuPd nanoparticles in a homogeneous alloy was also important for the catalytic performance. The stability of the catalysts was good for pure glycerol but had decreased for non-pure glycerol due to their deactivation caused by the leaching and also the agglomeration of the gold and palladium nanoparticles.

Dimitratos et al. (2016) prepared gold nanoparticles of various particle sizes by the reduction of the corresponding metal salts with sodium borohydride and polyvinylalcohol (PVA) polymer as stabilizers. The Au concentration seemed to have a notable influence on the preparation of the catalysts; for high Au concentration, the particle size obtained was larger. When the PVA amount diminished, the particles obtained were larger, except for the catalyst prepared with a big PVA amount; maybe due to the Au nanoparticles, immobilization over the support was more complicated. The catalysts were proved in the liquid phase glycerol oxidation, relating the particle size and the type of metal with the activity and selectivity of the process. The effect of the support was also studied for titania and carbon supports indicating that high values of TOF were obtained for all the catalysts except for the calcined varieties.

As previously shown, the glycerol oxidation in the liquid-phase has been extensively studied; an alternative to this process is the glycerol oxidation in the gas-phase that also has the advantage of reducing the glycerol viscosity. Thus, Kapkowski et al. (2015) investigated the gas phase glycerol dehydration over Au nanoparticles supported on silica and Cu, Fe, and Ni carriers by the sol-gel method.

They study different parameters to get the best reaction conditions; all the catalysts were selective to hydroxyacetone, with the bimetallic Cu supported Au nanoparticles as the most active catalyst (63% conversion and 70% selectivity).

Furthermore, electrochemical oxidation reactions of glycerol using electrodes modified with metal nanoparticles have also been reported. It is well known that Pd is cheaper than Au and Pt, but it is not as active for the glycerol oxidation in alkaline media as Au and Pt. Due to this, the preparation of Pd alloys using other metals has been tried for enhancing the activity in the glycerol oxidation reactions. Thus, Simoes et al. (2010) prepared nanocatalysts supporting on carbon Pd, Pt and Au nanoparticles and also PdAu and PdNi alloys of different composition to carry out the glycerol electrooxidation in alkaline media. The preparation of catalysts replacing the half of Pd atoms by Ni atoms improved its catalytic activity and reduced its cost. They observed that the reaction mechanism and the products obtained depended on the catalysts used. Thus, Pt/C and Pd/C nanocatalysts showed CO species absorbed on their surface while Au/C and $Pd_{0.3}Au_{0.7}$/C nanocatalysts did not show them maybe because they were not capable of breaking C-C bonds. The methodology proposed here allows for regulating the catalytic activity and selectivity, which is key in the field of fuel cells for high-added value chemicals and energy. Likewise, Lam et al. (2015) prepared a series of nanocatalysts of Pd-Ag and Pd-Au alloys at room temperature through a wet method. They also demonstrated that the preparation of Pt alloys with other metals allowed the researchers to reduce the overpotential for glycerol oxidation reactions in alkaline media (Lam et al. 2016). Thus, applying the same methodology, they prepared nanocatalysts loading PtAg alloys over the carbon black and tested them in the electrooxidation of glycerol in the alkaline medium. They proved that the activity of Pt in the glycerol oxidation was significantly improved when Pt was alloyed with Ag in the PtAg/carbon black electrodes.

Thia et al. (2016) studied that the electrodeposition of Cu species onto carbon-supported gold nanoparticles promoted the electrochemical oxidation of glycerol due to the interaction of gold and copper species. In particular, the interaction of Cu_2O species with Au led to the formation of Au^+ species that enhanced the selectivity to tartronate and glycerate products. In addition, the deposition of Cu or Cu_2O thick layers over Au/C avoided the dispersion of the intermediates from the surface of the catalyst when the oxidation ended. The optimal deposition time was of 90 min; when the deposition time extended beyond 120 min the selectivity decreased considerably. Also, González-Cobos et al. (2016) studied the electrooxidation of glycerol in alkaline media, in this case incorporating Bi (around 10 at%) onto PtPd nanocatalysts using a carbon support. It seems that the presence of bismuth avoided the carbon-carbon bond breaking, enhancing the selectivity to C3 compounds of great interest because of their high added value.

In the research conducted by Safavi et al. (2015), Pd nanoparticles were loaded on a hydroxyapatite/multiwalled carbon nanotube composite to use it as an electrocatalyst in the oxidation of aliphatic alcohols such as ethylene glycol, methanol or ethanol in alkaline media. They discovered that the use of this composite led to an electrocatalyst more efficient than would be created if hydroxyapatite was not present in the structure of the catalyst. It seems that the presence of hydroxyl groups on the surface promotes the oxidation of the aliphatic alcohols above the oxidation of the palladium nanoparticles even when large amounts of ethylene glycol were used. The presence of hydroxyapatite in the structure of the electrocatalyst favours the loading of Pd on the composite and avoids the aggregation of palladium nanoparticles.

Almeida et al. (2017) studied glycerol electrooxidation using a Pd nanocatalyst supported on mesoporous silica (Pd/SiO$_2$ NPs). The activity and stability study of the nanocatalyst prepared was carried out in alkaline media. The catalyst showed a structure with pores in the shape of a fountain pen and with a cage arrangement. They found that the well-ordered silica support played an important role trapping the reagents within the porous structure, increasing the probability that they collide with the Pd sites and resulting in more active catalysts than the commercial Pd/C. They also tested to see whether the nanocatalysts were stable after their use in full cycles and that the rapid removing of CO from the surface of the nanocatalyst improved its activity in the glycerol electrooxidation reaction.

Catalytic hydrogenolysis of glycerol

The glycerol hydrogenolysis reaction has aroused great commercial interest in obtaining glycols; for example, 1,3-propanediol (1,3-PDO), 1,2-propanediol (1,2-PDO), ethylene glycol (EG), 1-propanol (1-PrOH) and 2-propanol (2-PrOH) are products of high commercial value (Scheme 3) (Mota et al. 2017).

Huang et al. (2008) prepared Cu/SiO$_2$ catalysts by dispersing cooper nanoparticles onto silica using the precipitation-gel technique. The samples after calcination

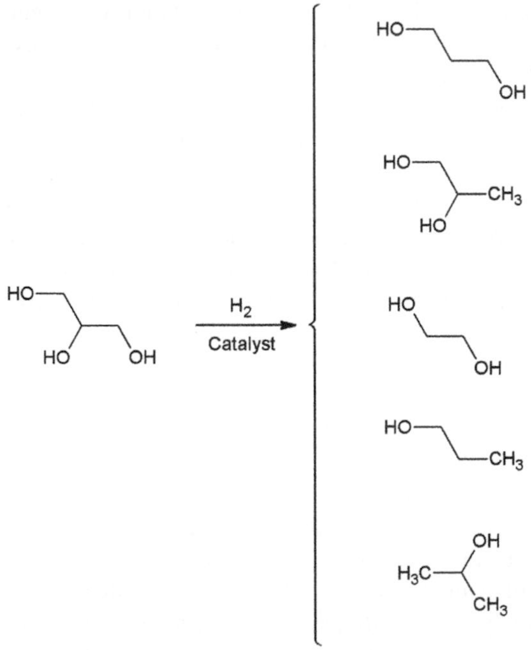

Scheme 3. Selective hydrogenolysis of glycerol to obtain glycols: from top to bottom, 1,3-PDO, 1,2-PDO, EG, 1-PrOH and 2-PrOH.

showed that CuO and Cu^{2+} species very well dispersed, with a strong interaction metal-support. These catalysts were compared to those prepared by the conventional impregnation method, showing in this case bulk and dispersed CuO species on the catalyst surface. Cu^0 and Cu^+ species coexisted on the catalyst surface when the samples were reduced in the presence of hydrogen in case of those samples prepared by the precipitation-gel method; however, the samples prepared by the impregnation method showed only Cu^0 species. Cu/SiO_2 catalysts prepared by both methods, impregnation and precipitation-gel methods, were very selective to 1,2-PDO in the glycerol hydrogenolysis. Catalysts prepared by the precipitation-gel method were more active and stable than the catalysts prepared by the impregnation method.

The hydrogenolysis of glycerol to 1,3-PDO in the aqueous phase in a batch reactor was performed by Nakagawa et al. (2010) using rhenium-oxide-modified iridium nanoparticles supported on silica catalysts ($Ir-ReO_x/SiO_2$). Under the optimized reaction conditions, the conversion of glycerol was 81% and the yield of 1,3-PDO was of 38% (36 h). These results were much better than those reported in the case of other systems such as when the nanoparticles were modified with rhodium, even though iridium has been less studied in hydrogenolysis reactions than other noble metals.

Li et al. (2011) reported the preparation of RuFe bimetallic nanoparticles supported on carbon nanotubes (RuFe/CNT) to catalyse selectively the hydrogenolysis of aqueous solutions of glycerol (20 wt %) in order to obtain 1,2-PDO and EG. It seems that monometallic catalysts such as Ru/CNT were very active for carbon-carbon cleavage, while also leading to the production of methane. However, the bimetallic RuFe/CNT catalyst was more active for carbon-oxygen cleavage, leading to the production of glycols with a selectivity of over 75% and full conversion. In addition, the reaction conditions employed were mild and the catalysts also showed a strong structure, being able to be reused successfully. It seems that depending on the amount of Fe, these species can form RuFe alloys or take the form of iron oxides such as FeO. In the last case, its presence is decisive for the stability of the bimetallic catalyst in successive runs; however, the presence of iron oxides in excess can block the catalytic surface, leading to a reduction of its activity. On the other hand, the presence of RuFe alloys caused a synergistic effect and their interaction with the iron oxides present on the carbon nanotube surfaces led to a very good performance. It seems that the RuFe nanoparticles promote the breaking of carbon-oxigen bond.

At the same time, Wu et al. (2011) supported Cu-Ru nanoparticles on carbon nanotubes to carry out the glycerol hydrogenolysis to 1,2-PDO. The characterization results showed the ruthenium formed clusters that were deposited on the outer surface of copper nanoparticles. The presence of ruthenium reduced the particle size of the copper and boosted the ability of Cu metal for the activation of hydrogen. The ruthenium clusters were not active in the glycerol hydrogenolysis but they were able to activate and produce active hydrogen that was moved to the surface of copper through the hydrogen spillover. Cu-Ru catalysts were so much more active and selective to 1,2-PDO than Cu metal due to the effect of the hydrogen spillover that was promoted by the presence of ruthenium clusters. Ruthenium and cooper-ruthenium did not catalyse the glycerol hydrogenolysis but promoted the hydrogen spillover.

Sun et al. (2011) prepared for the first time Pt@CNT/SiO$_2$ catalysts by controlling the mesoporous silica coating. The catalysts were very stable at high temperatures since the Pt nanoparticles were separated from each other by the framework of the hollow porous silica shell. This avoided the agglomeration of the metal nanoparticles or the sintering at high temperatures. The bimetallic CuRu@CNT/SiO$_2$ catalyst was also prepared following the same procedure. Both catalysts were very active in the hydrogenolysis of glycerol at 210°C; the CuRu@CNT/SiO$_2$ catalyst was reused up to four times without practically losing any activity.

Then, Jin et al. (2013) carried out the hydrogenolysis of glycerol and other biopolyols using bimetallic Cu nanocatalysts supported on reduced graphene oxide (Cu/rGO) (Jin et al. 2013). The catalysts were prepared by controlling the surface facets of the copper nanocrystals (Cu{111}) and incorporating Pd nanoparticles (CuPd/rGO) to improve the activity and also the stability of the samples due to the presence of a tandem synergistic effect where the hydrogen produced from polyols was consumed by consecutive hydrogenolysis of the glycerol.

Bimetallic Ru-Cu supported on TiO$_2$ catalysts were tested in the glycerol hydrogenolysis by Salazar et al. (2014). The incorporation of copper triggered a better selectivity to 1,2-PDO, mainly in the catalysts prepared with a 1:1 cooper/ruthenium mass ratio. These catalysts also displayed better activity due to the Ru-Cu interaction. Copper also prevented the deactivation of the catalyst containing ruthenium as occurred in the monometallic Ru catalyst. The best catalyst was 2.5Ru–2.5Cu/TiO$_2$, which led to the highest catalytic activity and selectivity.

Recently, Wang et al. (2016) carried out the hydrogenation of glycerol over Pt/metal oxide nanocomposites. They developed a new method to prepare Pt/metal oxide nanocomposites in a single step by regulating the Pt and Zinc precursor molar ratios and the amount of sodium bicarbonate in the absence of surfactants. This methodology allowed the researchers to control the size of the platinum nanoparticles and the platinum loading amount. The catalysts thus prepared showed very good activity; they tested different supports to load Pt nanoparticles, each of them showing a different catalytic behaviour in the hydrogenation of glycerol.

The hydrogenolysis of glycerol to 1,3-PDO over Pt/WO$_3$/ZrO$_2$ nanocatalysts was carried out by Zhou et al. (2016). In the catalysts prepared, WO$_3$ species were very well dispersed on a nano-ZrO$_2$ precursor and then Pt nanoparticles were uniformly dispersed. They demonstrated that the hydrogen spillover plays a significant role in the hydrogenolysis of glycerol since H/Pt and the conversion of glycerol keep a linear relationship. In particular, the role of both Pt(111) and WO$_3$ in the process of the spillover was deeply studied, finding a strong relationship among the Pt(111) terraces amount and the spillover capacity. The hydrogen spillover seemed to have an influence on the reaction rate but not on the strength and the total number of the acid centres.

Li et al. (2016) carried out the hydrogenolysis of glycerol in the aqueous phase using ZnPd/ZnO@Al$_2$O$_3$ catalysts. The catalysts were formed by an amorphous alumina layer and a crystalline satellite zinc oxide that confines the zinc and palladium nanoparticles. The catalysts prepared were very active and selective to 1,2-PDO (over 90% selectivity) and could be reused successfully up to five times even under extreme reaction conditions. The alumina layer prevented the erosion of zinc oxide in the presence of water and the formation of ZnPd aggregates.

Cu-ZnO nanocatalysts were used in the glycerol hydrogenolysis to obtain 1,2-PDO (Du et al. 2016). Du et al. discovered that the loss of the catalytic activity of these catalysts was due to their structure and also that the morphology of the catalysts changed after several cycles of the reaction; first into a spherical shape, then into layers in the form of lamellae and lastly, into a rod-like structure. This fact caused the formation of Cu-ZnO aggregates, the lack of uniformity in the copper and zinc distribution and the loss of copper on the catalyst surface.

Pandhare et al. (2016) synthesized Cu/MgO catalysts to perform the gas phase glycerol hydrogenolysis to obtain 1,2-PDO. They studied the effect of different parameters on the glycerol conversion and the selectivity to 1,2-PDO; the parameters studied were pressure, temperature, metal loading and the weight hourly space velocity. Metal loadings higher than 10 wt % caused a diminishing in the 1,2-PDO yield as a result of the formation of agglomerates and a decrease of the metallic surface area. It seemed that Cu/MgO catalysts endorsed the hydrogenolysis of carbon-oxygen bonds while they blocked the carbon-carbon bond cleavage that increased the selectivity to other by-products. When 10 wt % of the Cu/MgO catalyst was used, yields of up to 95.5% to 1,2-PDO were reached with full glycerol conversion. The presence of acid-base pairs and a synergism between the support and the Cu nanoparticles, as well as the small size of particle seemed to be the reason for such great performe (total glycerol conversion and yield values of near 96 % of 1,2-PDO over the 10 wt % Cu/MgO catalyst). The catalysts were also very stable, leading to longer reaction times.

More recently, Edake et al. (2017) carried out the hydrogenolysis of glycerol in the gas phase to obtain 1,3-PDO and 1,2-PDO using $Pt/WO_3/Al_2O_3$ catalysts deposited in a fluidized bed working at high temperatures no lower than 240°C and atmospheric pressure. The use of fluidized beds allows the researchers to reach heat transfer rates high enough to vaporise the glycerol and diminish the glycerol combustion or its thermal decomposition. Beside 1,3-PDO, other co-products such as 1,2-PDO, 1-PrOH and 2-PrOH appeared. The higher temperatures led to an increased production of the by-products. The authors proposed a mechanism for the reaction; first, glycerol will convert to acrolein by dehydration, then, it may be rehydrated to obtain 3-hydroxypropanal and finally, hydrogenated to obtain 1,3-PDO. The selectivity to 1,3-PDO could increase if the reaction that produces simultaneously 1,2-PDO could be avoided.

Dehydration of glycerol

The dehydration of glycerol is a renewable route used in the production of acrolein, an important intermediate used in the synthesis of high added value products such as acrylic acid and methionine (Rahmat et al. 2010; Katryniok et al. 2009). This reaction is usually carried out in the gas or liquid phase using different acid catalysts; for example, mineral acids, metal oxides, heteropoly acids, zeolites, etc. (Katryniok et al. 2013).

Talebian-Kiakalaieh et al. (2016) carried out the dehydration of glycerol in the gas-phase using gamma-alumina nanoparticles as a catalyst. The reaction was studied using varying different reaction parameters such as reaction temperature, from 280 to 320°C, the concentration of the glycerol feed (from 0,5 to 20 weight%) and also the catalyst loading (from 0,1 to 1 g). It seems that the use of gamma-alumina nanoparticles as support enhanced the selectivity to acrolein when they compared these results with those obtained using ZrO_2 supported on silicotungstic acid. The results showed that gamma-alumina led to an acrolein yield of 30,2% compared to 24,8% for ZrO_2 obtained under the same reaction conditions (300°C, 10wt% of glycerol concentration and 0,5 g of catalyst). The reasons were the very large surface area and mesoporosity of gamma-alumina nanoparticles, the extremely high thermal stability and the fact that the acidity was tuned more precisely, getting a better dispersion of the active phase. The best results were obtained for the catalyst prepared by impregnation loading 30 wt % of HSiW on 20 wt % of alumina nanoparticles (30HZ-20A catalyst), reaching values of selectivity to acrolein of 88,5% at conversion values of 97% (300°C of temperature, 3 h of reaction time, 0,5 g of catalyst, 10 wt % glycerol feed concentration). Moreover, this catalyst showed great stability even after longer reaction times, that is, more than 40 hours of reaction time.

Glycerol steam reforming

Glycerol steam reforming is carried out using Pt, Ru, Co, Ni, Rh and Co catalysts for the production of syngas for sustainable technologies based on hydrogen and for fuel cells (Gandia et al. 2013; Dou et al. 2014). The optimal temperature to carry out the reaction has been established at 550–750°C; however, it is difficult for the catalyst to maintain its hydrothermal stability at this temperature. Therefore, many attempts have been made to strength the metal-support interactions and to achieve glycerol reforming at lower temperatures than those used in conventional gasification while the reaction selectivity and the stability of the catalysts are preserved.

Nanocatalysts' morphology can be tuned with the purpose of prepared heterogeneous catalysts with a very high performance by selectivity displaying the facets with high activity to the reagents. In this way, it is possible to generate numerous active sites of different types in a solid catalyst, taking into account the location (e.g., edge, corners) and the surface energy. Huang et al. (2015) supported iridium nanoparticles on rod-shaped $La_2O_2CO_3$ in order to obtain stable catalysts with very high activity and selectivity in the steam reforming of glycerol and the oxidative steam reforming of glycerol at a high reaction temperature, 650°C. It seems that the very high performance of these catalysts is due to their strong basicity and the medium strength interaction between Ir nanoparticles and the surface facets of the surface of the support, in particular {110} facets. The authors demonstrated that the shape effect of facets present in catalysts prepared supporting Ir nanoparticles onto $La_2O_2CO_3$ nanorods considerably increases the catalytic activity in the steam reforming of glycerol at high temperatures. The best catalytic activity of nanorods catalysts was due to the development of hexagonal $La_2O_2CO_3$ nanorods where {110} surfaces strongly interrelate with Ir. This facet had also significance importance in the catalytic activity allowing a stable production of hydrogen for 100 h without significant loss of activity. Through this methodology, it was possible to avoid the sintering of Ir nanoparticles at high temperatures, the coking of the catalysts and the easy regeneration of the support.

In a later study, Touri et al. (2016) prepared Pt/SiO$_2$ nanocatalysts by the sol-gel technique to obtain hydrogen from steam reforming of glycerol at 300–400°C and at atmospheric pressure using a microchannel reactor (Touri et al. 2016). The catalysts were very active and selective to hydrogen, leading to a low selectivity to carbon monoxide. It seems that the use of a reactor with a smaller volume in which a lower amount of catalyst was loaded led to a lower selectivity to carbon monoxide and a better yield to hydrogen. The same researchers also found that a low flow rate (3 mL/h) and a temperature of 400°C were the best reaction conditions for the reaction.

Senseni et al. (2016) carried out the steam reforming of glycerol using Ni/Al$_2$O$_3$ catalysts varying the reaction conditions. They also studied the abovementioned reaction by using a response surface methodology comparing these results with those obtained experimentally. The catalyst was prepared by the impregnation method and different reaction parameters were adjusted (temperature, feed ratio and gas hourly space velocity), obtaining catalysts more useful for glycerol steam reforming process in industries. The best reaction conditions were: temperature around 672°C, feed ratio of 9 and gas hourly space velocity of around 37 mL/g h, leading to excellent results, conversion values of glycerol of almost 98% and 74,4% selectivity to hydrogen. At the same time, Subramanian et al. (2016) supported different metals onto gamma-alumina to catalyse the production of hydrogen from glycerol in the aqueous phase, following a different reaction pathway from the one followed in the gas phase. The best catalyst was Pt/Al$_2$O$_3$ as it showed the better catalytic activity (18% glycerol conversion and 17% yield to hydrogen) at an optimized reaction conditions (240°C, 42 bar, 10 wt % glycerol feed and $\geq$ 4100 substrate:metal molar ratio) compared to Pd/Al2O3, Rh/Al2O3, and Au/Al2O3 catalysts tested in this study. The next works of the authors will be to study the reaction in continuous flow to investigate more deeply the kinetic of the reaction, the stability of the catalysts and the possibility to reuse them in a less tedious way than in a batch reactor.

Other glycerol transformations

González-Arellano et al. (2014) carried out the synthesis of cyclic acetals by glycerol acetalization, using different aldehydes and ketones, and the synthesis of monoacetylglycerides, diacetylglycerides and triacetylglycerides by glycerol esterification, using levulinic acid (Scheme 4). As catalysts, they used iron oxide nanoparticles supported on aluminosilicate (Fe/Al-SBA-15). These catalysts were highly active and selective, optimizing different reaction parameters. The catalysts could be reused without loss of activity and stability. Also, Kapkowski et al. (2017) carried out the synthesis of Re, Ru, Ir, Rh nanoparticles supported on nano-silica catalysts to be applied in the acetalization of glycerol using two different ketones, acetone and butanone, at room temperature (Scheme 4). Two different products could be obtained, cyclic acetals with five-membered or cyclic acetals with six-membered.

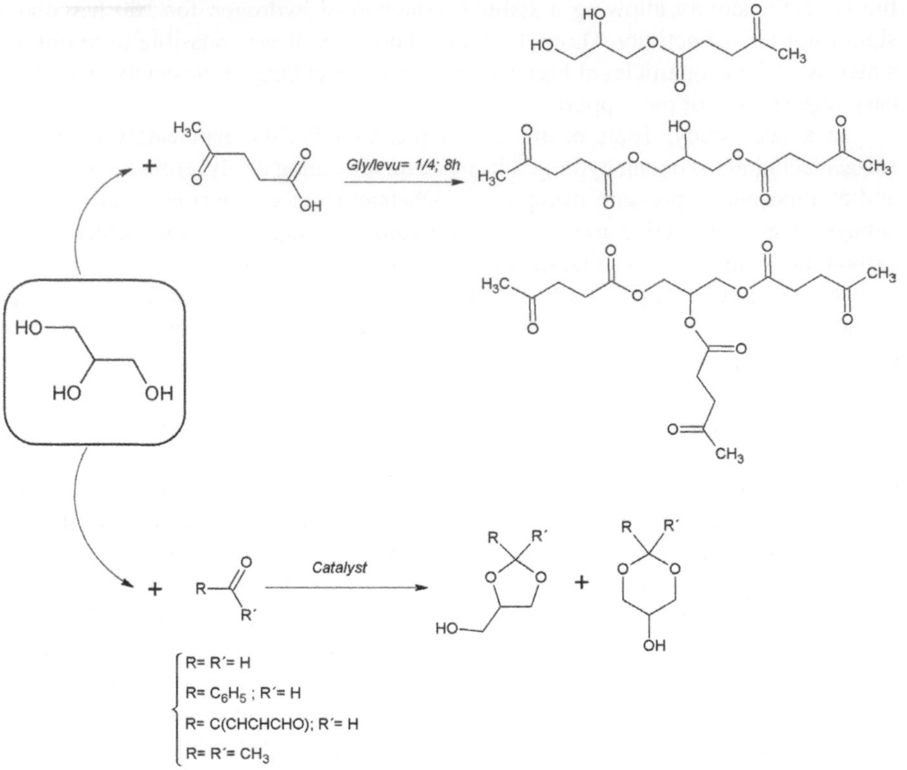

Scheme 4. Esterification of glycerol using levulinic acid and Fe/Al-SBA-15 as catalyst (top) (González-Arellano et al. 2014). Condensation of glycerol with aldehydes or acetone to obtain cyclic acetals (bottom) (Kapkowski et al. 2017).

They prepared a series of catalysts containing one, two or three metal nanoparticles, in order to study the presence of synergy effects. The most selective catalyst to obtain cycic acetals of five-membered seemed to be those containing Re. Kapkowski et al. also demonstrated the significance of the influence of the silica support on the acetalization of glycerol since it provides the acidity necessary to successfully perform the reaction. In addition, the same researchers reported improved data of TON compared with other data described in the literature using other catalysts (Re/nano-SiO$_2$, TON = 1724 mol/mol versus H$_3$PW$_{12}$O$_{40}$, TON = 6,2 mol/mol) even on a large-scale production (30°C, 100% conversion and 98% selectivity to solketal, the amount of catalyst used minimized up to 60% compared with the small-scale system).

Jin et al. (2013) synthesized nanocatalysts based on copper loaded on reduced graphene oxide (Cu/rGO) to be applied in reactions of conversion of biopolyols such as glycerol, sorbitol and xylitol to high added value products in aqueous phase and at room temperature (Jin et al. 2013). It seems that the {111} surface facet of copper was the one that showed the best catalytic activity in reactions of dehydrogenation/hydrogenation. Cu-graphene catalysts were very active in obtaining products such as lactic acid and linear alcohols among others (values of TOF up to 114 mol/g atom Cu/h). They also tested bimetallic catalysts based on the combination of palladium (traces) and cooper over reduced graphene oxide. The bimetallic catalyst (CuPd/rGO) showed a synergism that allowed researchers to produce hydrogen from polyols to be used in hydrogenolysis reactions. The incorporation of palladium into the Cu-graphene system also provided more catalytic stability. The activity, selectivity and stability showed by these catalysts allow them to be applied in different catalytic reactions.

Oberhauser et al. (2016) supported Pt nanoparticles onto Ketjenblack carbon by means of the metal vapour synthesis method to convert glycerol to lactate (95% chemoselectivity during three successive catalytic cycles and the catalytic activity held for 780 h^{-1}). They also changed the hydrogen acceptor, used during the reaction ethylene instead of oxygen to avoid the formation of products derived from the oxidation of glycerol. It seems that the small size of Pt nanoparticles (1,5 nm) and the high partial pressure of ethylene avoided its aggregation and promotes its catalytic activity in the conversion of glycerol to lactate. The role of the ethylene was also to stabilize the catalyst during the conversion of the glycerol to the lactate.

Conclusions and future outlook

The transformation of glycerol into high added value products using nanocatalysts has aroused great interest in the last decade. Monometallic, bimetallic and even trimetallic nanoparticles supported catalysts with particular geometric and electronic properties have been prepared adjusting their morphology; shape and size of their nanoparticles. It has been well studied that the interaction of the nanometal species with the support plays an important role in catalysis, leading to a series of considerable enhancements, not only as far as activity and selectivity are concerned, but also in terms of performance during their whole cycle of lifetime. To finish, It would be very interesting in the future to study the connection between structure/activity/selectivity using *in situ* techniques that led researchers to monitor in real-time the behaviour of the nanocatalyst during its performance in the reaction in order to prepare efficient nanocatalysts with improved properties.

Reference cited

Almeida, T.S.D., K.-E. Guima, R.M. Silveira, G.C. da Silva, M.A.U. Martines and C.A. Martins. 2017. A Pd nanocatalyst supported on multifaceted mesoporous silica with enhanced activity and stability for glycerol electrooxidation. RSC Adv. 7: 12006–12016.

Astruc, D. 2008. Nanoparticles and Catalysis. Wiley-VCH.

Bert, S. and M. Van de Voorde. 2017. Nanotechnology in catalysis: Applications in the Chemical Industry, Energy Development and Environment Protection. Wiley-VCH.

Carrettin, S., P. McMorn, P. Johnston, K. Griffin and G.J. Hutchings. 2002. Selective oxidation of glycerol to glyceric acid using a gold catalyst in aqueous sodium hydroxide. Chem. Commun. 7: 696–697.

Chan-Thaw, C.E., S. Campisi, D. Wang, L. Prati and A. Villa. 2015. Selective oxidation of raw glycerol using supported AuPd nanoparticles. Catalysts 5: 131–144.

Corma, A. and H. Garcia. 2008. Supported gold nanoparticles as catalysts for organic reactions. Chem. Soc. Rev. 37: 2096–2126.

Dimitratos, N., A. Villa, L. Prati, C. Hammond, C.E. Chan-Thaw, J. Cookson and P.T. Bishop. 2016. Effect of the preparation method of supported Au nanoparticles in the liquid phase oxidation of glycerol. Appl. Catal. A Gen. 514: 267–275.

Dou, B., Y. Song, C. Wang, H. Chen and Y. Xu. 2014. Hydrogen production from catalytic steam reforming of biodiesel byproduct glycerol: Issues and challenges. Renew. Sustain. Energy Rev. 30: 950–960.

Du, Y., C. Wang, H. Jiang, C. Chen and R. Chen. 2016. Insights into deactivation mechanism of Cu–ZnO catalyst in hydrogenolysis of glycerol to 1,2-propanediol. J. Ind. Engin. Chem. 35: 262–267.

Edake, M., M. Dalil, M.J.D. Mahboub, J.-L. Dubois and G.S. Patience. 2017. Catalytic glycerol hydrogenolysis to 1,3-propanediol in a gas–solid fluidized bed. RSC Adv. 7: 3853–3860.

Gandia, L.M., G. Arzamedi and P.M. Dieguez. 2013. Renewable Hydrogen Technologies: Production, Purification, Storage, Applications and Safety. Elsevier.

González-Arellano, C., S. De and R. Luque. 2014. Selective glycerol transformations to high value-added products catalysed by aluminosilicate-supported iron oxide nanoparticles. Catal. Sci. Technol. 4: 4242–4249.

González-Cobos, J., S. Baranton and C. Coutanceau. 2016. Development of Bismuth-modified PtPd nanocatalysts for the electrochemical reforming of polyols into hydrogen and value-added chemicals. ChemElectroChem. 3: 1694–1704.

Heiz, U. and U. Landman. 2008. Nanocatalysis. Springer.

Hong, C., X. Jin, J. Totleben, J. Lohrman, E. Harak, B. Subramaniam, R.V. Chaudhari and S. Ren. 2014. Graphene oxide stabilized Cu_2O for shape selective nanocatalysis. Mater Chem. A 2: 7147–7151.

Huang, Z., F. Cui, H. Kang, J. Chen, X. Zhang and C. Xia. 2008. Highly dispersed silica-supported copper nanoparticles prepared by precipitation-gel method: a simple but efficient and stable catalyst for glycerol hydrogenolysis. Chem. Mater. 20: 5090–5099.

Huang, X., C. Dang, H. Yu, H. Wang and F. Peng. 2015. Morphology effect of $Ir/La_2O_2CO_3$ nanorods with selectively exposed {110} facets in catalytic steam reforming of glycerol. ACS Catal. 5: 1155–1163.

Jin, X., L. Dang, J. Lohrman, B. Subramaniam, S. Ren and R.V. Chaudhari. 2013. Lattice-matched bimetallic CuPd-graphene nanocatalysts for facile conversion of biomass-derived polyols to chemicals. ACS Nano 7: 1309–1316.

Kapkowski, M., P. Bartczak, M. Korzec, R. Sitko, J. Szade, K. Balin, J. Lelako and J. Polanski. 2014. SiO_2-, Cu-, and Ni-supported Au nanoparticles for selective glycerol oxidation in the liquid phase. J. Catal. 319: 110–118.

Kapkowski, M., T. Siudyga, R. Sitko, J. Lelątko, J. Szade, K. Balin, J. Klimontko, P. Bartczak and J. Polanski. 2015. Catalytic gas-phase glycerol processing over SiO_2-, Cu-, Ni- and Fe-supported Au nanoparticles. PLoS ONE (in press).

Kapkowski, M., W. Ambrozkiewicz, T. Siudyga, R. Sitko, J. Szade, J. Klimontko, K. Balin, J. Lelątko and J. Polanski. 2017. Nano silica and molybdenum supported Re, Rh, Ru or Ir nanoparticles for selective solvent-free glycerol conversion to cyclic acetals with propanone and butanone under mild conditions. Appl. Catal. B Environ. 202: 335–345.

Katryniok, B., S. Paul, M. Capron and F. Dumeignil. 2009. Towards the sustainable production of Acrolein by glycerol dehydration. ChemSusChem. 2: 719–730.

Katryniok, B., H. Kimura, E. Skrzynska, J.S. Girardon, P. Fongarland, M. Capron, R. Ducoulombier, N. Mimura, S. Paul and F. Dumeignil. 2011. Selective catalytic oxidation of glycerol: perspectives for high value chemicals. Green Chem. 13: 1960–1979.

Katryniok, B., S. Paul and F. Dumeignil. 2013. Recent developments in the field of catalytic dehydration of glycerol to Acrolein. ACS Catal. 3: 1819–1834.

Lam, B.T.X., M. Chiku, E. Higuchi and H. Inoue. 2015. Preparation of PdAg and PdAu nanoparticle-loaded carbon black catalysts and their electrocatalytic activity for the glycerol oxidation reaction in alkaline medium. J. Power Sources 297: 149–157.

Lam, B.T.X., M. Chiku, E. Higuchi and H. Inoue. 2016. PtAg nanoparticle electrocatalysts for the glycerol oxidation reaction in alkaline medium. Advances in Nanoparticles 5: 167–175.

Li, B., J. Wang, Y. Yuan, H. Ariga, S. Takakusagi and K. Asakura. 2011. Carbon nanotube-supported RuFe bimetallic nanoparticles as efficient and robust catalysts for aqueous-phase selective hydrogenolysis of glycerol to glycols. ACS Catal. 1: 1521–1528.

Li, X., B. Zhang, Q. Wu, C. Zhang, C. Yu, Y. Li, W. Lin, H. Cheng and F. Zhao. 2016. A facile strategy for confining ZnPd nanoparticles into a $ZnO@Al_2O_3$ support: A stable catalyst for glycerol hydrogenolysis. J. Catal. 337: 284–292.

Mimura, N., N. Hiyoshi, T. Fujitani and F. Dumeignil. 2014. Liquid phase oxidation of glycerol in batch and flow-type reactors with oxygen over Au–Pd nanoparticles stabilized in anion-exchange resin. RSC Adv. 4: 33416–33423.

Mota, C.J.A., B. Peres Pinto and A.L. de Lima. 2017. Glycerol: A Versatile Renewable Feedstock for the Chemical Industry. Springer.

Nakagawa, Y., Y. Shinmi, S. Koso and K. Tomishige. 2010. Direct hydrogenolysis of glycerol into 1,3-propanediol over rhenium-modified iridium catalyst. J. Catal. 272: 191–194.

Oberhauser, W., C. Evangelisti, C. Tiozzo, F. Vizza and R. Psaro. 2016. Lactic acid from glycerol by ethylene-stabilized platinum-nanoparticles. ACS Catal. 6: 1671–1674.

Pandhare, N.N., S. Murty Pudi, P. Biswas and S. Sinha. 2016. Selective hydrogenolysis of glycerol to 1,2-propanediol over highly active and stable Cu/MgO catalyst in the vapor phase. Org. Process Res. Dev. 20: 1059–1067.

Philippot, K. and P. Serp. 2013. Nanomaterials in Catalysis. Wiley-VCH.

Polshettiwar, V. and T. Asefa. 2013. Nanocatalysis: Synthesis and Applications. John Wiley & Sons.

Rahmat, N., A.Z. Abdullah and A.R. Mohamed. 2010. Recent progress on innovative and potential technologies for glycerol transformation into fuel additives: A critical review. 14: 987–1000.

Ricciardi, R., J. Huskens and W. Verboom. 2015. Nanocatalysis in flow. Renew. Sust. Energ. Rev. 8: 2586–2605.

Rogers, S.M., C.R.A. Catlow, C.E. Chan-Thaw, D. Gianolio, E. Gibson, A.L. Gould, N.. Jian, A.J. Logsdail, R.E. Palmer, L. Prati, N. Dimitratos, A. Villa and P.P. Wells. 2015. Tailoring gold nanoparticle characteristics and the impact on aqueous-phase oxidation of glycerol. ACS Catal. 5: 4377–4384.

Safavi, A., A. Abbaspour and M. Sorouri. 2015. Hydroxyapatite wrapped multiwalled carbon nanotubes composite, a highly efficient template for palladium loading for electrooxidation of alcohols. J. Power Sources 287: 458–464.

Salazar, J.B., D.D. Falcone, H.N. Pham, A.K. Datye, F.B. Passosa and R.J. Davis. 2014. Selective production of 1,2-propanediol by hydrogenolysis of glycerol over bimetallic Ru–Cu nanoparticles supported on TiO$_2$. Appl. Catal. A Gen. 482: 137–144.

Sankar, M., N. Dimitratos, D.W. Knight, A.F. Carley, R. Tiruvalam, C.J. Kiely, D. Thomas and G.J. Hutchings. 2009. Oxidation of glycerol to glycolate by using supported gold and palladium nanoparticles. ChemSusChem. 2: 1145–1151.

Schünemann, S., G. Dodekatos and H. Tüysüz. 2015. Mesoporous silica supported Au and AuCu nanoparticles for surface plasmon driven glycerol oxidation. Chem. Mater. 27: 7743–7750.

Senseni, A.Z., S.M.S. Fattahi, M. Rezaei and F. Meshkani. 2016. A comparative study of experimental investigation and response surface optimization of steam reforming of glycerol over nickel nano-catalysts. Int. J. Hydrog. Energy 41: 10178–10192.

Sharma, S.K. and A. Mudhoo. 2011. Green Chemistry for Environmental Sustainability. CRC Press, Taylor & Francis Group.

Simoes, M., S. Baranton and C. Coutanceau. 2010. Electro-oxidation of glycerol at Pd based nano-catalysts for an application in alkaline fuel cells for chemicals and energy cogeneration. Appl. Catal. B Environ. 93: 354–362.

Subramanian, N.D., J. Callison, C.R.A. Catlow, P.P. Well and N. Dimitratos. 2016. Optimized hydrogen production by aqueous phase reforming of glycerol on Pt/Al$_2$O$_3$. Int. J. Hydrog. Energy 41: 18441–18450.

Sun, Z., H. Zhang, Y. Zhao, C. Huang, R. Tao, Z. Liu and Z. Wu. 2011. Thermal-stable carbon nanotube-supported metal nanocatalysts by mesoporous silica coating. Langmuir 27: 6244–6251.

Talebian-Kiakalaieh, A., N.A.S. Amin and Z.Y. Zakaria. 2016. Gas phase selective conversion of glycerol to acrolein over supported silicotungstic acid catalyst. J. Ind. Engin. Chem. 34: 300–312.

Thia, L., M. Xie, Z. Liu, X. Ge, Y. Lu, W.E. Fong and X. Wang. 2016. Copper-modified gold nanoparticles as highly selective catalysts for glycerol electro-oxidation in alkaline solution. ChemCatChem. 8: 1–8.

Touri, A.E. and M. Taghizadeh. 2016. Hydrogen production via glycerol reforming over Pt/SiO$_2$ nanocatalyst in a spiral-shaped microchannel reactor. Int. J. Chem. React. Eng. 14: 1059–1068.

Wang, D., A. Villa, D. Su, L. Prati and R. Schlögl. 2013. Carbon-supported gold nanocatalysts: Shape effect in the selective glycerol oxidation. ChemCatChem. 5: 2717–2723.

Wang, H., Z. Lu, C. Li, D. Lu and X. Wang. 2016. Controllable preparation and catalytic performance of Pt/metal oxide nanocomposites for glycerol hydrogenation. Mater. Res. Innov. 21: 21–26.

Wu, Z., Y. Mao, X. Wang and M. Zhang. 2011. Preparation of a Cu–Ru/carbon nanotube catalyst for hydrogenolysis of glycerol to 1,2-propanediol via hydrogen spillover. Green Chem. 13: 1311–1316.

Yuan, Z., Z. Gao and B.-Q. Xu. 2015. Acid-base property of the supporting material controls the selectivity of Au catalyst for glycerol oxidation in base-free water. Chin. J. Catal. 36: 1543–1551.

Zhao, Z., J. Arentz, L.A. Pretzer, P. Limpornpipat, J.M. Clomburg, R. Gonzalez, N.M. Schweitzer, T. Wu, J.T. Miller and M.S. Wong. 2014. Volcano-shape glycerol oxidation activity of palladium-decorated gold nanoparticles. Chem. Sci. 5: 3715–3728.

Zhou, W., Y. Zhao, Y. Wang, S. Wang and X. Ma. 2016. Glycerol hydrogenolysis to 1,3-propanediol on Pt/WO$_3$/ZrO$_2$: Hydrogen spillover facilitated by Pt(111) formation. ChemCatChem. 8: 3663–3671.

Part III
Energy and Biomass

Chapter 4

Producing Fuels and Fine Chemicals from Biomass using Nanomagnetic Materials

Alessio Zuliani and *Rafael Luque**

INTRODUCTION

In 2015, the world chemicals turnover reached an amount of 3500 billion euros, representing 5% of the global Gross World Product. More than two third of the world's chemicals' raw materials were derived from mineral oil (Okkerse and van Bekkum 1999; Central Intelligence Acengy 2016).

In parallel, in 2015 the World Energy Consumption amounted to 13 billion metric tons of oil equivalent. Of these, 11 billion metric tons were derived from oil, coal and natural gas. As economic growth is pushing countries to increase their material and energetic demand, these amounts will upsurge in the future (Krausmann et al. 2017). Although the geological reserves can provide these resources per market request for some decades, they are limited and thus have to be substituted before we freeze in the dark. Moreover, the exploitation of nonrenewable resources implies the progression of considerable harms such as geopolitical disequilibrium, unexpected stock market fluctuations—and thus, possible economic crises—and, above all, increase in the release of CO_2 into the atmosphere (Carraretto et al. 2004; Kjarstad and Johnsson 2009; Bentham 2014).

One of the most captivating replies to this challenge is the employment of renewable bioresources through new sustainable processes (Ragauskas et al. 2006). Indeed, the exploitation of biomass can potentially satisfy a variety of needs, including generating electricity, heating homes, fueling vehicles and providing raw

Departamento de Quimica Organica, Universidad de Cordoba, Edificio Marie-Curie (C-3), Ctra Nnal IV, Km 396, Cordoba, Spain.
Email: z62zuzua@uco.es
* Corresponding author: q62alsor@uco.es

materials (Bozell 2010; Osatiashtiani et al. 2015). Moreover, as shown in Fig. 1, the treatment of feedstock derived from biomass don't add carbon dioxide to the environment, unlike fossil derivatives, which store carbon dioxide from eons of the past (Twidell and Brice 1992).

In term of efficiency, the use of an appropriate catalyst plays a crucial rule for the conversion of biomass into chemicals and fuels (Dethlefsen et al. 2015; Shi and Vohs 2016). Homogeneous catalysts are higly active, but their recovery and reuse is difficult. Heterogeneous catalysts are reusable, but have less catalytic activity. Nanocatalysts have been developed as intermediate catalysts between heterogeneous and homogeneous systems, that is to say, they are recoverable catalysts, with high activity, due to their elevated surface area and their nanosize (Gawande et al. 2014). To optimize the recovery of the nanocatalysts, magnetic nanoparticles (MNPs) are posited as relevant alternatives to conventional nanoparticles supported on inert materials. In fact, as shown in Fig. 2, the MNPs can be easily separated from the reaction mixture with an external magnetic field, conferring recycling properties to the MNPs. Nowadays, some MNPs are used in different types of reactions in a few biorefineries for the synthesis of chemicals (biofactories) and fuels, but they are still poorly employed in the majority of world's bioindustries (Shylesh and Thiel 2011; Baig and Varma 2014; Baig et al. 2015; Baig et al. 2015; Liu and Zhang 2016). Therefore, the short-term goal is to achieve the large scale production of high cost good quality chemicals and high energy density biodiesel through the exploitation of the reusable properties of the MNPs catalysts. Recent research projects on MNPs have focussed on different reactions such as transesterification, oxidation and hydrogenation reactions, carbon-carbon coupling reactions and other important

Fig. 1. Schematic representation of CO_2 cycle in petroleum industry vs. bio-based industry.

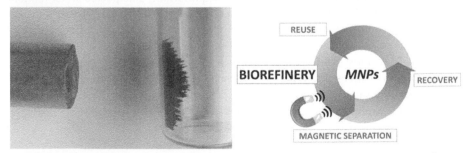

Fig. 2. Picture of MNPs attracted by a magnet and concept of the reutilization of MNPs in biorefinery.

reactions such as hydroformylation and olefin metathesis. Besides the optimization of the current applications, new technologies such as photo- and bio-catalysis are drawing a promising path for the future of MNPs (Polshettiwar et al. 2011).

Biorefinery: the key for a greener future

Biomass is defined as the entirety of the Earth's living matter. Most commonly, it is the term related to the material derived from growing plants or from animals. Thus, it describes wood and wood wastes, agricultural crops and their byproducts, animal manure and wastes, waste from food processing, municipal solid waste (MSW) and aquatic plants and algae. The major components of biomass, the carbohydrates and the lipids, are considered to be ideal feedstock for production of both biofuels and platform chemicals. From a strict point of view, hydrocarbons could be considered biomass as well, as they are derived from ancient animals and plants. However, biomass refers to materials directly derived from animals and plants, excluding long time geologic transformation.

Until the middle of the twentieth century, many industrial materials such as solvents, dyes and synthetic fibres were made from trees and agricultural crops (Ragauskas et al. 2006). Besides food provision, our history and civilizations are overflowing with interesting examples of biomaterials productions; for example, more than 2500 years ago, the Phoenicians were successful sellers of tyrian purple, a dye naturally extracted from the secretions of the sea snail muricidae (Ziderman 1986).

By the late 1960s, many bio-based chemical products had been displaced by petroleum derivatives which allowed for the overcoming the difficulties of the exploitation of renewable biomass, such as low sources quantities or high production price. However, a renewed interest in the synthesis of fuels and materials from bioresources was already aroused during the 1970s petroleum crisis (Wyk 2001).

At the present, societal growth is pushed both by economical and ethical issues, focusing on the meaningful aim of a greener future for the future generations. In fact, as petroleum-based industries use and produce hazardous substances, a shift to a sustainable chemical industry is needed.

In 1998, Anastas and Warner defined the twelve principles of the green chemistry. The principles state the outline for minimizing the environmental and health impact of a chemical process. The entrepreneurship translation of the twelve principle is the so-called biorefinery, defined by the *International Energy Agency* Task 42 as a facility that integrates biomass sustainable conversion and equipment to produce power, heat and chemicals from biomass (Jungmeier et al. 2014).

It may sound strange but the concept of biorefinery is quite old. For example, during the Roman Empire, human urine was treated to extract ammonia for the tanning of leather. In addition, the Emperor Vespasian imposed a tax on those whom didn't use public bath as some urine could get lost with the precious ammonia—This is where the famous saying *"Pecunia non olet"* (lit. "Money doesn't smell", meaning that urine could have a bad smell, but money doesn't) comes from. In this way a biorefinery simultaneously solved two problems: waste disposal and the provision of raw materials. Similarly, nowadays, biogas is produced in many countries from municipal solid wastes or farm manure, solving the problem of waste disposal as this waste produces gas that can be upgraded, as almost pure methane, in the line (De Baere 2006).

The biomass conversion of starch, sucrose and vegetable oils to bioproducts have been operating well since many years. Million tons per year of corn, wheat and potato starch have been processed to produce different products. Cognis Company (owned from 2010 by BASF) produced hundreds of different wellness, nutritional and functional products from vegetable oil. Roquette Company converts 6 M tons per year of starch to produce over 700 different products. Lots of startups are developing products produced via innovative processes involving microalgae or sea weeds (Manzer 2010).

Considering biorefineries productions, Bozell and Peterson reported two different strategies for biomass conversion to chemicals (Bozell 2008; Bozell and Petersen 2010):

- Target driven approach: utilizes analysis methodologies to discover the most efficient synthesis routes to produce a given chemical from a well-identified platform molecule. This methodology is generally uneconomical in respect to conventional synthesis routes from hydrocarbons.

- Process driven approach: biomass is converted into a family of valuable products by one or more catalytic processes (hydrogenation, oxidation, etc.). This approach is not aimed to duplicate chemicals produced from fossil fuels; therefore, it can lead to the development new valuable bioproducts. The starting feedstock are mixture of homogeneous species that are converted to a mixture of molecules with similar functionalities that can be used without further separation for the manufacture of paper, resins, pains, cosmetics, etc.

A key aspect of a biorefinery is the correct ratio between commodity chemicals and fuels produced. As the biorefinery is supposed to substitute petroleum refinery, 5% of the total output should go to chemical products, while the rest should be directed to transportation fuels and energy (Kamm and Kamm 2004).

The peculiarities of magnetic nanoparticles for the biorefineries

Green synthetic protocols target to increase the efficiency of the production of chemicals avoiding the use of volatile organic compounds (VOCs) and toxics reagents, the hazards reaction conditions, the high energy consumption and the waste accumulation of conventional chemical production (Clark and Macquarrie 2002; Lancaster 2002; Poliakoff et al. 2002; Polshettiwar and Varma 2008).

Aiming at this goal, the research of new catalytic systems, characterized by environmental sustainability and high efficiency, has become critical. This is practically possible though the exploitation of reusable and low/nontoxic catalysts (Anastas and Kirchhoff 2002; Dalpozzo 2015).

All catalysts can be divided into two classes: homogeneous and heterogeneous. Homogeneous catalysts are found in the same phase of the reactants, permitting easier interaction between the components. The direct result of the easier interactions is a sensibly increased activity. Moreover, it's possible to control the chemo-, region- and enantio-selectivity of the products by modifying the structure of the catalytic molecules. However, the final separation of the mixture is difficult. Actually, it's unfortunately common to find trace amount of the catalyst in the final product (Gawande et al. 2014).

Heterogeneous catalysts (in a different phase from the reactant) are employed in order to simplify the separation procedures. These heterogeneous catalysts are commonly synthesized by grafting active molecules on the surfaces or into the pores of a solid support such as silica, alumina, ceria, etc. Although these systems are easily separated, the active sites are less accessible, resulting in a decreased activity compared to homogeneous catalysts (Sheldon 2005).

As both the two classes of catalysts show advantages and weakness, the ideal catalyst should possess intermediate characteristics.

The nanodimension catalysts bind together the homogeneous catalysts' activity with the heterogeneous system's separation characteristics (Corma and Garcia 2003). Furthermore, the nanocatalysts possess extremely interesting properties such as high surface area, high surface/volume ratio and novel properties that are not found at the micro scale (Cuenya 2010). However, the isolation and the recovery of nanocatalysts is not so economically convenient: easy recovering techniques are inefficient because of the nanodimension of the materials; therefore, long and complicated separation techniques are needed, such as cumbersome filtration and centrifugation (Weir et al. 2012). A really smart solution to this limitation is the utilization of magnetic nanoparticles. Essentially, the MNPs' insoluble and paramagnetic nature allows an easy and efficient separation of the nanocatalysts from the products just due to the utilization of an external magnet. This characteristic is extremely useful and important when the substances treated have high viscosity and are full of impurities, such as in the case of biomass. These type of nanocatalysts, called magnetically separable nanoparticles (MSNPs or MNPs), show the peculiars characteristics of nanocatalysts joint together with the properties of easy recovery and reuse. These characteristics directly result in reduced energy consumption, limited catalysts losses, restricted processing time and finally, an increased economic gain. These nanocatalysts can be hence classified as forefront catalysts with high accessibility and increased reusability (Astruc et al. 2005; Shylesh et al. 2010). Furthermore, the economic characteristics of MNPs catalysts are of primary importance for the treatment of biomass; in fact, biomass have high prices compared to petroleum substances. Therefore, to compete with petroleum industry, the production of MNPs should be economically accessible and should balance the high costs of biomass.

Newest achievements in the development of new supporting materials and coatings have led to widely expand the potentiality of MNPs. In fact, adopting innovative strategies such as surface modification, grafting, self-assembly and nanocasting has made it possible to find sensible new applications including complex multistep catalytic transformations carried out in one pot reactions; enzimatic transformations and applications of hyperthermic capabilities (Govan and Gun'ko 2014).

Recently, several types of MNPs have been developed for important catalytic applications. Among them, metal nanoparticles (Pd, Ni, Cu, Ni, Co, Au and others), magnetite-supported organocatalysts, NHC and chiral catalysts have been investigated in various organic transformations suitable for biomass conversion. Specifically, magnetic metals and NHC nanocatalysts have been exploited in reduction, oxidation, coupling, hydration, hydrogenation and alkylation reactions. Elsewhere, magnetic organocatalysts exhibit good activity in more complex reactions such as Sonogashira, Stille, Suzuki couplings and Paal-Knorr, Aza-Michael and Mannich reactions (Gawande et al. 2014).

Magnetic nanoparticles synthesis: recent developments

The interactions between magnetic fields and substances have helped in segregating different classes of materials. Materials which show persistent magnetic moments are defined as permanent magnets. Most materials do not have permanent moments, therefore they can be attracted, repulsed or not/negligibly affected by a magnetic field, thus defining, respectively, paramagnetic, diamagnetic and non-magnetic materials. Other materials show complex behaviours such as spin glass behaviour and antiferromagnetism. The magnetic state of a material depends on the temperature, pressure, applied magnetic field and other variables. Consequently, the materials can show more than one form of magnetism as these variables changes; for example, pure oxygen shows magnetic properties in its liquid state.

Magnetic nanocatalysts are mainly produced by the use of a permanent magnetic material where the active site is usually immobilized. The literature reports different methods for the preparation of magnetic supports, such as the microemulsion technique, the co-precipitation method, the sol-gel method, the hydrothermal reaction method and thermal decomposition (Laurent et al. 2008; Zhang et al. 2012; Varma 2014). Magnetic supports mainly include metals (mainly Fe, Co and Ni), iron oxides (FeO, Fe_2O_3, Fe_3O_4), alloys (FePt, CoPt) or spinel ferrites MFe_2O_4 (M = Co, Mn, Cu, Zn) to name a few. Among them, magnetite, Fe_3O_4, is the most used support because of its low cost and easy preparation which make it one of the most interesting magnetic material to prepare MNPs for biomass conversion.

Recent trends and developments in magnetic nanoparticles preparation include template assisted synthesis, the sonochemical method, the extended LaMer Approach, flow and microwave preparations, bio-inspired co-precipitation and the biomass-derived method (Sharma et al. 2016). Among them, the methods based on biomass derivation are most interesting, as they are derived from biological recyclable material (Polshettiwar et al. 2011).

The template-assisted synthesis aims to prepare nanocatalysts with well-defined structures with a simple and versatile methodology. In this methodology a sacrificial template of metal, biological, polymeric or chemical origin is used to define the original shape and is then totally or partially removed with an acid or basic solution to produce free standing nanostructures (Wang et al. 2008; Dong et al. 2015; Guignard and Lattuada 2015).

The innovative sonochemical method for the synthesis of MNPs is still nowadays poorly reported in literature (Aliramaji et al. 2015; Soltani 2017). The technique can provide precise control over the size, morphology, composition, magnetic and surface properties of the MNPs. This is achieved with a short, extremely high temperature and a pressure reaction unattainable in traditional energy reactions.

Vreeland et al. (2015) purposed the so-called "Extended LaMer approach" (Lamer and Dinegar 1950) to finely regulate the size of nanoparticles. The method allows the preparation of highly crystalline MNPs with reproducible dimensions. This is achieved through the continuous and controlled addition of the precursor, which enables a uniform rate of growth of the MNPs. Specifically, Vreeland and co-workers synthesized magnetite nanoparticles by the addition of iron oleate precursor to a heated flask (Vreeland et al. 2015). The variation of the reaction duration and the volume of the added precursor yields high crystalline nanoparticles with sub-nanometer precision.

In the past decade, the development of industrial processes for the preparation of MNPs has focused on microstructured devices for flow chemistry. The synthesis through flow devices provides several advantages compared to conventional reaction systems. In fact, flow chemistry promotes efficient mixing and rapid chemical reactions, allowing better control over the synthetic parameters which define nanoparticles size and morphology. Thanks to these characteristics, Thomas et al. (2016) recently developed a method to produce functionalized magnetite nanocatalysts with nanoflower morphology, with a rapid one-step hydrothermal continuous process. The nanoparticles were functionalized with 3,4-dihydroxy-L-phenylalanine (LDOPA) or 3,4-dihydroxyhydrocinnamic acid (DHCA) and tested as a magnetic resonance imaging contrast agent. The results demonstrated that the fine morphology control of the Fe_3O_4-LDOPA nanomaterials resulted in a twice transverse relativity compared to that of commercial Feridex®/Endorem®.

In addition, flow technology provides possibilities for using alternatives energy sources, such as microwave or ultrasounds.

The irradiation of dielectric material with microwaves generate various phenomena. For example, the microwave irradiaiton of a solution of an electrolyte in water exhibit dielectric heating. This event rises from the orientation polarization of water and the resistance heating (Joule processes) between the electrolyte and the water. The exploitation of these phenomena led to the development of a new synthesis procedure for MNPs, characterized by uniform heating, low energy dispersion and very high purity and yield of the products (Kappe 2004).

For example, Makridis et al. (2016) reported a facile microwave assisted hydrothermal synthesis of Cobalt ferrite and Manganese ferrite. The nanoparticles were developed as a heating mediator for magnetic hypothermia. Interestingly, Makridis and co-workes observed that as the reaction time increased, the nanomaterials crystallinity improved, while the size fairly remained the same. These reults permitted the evaluation of the crystallinity influence on magnetic hypothermia. In this way, it has been proved that high crystallinity magnetic nanoparticles showed exceptional manetic heating properties.

The production of MNPs via biomass valourization is one of the most captivating challenges for a sustainable development. In fact, MNPs that are biomass-derived express their greener applications in the biorefinery, where themselves are produced, representing an extraordinary example of a circular economy product. These kind of synthesis include bio-inspired co-precipitation methodology and biogenic methods.

Some living organisms such as chitons and magnetotactic bacteria are capable of mineralizing magnetic crystals with controlled sizes and shapes in aqueous solutions and under ambient conditions. The bacteria control the crystal growth and nucleation by the interaction with bimolecular templates and additives (Luo et al. 2015). This natural phenomena gave the inspiration to design a room temperature co-precipitation approach using biologically derived peptide additives as nucleation controllers. As the biological mineralization is extremely complex, Lenders et al. (2017) have recently developed an automated dispensing system to prepare and rapidly screen hundreds of mineralization reactions with randomized conditions. With this study, it has been proven that the nanoparticles size, size distribution and shape can be tuned by the concentration and compositions of the copolypeptides. Furthermore, it has been well demonstrated that the pH is the most important factor in controlling the crystalline phase.

The literature reports some green biogenic methods for the synthesis of biomass-derived MNPs, with low costs of synthesis and low/nontoxicity (Mahdavi et al. 2013; Virkutyte and Varma 2013). Recently, Prasad et al. (2016) reported the preparation of Fe_3O_4 magnetic nanoparticles using a watermelon rind extract. The agricultural waste was added to $FeCl_3 \cdot 6H_2O$ and sodium acetate and stirred for 3 hr at 80°C. The watermelon extract contained polyphenols that acted as reducing agents. The obtained MNPs had 2–20 nm size, exhibit high magnetic response behaviour, and high activity through the synthesis of 2-oxo-1,2,3,4-tetrahydropyrimidine derivatives, with 94% yield. Venkateswarlu et al. (2013) synthesized magnetite nanoparticles with a plantain peel extract, with dimensions below 50 nm. Very recently, the same procedure was adapted for the synthesis of MNPs with *Pisum Sativum* peels as reducing agents (Prasad et al. 2017).

Despite the different synthesis techniques, MNPs themselves are not so stable. Firstly, they are sensitive to oxidation, which can form a thin oxide layer on the surface and sensibly change the MNPs properties (Robinson et al. 2010). Furthermore, they tend to agglomerate because of their magnetic nature and the strong magnetic dipole interactions, limiting the use of MNPs in various applications. To increase MNPs' stability, a modification using stabilizing ligands or coating materials, such as carbon, inorganic components like silica, polymers and metals has been proved to be the best solution so far (Sharma et al. 2016). In addition, the coating can also provide large surface area and reaction sites or an active group for grafting the active sites on the MNPs.

The production of MNPs coated with silica normally involve sol-gel or microemulsion preparation methods. The most used precursors include TEOS or sodium silicate. The silica coating can provide several functionalized sites for covalent attachments. Qiu et al. (2015) simplify a literature sol-gel method (Sun et al. 2013) to obtain a nanocatalyst with a magnetite core and a mesoporous silica shell. The formation of Fe_3O_4@MSN-CBA magnetic nanoparticles was obtained with the coating of an ultrasmall superparamagnetic magnetite nanoparticle with (3-Aminopropyl) triethoxysilane (APTES) with a facile synthesis protocol. The nanomagnetic catalysts were demonstrated to have good stability, uniform size, excellent dispersity, ordered mesoporosity and water solubility.

Carbon coatings are obtained through hydrothermal, sonochemical and flame spray pyrolysis techniques, which allow for the production of highly chemical, thermal stable and biocompatible compounds. Although standard precursors are glucose and cellulose, [...] other interesting precursors such as polydopamine have been recently studied. Jiang et al. (2015) reported the synthesis of a novel hierarchical nanostructure composed of carbon coated Fe_3O_4 nanoparticles supported on graphene (G/Fe_3O_4@C). The MNPs were obtained through the phase transformation of β–FeOOH to Fe_3O_4 and the simultaneous carbonization of a polydopamine nanocoating through a thermal annealing at 500°C. The carbon layer was demonstrated to prevent the agglomeration of Fe_3O_4 nanoparticles. Moreover, Fan et al. (2015) purposed the one pot synthesis of the MNPs G/Fe_3O_4@C, simplifying the production process.

Polymer coating enhances the colloidal stability of MNPs. The preparation is usually achieved via inverted microemulsion or oxidative polymerization techniques. The precursor are polyesters, polyaniline, polyethylene glycol and other polymers.

A part from the above-mentioned classical coatings, also biomass derived coatings have been studied aiming to the ambitius circular economy. For example, Baig et al. (2014a) prepared carbon coated MNPs with the most abundant polymer on Earth. The magnetic carbon-supported Pd catalyst was in fact easily synthesized by the *in situ* generation of nanoferrites and the incorporation of carbon from cellulose. The cellulose and the $PdCl_2$ were added to a solution of $FeSO_4 \cdot 7H_2O$ and $Fe_2(SO_4)_3$ and calcinated at 450°C. The obtained Fe_3O_4@CPd showed complete hydrogenation of nitro benzene in 40 min, while commercial Pd/C required 5 hr.

In addition, some biopolymers have been successfully employed as coating for MNPs. Among the various biopolymers, chitosan (CS) contains NH_2 and OH functional groups which can capture and stabilize the metal nanoparticles (Guibal 2004; Kramareva et al. 2004; Baudoux et al. 2007). Additionally, the polymer is obtained from chitin, the second most abundant natural polysaccharide, derived from shrimp and crab shells by deacetylation. The chitosan is also optically active and a good stabilizer of Pd metal nanoparticles (Hardy et al. 2004). Yin et al. (1999) reported an easy preparation of the silica-supported chitosan-palladium complex. The silica was added to an acid solution of chitosan, which was then neutralized to obtain the SiO_2-CS. Secondly, the Pd was added in the form of $PdCl_2 \cdot 2H_2O$ in an ethanol solution to prepare the final complex. The obtained nanocatalyst was used in the asymmetric hydrogenation of some ketones at room temperature and under 1 atm H_2, showing an excellent result as compared to the other literature reported asymmetric catalysts (Sibikinap et al. 2009).

MNPs are characterized with several standard physicochemical techniques. Characterization by Transmission Electron Microscopy (TEM) and Scanning Electron Microscopy helps to determine the size, the shape and the morphology of the nanoparticles. Fourier Transform Infrared Spectroscopy (FTIR) is important for the analysis of the functional group and is also used to confirm the surface modification of the MNPs. Powder X-ray diffraction analysis (XRD) gives information about the crystallographic structure and the chemical composition of the nanomaterials. X-ray photoelectron spectroscopy (XPS) is useful to determine the oxidation state of the metal present in the final catalyst. Thermo-gravimetric Analysis (TGA) is important to investigate the thermal stability of the MNPs. Vibrating sample magnetometry (VSM) helps to examine the magnetic properties of the modified and unmodified MNPs.

Biofuel from MNPs

Introduction

The photosynthesis allows plants to absorb solar light and transfer low energy CO_2 into energy rich organic compounds. These energy rich compounds enable the plants to live. Thus, the solar energy is stored in the form of chemical energy. Animals eat plants or other animals and access the biomass energy. The conversion of this energy into fuels through biomass valorisation processes leads to different classes of biofuels: bioalcohols, biodiesels, bioethers and biogas. MNPs are mainly exploited for making biodiesels, and very recently they have been proposed as catalysts in the anaerobic digestion process for the production of biogas.

Biodiesel

Biodiesel, a biodegradable, nontoxic, low emission fuel, is produced from vegetables oils or animal fats and can be used as a fuel or additive (Ma and Hanna 1999). More than 100 years ago, Rudolph Diesel tested for the first time a biodiesel, merely vegetable oil, in his engine (Griffin 1993). However, the advent of petroleum as a fuel led to the evolution of a diesel engine strictly related to mineral oil, leaving the utilization of vegetable derivatives in emergency situations during the 1930s and the 1940s (Ma and Hanna 1999). Nowadays, the utilization of biodiesel as fuel is posited as a green answer to the request for sustainable development, for example, the Renewable Energy Directive states that all the EU members target raising the share of renewables in the energy consumption by 2020 to a 20% average (doubling the 9,8% of 2010) while the U.S. Department of Energy has set the aim to replace 30% of the petroleum derived transportation fuel with biofuels by 2025 (Bachmann 2005).

The direct use of oils and fats as diesel is hindered by viscosity and carbon deposition in the engine. Therefore, oils and fats are processed to be compatible with existing engine. There are three primary conversion methodologies for producing biodiesels: microemulsion, pyrolysis and transesterification. While pyrolysis is a thermal treatment and microemulsion is a mechanical process, the catalyst plays a crucial rule in the transesterification reaction.

The reaction of transesterification, or alcoholysis, is the reaction of a fat or oil with an alcohol to form esters and glycerol. To improve the reaction rate and yield, a catalyst is usually needed. The production of biodiesel via transesterification start with the extraction of natural vegetable oils and animal fats to obtain crude oil or fat. The crude oil is not pure, as it contains free fatty acids, water, sterols, phospholipids, odorants and other impurities. The content of water and fatty acids has significant effects on the catalytic transesterification process. It also interferes with the separation of the esters.

The crude fat or oil, mainly composed by trygliceride, is treated with methanol in the presence of a catalyst to obtain the fatty acid esters and glycerol. As the reaction is reversible, an excess amount of alcohol is used to shift the equilibrium. Different alcohols can be used, such as methanol, ethanol, propanol, butanol and amyl alcohol. Methanol is the most common alcohol used due to its low cost.

The reaction can be catalysed by a variety of base and acids catalysts or enzymes lipases (Cai et al. 2015; Fu et al. 2015; Xie et al. 2015). Conventionally, homogeneous basic or acid catalysts, such as NaOH or H_2SO_4, have been used to produce biodiesel at a large scale (Boon-anuwat et al. 2015; Gu et al. 2015). However, the difficult recovery of the catalysts from the reaction solution, the soap formation and the

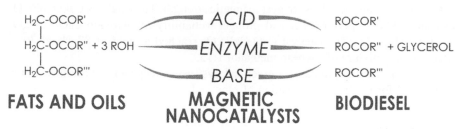

Fig. 3. Scheme of the transesterificition reaction of oil and fats catalysed by alkali, acid and enzymes recoverable catalyst.

corrosive nature of the substances fix important limits to these kinds of technology. The utilization of nanomagnetic materials offers a valid alternative for a high performance catalyst of the transesterification process due to the heterogeneous-homogeneous and recovery characteristics.

During recent years, many kinds of magnetic base catalysts have been studied for an efficient synthesis of biodiesel via the transesterification reaction, producing alkali, acid and enzymes recoverable catalyst, as summarized in Fig. 3.

Alkalis MNPs

Many kinds of magnetic alkali nanoparticles have been reported as efficient transesterification catalysts. Among them, the employment of an active compound of CaO has received more attention (Schachter and Pines 1968; Peterson and Scarrah 1984; Kabashima and Hattori 1998; Ono 2003). In fact, the utilization of calcium-oxide as a base catalyst offer several advantages, including higher activity, long catalyst lifetimes, low cost and mild reaction conditions (Liu et al. 2008). However, the recovery of the catalyst poses some limitations. It has been demonstrated that during the transesterification process, lattice oxygen species form hydrogen bonds with methanol and glycerin, increasing the viscosity of glycerin and forming suspensolid with CaO. Therefore, the CaO separation is hardly carried out. The combination of calcium-oxide with magnetic nanoparticles overcome these separation limits (Kouzu and Hidaka 2012).

Liu et al. (2010) synthesized a simple magnetic solid base catalyst of CaO/Fe_3O_4 by loading calcium-oxide on Fe_3O_4 using sodium carbonate and sodium hydride as precipitators. The catalyst was tested in the transesterification reaction of *Jatropha curcas* oil. It was observed that the proportion between Ca^{2+} and Fe_3O_4 affected the catalytic performance, reaching the highest activity when the proportion of Ca^{2+} to Fe_3O_4 was 7:1. The reaction showed 95% yield at 70°C after 80 min, with a catalyst dosage of 2% and methanol/oil molar ratio of 15:1. The catalysts also exhibit a recovery rate of 91,45%, which was sensibly higher compared to the recovery rate of 55,95% of CaO. Moreover, the conversion rate of the transesterification reaction was up to 90% after five cycles, and above 70% after 10 cycles.

Fan et al. (2014) prepared a series of magnetic CaO/MFe_2O_4 (M^{2+}) catalysts, by a hydrothermal method, with MFe_2O_4 as the magnetic core. The MNPs catalysed the transesterification reaction of soybean oil for the biodiesel production. The catalyst $CaO/CoFe_2O_4$ exhibited the strongest magnetic strength. The catalysts $CaO/CoFe_2O_4$ also showed better performance and stronger wettability compared to $CaO/ZnFe_2O_4$ and $CaO/MnFe_2O_4$. The catalyst $CaO/CoFe_2O_4$ reached a yield of 87,4% at 70°C after 5 hr in a mixture of soybean:methanol 1:15.

Lin et al. (2013) produced a green catalyst using recycled CaO hollow fibres (CaO/α-Fe) for the transesterification of rapeseed oil. The catalyst was synthesized by an organic gel-thermal decomposition method. Under the optimal condition of methanol/oil ratio 12:1, reaction time 2 hr, catalyst amount 5%, and temperature 60°C, biodiesel yield was of 95,7%. The catalyst was recovered and recycled up to 20 cycles and 85,2% yield was still obtained. The high catalytic activity of the CaO/α-Fe hollow fibres was attributed to the excellent magnetic properties of the catalyst and the hollow microstructure. While the hollow microstructure increased the catalytic activity due to the higher surface area, the high magnetic properties allowed a rapid separation, avoiding the dissolution of CaO and therefore, the deactivation of the catalyst. The combination of high catalytic activity, long lifetime and good magnetic properties create a potential usage of this recyclable catalyst in the biodiesel industry.

To increase the CaO activity, Zhang et al. (2016) used SrO to produce a magnetic $CaO@(Sr_2Fe_2O_5-Fe_2O_3)$ catalyst. The catalyst was synthesized by a co-precipitated method. The catalyst was applied to the transesterification of soybean into biodiesel, reaching, after 2 hr of reaction, a maximum yield of 94,9% at 70°C with 0,5% catalyst dosage and a methanol:soybean oil ratio of 12:1. The catalyst showed high efficiency and high stability upon five repeated runs.

Similarly Liu et al. (2016) produced an $MgFe_2O_4@CaO$ catalyst, which exhibited a high activity due to the synergy between calcium and magnesium. The catalyst was synthesized with an alkali precipitation with the presence of sodium dodecylbenzensulfonate (SDBS) and was tested in the transesterification of soybean oil. The highest biodiesel yield of 98,3% was obtained with a catalyst dosage of 1 wt. %, methanol:soybean oil ratio of 12:1; reaction temperature of 70°C and reaction time of 3 hr. The catalysts showed strong magnetic strength and was easily recovered after every cycle. The yield of biodiesel still remained above 89% after five times. The catalysts also showed improved acid-resistance and water-resistance properties compared to CaO.

With respect to biorefinery, the use of CaO offers the possibility of double biomass valorisation. In fact, CaO is a biomass catalyst, but also a biomass derivable product. Anr et al. (2016) synthesized an active heterogeneous CaO from the tropical biodiversity seashell *Anadara granos*. The catalyst was prepared via a "Calcination–hydration–dehydration" protocol and showed high activity for the transesterification process of *Jatropha curcas* oil, yielding 78.4% after the third reuse cycle.

Na_2O-SiO_2/Fe_3O_4 is another type of basic magnetic nanocatalyst that has been recently purposed for biodiesel production. Huang et al. (2012) prepared a catalyst by loading Na_2SiO_3 on Fe_3O_4 nano-particles with $Na_2O\cdot3SiO_2$ and NaOH as precipitators. The catalyst showed best catalytic activity when the Si/F molar ratio was 2,5 and the calcination temperature was 350°C. The catalyst was tested in the transesterification of cottonseed oil to FAME and the yield was strictly related to the methanol/oil molar ratio, with a maximum at 7:1. The best performance was obtained with 5% catalyst dosage, 60°C reaction temperature and 100 min reaction time, reaching a yield of 99.6%.

Acids MNPs

The biggest lack of base-catalyzed biodiesel production is an economic one. In fact, despite the fact that base-catalysed transesterification is a facile process, it requires high cost virgin oil as feedstock (Zhang et al. 2003a). An alternative to base catalysts, acid catalysts promote both esterification of free fatty acids and transesterification of triglycerides simultaneously, therefore, the waste cooking oil can be classified as feedstock for biodiesel production (Canakci and Van Gerpen 2001; Zhang et al. 2003b).

Wang et al. (2015) reported the synthesis of acids catalyst for biodiesel production in the form of sulfamic acid and sulfonic acid functionalized silica coated crystalline Fe/Fe_3O_4 core/shell magnetic nanoparticles (MNPs). These MNPs have been demonstrated to be efficient recoverable catalysts for the biodiesel production from oils and fats containing a high level of free fatty acids, like waste cooking oil. The synthesis of the catalysts consisted of three steps. First, magnetic crystalline high magnetic nanoparticles were prepared by thermal decomposition of iron pentacarbonyl, in the second step, the particles were covered by silica by hydrolyzing tetraethoxysilane (TEOS) on the surface of MNPs under basic conditions and in the last step, the particles were functionalized by silanation with (3-aminopropyl) triethoxysilane (APTES) and (3-mercaptopropyl) trimethoxysilane (MPTMS), producing NH_2-SiO_2-functionalized-MNPs. The treatment with sulfuric acid or chlorosulforic acid created two different types of acid MNPs, sulfonic acid functionalized MNPs and sulfamic acid functionalized MNPs. The catalysts were tested in the transesterification of glyceryl trioleate and in the esterification of oleic acid in methanol. The catalysts were tested varying the temperature from 60 to 100°C and showed better performances at a higher temperature. In 20 hours of transesterification reaction at 100°C, sulfonic acid functionalized MNPs and sulfuric acid functionalized MNPs reached respectively 88% and 100% of conversion. The esterification of oleic acid was concluded within four hours with 100% of conversion for both the catalysts at 70°C in methanol.

While the sulfonic acid functionalized MNPs had low reusability, and the conversion dropped to 62% at the fifth run, sulfamic acid functionalized MNPs maintained 95% conversion throughout five reaction cycles.

Enzymes MNPs

Lipases, enzymes naturally found in animals, plants and microorganisms, are hydrolases which take part in the deposition, transfer and metabolism of lipids (Villeneuve et al. 2000). As a matter of fact, "Lipases" derived from the ancient Greek "lipsos", which meant fat.

Lipases can convert triglycerides to FAAE, thus, they have gained attention in the production of biodiesel. Specifically, lipases act on carboxyl esters bonds to yield fatty acids and glycerol at the lipid-water interface (Pessoa et al. 2009). The structure of lipases has a central L-sheet with an active site of serine on a nucleophilic elbow placed in a channel covered by a peptide lid: when the lipase is in contact with a lipid-water interface, the lid undergoes conformational changes, making the active site accessible for the acyl group. The most commonly used lipases for biodiesel production include lipase from *Candida antartica*, *Candida cylindracea*, *Candida rugosa*, *Pseudomonas cepacia*, *Pseudomonas fluorescens*, *Rhizopus oryzae*, *Aspergillus niger*, *Thermomyces lanuginosus*, *Rhizopus delema*, and *Rhizomucor miehei*.

The biocatalysis of oil feedstocks exhibits many advantages such as mild reaction conditions, environmental sustainability, board adaptability to crude materials and ease of down-stream processing (Zhao et al. 2015). The reaction is strictly dependent on various factors such as the source of lipase, the application technique, the temperature, the presence of solvent, the alcohol to molar ratio, the water content and the presence of an immobilizing material which the lipase is supported on (Guldhe et al. 2015). In fact, when the optimum temperature is overpassed, the enzyme starts to denaturalize becoming inactive, despite an increasing of the activity can be initially being observed (Fjerbaek et al. 2009). The utilization of methanol, as acyl acceptor, should also be strictly controlled. In fact, it has been demonstrated that an excess amount of alcohol form droplets in the oil which causes the denaturation of the enzyme. Various studies have optimized the methanol to oil molar ratio in the range of 3:1 to 4:1 (Fjerbaek et al. 2009; Kumari et al. 2009; Lu et al. 2009), with the exception of some lipases, which show the optimum at an higher ratio, that is, around 7:1 (Li et al. 2011; Meunier and Legge 2012). To overcome this methanol inhibition, some alternatives have been proposed, including the stepwise addition of methanol (Shimada et al. 1999); the use of other acyl acceptors, such as methyl acetate (Du et al. 2004), or ethyl acetate (Modi et al. 2007); the use of solvent to increase alcohol solubility (Iso et al. 2001) and the use of methanol-tolerant lipase (Yang et al. 2009).

The utilization of lipases enzyme for the transesterification of oils can be carried out directly employing cells containing the lipase (intracellular), or using the lipase extracted from the cells (extracellular). Extracellular lipases have been frequently used for biodiesel production, preferentially in the immobilized form. Normal feedstock for the extracellular lipases are Sunflower oil, Palm oil, Soybean oil, *Jatropha* oil, Cotton oil and Olive oil. As the cost of the extraction of the lipase is the limiting factor for the industrial application, the intracellular lipase is gaining interest among the researchers.

However, the major limitation in the biocatalysis for biodiesel production is the high cost of the enzymes. One attractive solution is the utilization of an immobilizing material, to reduce the losses during the recovery processes. Therefore, the price of the lipase could be amortized in several cycles. It has been demonstrated that the *Candida Antartica* lipase can be reused without loss of its activity for up to 50 cycles (Iso et al. 2001). Modifying other process parameters such as acyl acceptor and solvent, it has been demonstrated a complete reusability of the catalyst after 100 and 200 cycles (Du et al. 2004). Furthermore, immobilized lipases have shown higher and faster activity compared to free lipases (Iso et al. 2001). The most used techniques for the immobilization of lipase include adsorption, entrapment, encapsulation, and cross-linking, while between the newest immobilization techniques, the utilization of nanomagnetic carriers is gaining more attention.

In fact, the immobilization on nanomagnetic materials have the advantage of an easy separation and the possibility to concentrate the catalyst at specific sites in a reactor by applying an external magnetic field. This is an important characteristic when it's necessary to treat inhomogeneous biomass substances. Furthermore, the nanodimension gave to the catalysts a high surface active area. For example, multi walled carbon nanotubes (CNTs) has been purposed as a good immobilizing agent for the lipase as they can be easily functionalized. Furthermore, they are made magnetically active with simple processes (Georgakilas et al. 2005). Recently, Fan et al. (2017) reported the preparation of recoverable carbon nanotubes filled with magnetic iron oxide functionalized with polyamidoamine (PAMAM) dendrimers. The dendrimers were capable of immobilizing *Burkholderia Cepacia* lipase for the biodiesel production from oils. The nanoparticles were prepared through a facile approach of some steps. Firstly, the amino-functionalized Multi Wall Carbon NanoTubes (MWCNTs-NH_2) were synthesized. Then, the iron oxide nanoparticles were encapsulated in the MWCNTs-NH_2. The MWCNTs-NH_2 were sonicated in a water-hydrazine hydrate solution containing ammonium iron(II) sulfate, adjusting the pH to 11–13. The solution was then refluxed at 130°C for 2 hrs, washed with N,N'-dimethylformamide, ethanol, water, and HCl, and finally dried. The magnetic MWCNTs-NH_2 (defined as mMWCNTs-NH_2) was then processed with a multistep repetition of room temperature treatments with methyl acrylate, ethylendiamine and methanol to obtain the PAMAM dendrimers. Finally, the lipase was immobilized on the mMWCNTs by simply mixing the MNPs and the lipase in a phosphate-buffered solution. The MNPs were tested in the transesterification of soybean oil with methanol, showing interesting results. It has been found that the water content influence the reaction yield with an optimum at 2%. These results give positive possibilities to the utilization of low grade oil feed-stocks containing water, which gives saponification in standard transesterification reactions with NaOH. Additionally, it has been found that the temperature for the highest yield of 93,1% was carried out in the mild conditions of 35°C, while the activity decreased at a higher temperature. The reusability of the catalysts was investigated by recovering the MNPs with a magnet and by washing them with *t*-butanol. After 20 cycles the particles retained 89,6% yield.

Clearly, the nature of the enzyme strongly influences the reaction yield. Aiming to develop a recoverable biocatalysts for the treatment of waste cooking oils, Mehrasbi et al. (2017) immobilized lipase from Candida Antarctica lipase (CALB) on silica core shell magnetic nanoparticles ($Fe_3O_4@SiO_2$). This lipase is in fact not effective for the transesterification of triglycerides but is extremely active for the esterification of free fatty acids, which are a major component in waste cooking oil. The MNPs were prepared with a simple sol-gel process to obtain $Fe_3O_4@SiO_2$. The silica coated magnetic nanoparticles were then functionalized with (3-glycidyloxypropyl) trimethoxysilane (GPTMS) and Et_3N, obtaining epoxy-functionalized silica-MNPs. The immobilization of lipase was achieved by adding the MNPs to a sodium phosphate buffer containing the enzyme. The author and co-workers tested the MNPs in the esterification of waste cooking oil using molecular sieves as a water adsorbing agent, as reported in the literature (Hsu et al. 2002), obtaining 100% yield and 100% reusability after five cycles.

Biogas and MNPs: a possible future?

Biogas is a mixture of gases produced by the anaerobic digestion (AD) of biomass. Varying with the biomass nature, the two major components of the biogas, methane and carbon dioxide, are normally found in a 3:1 volume ratio, representing together 90% of the biogas mixture. The biogas is directly combusted to produce heating and electric energy, or can be up-graded and purified for its direct utilization in pipelines, or for functioning as a biofuel for vehicles. The digested biomass is valorised as fertilizer or as compost after an aerobic treatment. The AD consists of a bacteria biologic processes which result in an almost complete degradation of the total organic content of the biomass. The AD digestion process consists of the sequential concatenation of hydrolysis, acidogenesis, acetogenesis and methanegenesis. These phases depends on the correct growth and behaviour of the bacteria. The bacteria growth is more sensitive to several parameters step-by-step. The parameters that can strongly influence the process are pH, temperature fluctuations, C/N ratio, etc., and the temperature has a major influence on the AD process. The process can occur under psychrophilic ($\sim 20°C$), mesophilic (25–40°C) or thermophilic (50–65°C) conditions. Mesophilic conditions are the most common. Normal feed stocks for the biogas production include municipal organic solid waste, dedicated culture (corn, *Triticum*, etc.) and animal effluents. Different experiments have aimed to stimulate the microbial activity and increase biogas production (Yadvika et al. 2004). Some experiments has led researchers to discover that the addition of heavy metal ions can be used as additive to improve the biogas production (Luna-del Risco et al. 2011). Specifically, it has been established that the growth of methanogenic bacteria depends on the presence of Fe^0, Co^0 and Ni^0 during the enzyme synthesis. Moreover, Zhang et al. (2011) proved that specifically zerovalent iron not only serves as an electron donor, but also enhances the AD. However, the accumulation of the metals in the reactor reach toxic concentrations for the anaerobic bacteria (Chen et al. 2008). Recently, Abdelsalam et al. (2017), hypothesized that iron magnetic nanoparticles enhance the production of biogas in the AD process and, due to their recoverability, they avoid the accumulation of metals in the reactor. The MNPs were easily synthesized, dissolving iron chloride in a solution of $NaCO_3$ where secondary ascorbic acid was added. The effect of iron magnetic nanoparticles on biogas production was investigated by the anaerobic digestion of fresh manure (faeces and urine). Fe_3O_4 MNPs exhibited the best activity at the concentration of 20 mg/L manure. It has been demonstrated that the magnetic particles afflicted not only the production of biogas, but also the methane/CO_2 ratio. Specifically, the biogas production increased by 59.4% while the methane production increased by 66%.

Biochemicals from MNPs

Introduction

Besides the production of energy, magnetically recoverable nanoparticles are intensely investigated for the production of chemicals from biomass. In fact, using MNPs it's possible to convert renewable sources into high value materials exploiting the recovery characteristics of the catalysts. This process could be competitive to the petroleum production as the easy reutilization of the catalysts lowers the production costs. In addition, possible chemicals derived from biomass have been already investigated for decades and have attracted interest from government authorities. For example, according to the *US Department of Energy* (DOE), there are twelve kinds of top valuable building blocks derived from carbohydrates, including 1,4-diacids (succinic, fumaric and malic), 2,5-furandicarboxylic acid, 3-hydroxy propionic acid, aspartic acid, glucaric acid, glutamic acid, itaconic acid, levulinic acid, 3-hydroxybutyrolactone, glycerol, sorbito and xylitol/arabinitol.

The most studied reactions of the conversion of biomass into chemicals with magnetic nanoparticles include:

*   The hydrolysis of lignocellulose into glucose;
*   The dehydration of fructose into HMF (5-hydroxymethylfurfural);
*   The derivatization of HMF.

Hydrolysis of lignocelluloses

Lignocellulose, a term used to refer to plant dry matter, is the most abundant biomass in the Earth, therefore, its utilization as a chemical resource is extremely exciting from both environmental and economical points of view. Normally, lignocellulose is composed by lignin (15–25%), hemicellulose (23–32%) and cellulose (38–50%), in some cases the content of cellulose represent more than 90% of the lignocellulose, such as in the case of cotton fibre. As the cellulose is the major component in the lignocellulose biomass, it has gained attention as sustainable feedstock for the production of chemicals.

"Cellulose" was first coined by French chemist Anselme Payen in 1835, as an adjective referring to "consisting of cells". In turn, the term "cell" was derived from the Latin "cellula", which meant "small simple chamber". If you think on a small simple chamber, it's quite easy to notice the walls above everything, so that a "cellula" seems to consist of walls. Actually, the cellulose can be mainly found in the primary cell wall of green plant, many forms of algae and also in oomycetes, therefore, the "consisting of cells walls" it's extremely right.

More properly, cellulose refers to a linear polymer composed of glucose units repeatedly connected through β-1,4-glycosidic linkages, as shown in Fig. 4.

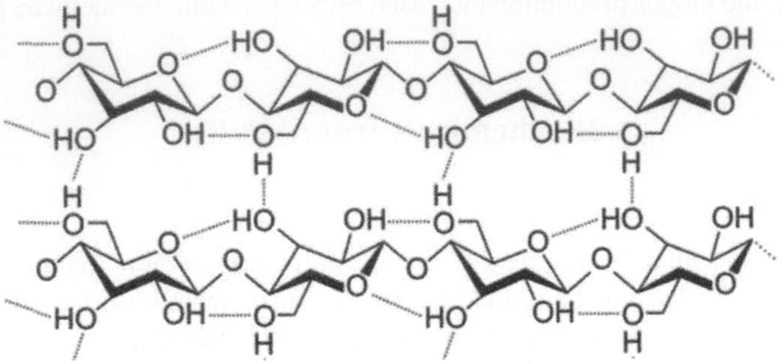

Fig. 4. Schematic representation of cellulose structure.

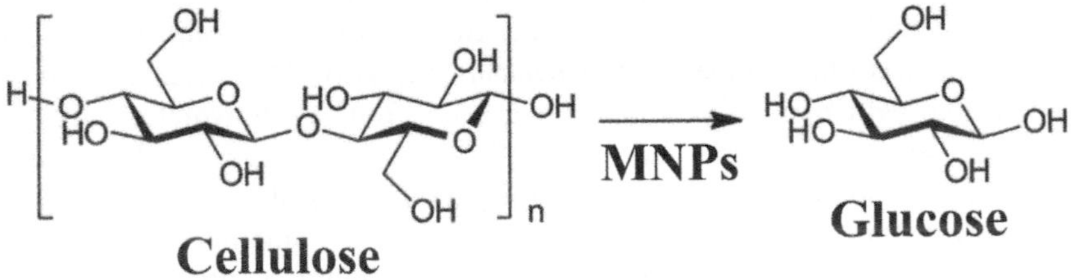

Fig. 5. Reaction scheme of the destructuration reaction of cellulose into glucose catalysed by MNPs.

The most important intramolecular interactions of the cellulose are the hydrogen bonds that form a cross-linked structure between the cellulose chains. The structure is therefore highly compacted. As a result, cellulose is insoluble in water and in common organic solvents, limiting its utilization. Thus, a destructuration of the cellulose in the simple monomer glucose, showed in Fig. 5, is an essential step for the production of valuable chemicals.

Magnetic nanoparticles have been successfully employed for this destructurization, through the reaction of hydrolysis.

Acid MNPs

Mineral acids have been employed as the first catalysts for the hydrolysis of cellulose. In fact, the free H^+ could easily attack the β-1,4-glycosidic linkages. However, the utilization of mineral acids has some drawbacks, such as poor recyclability, reactor corrosion, waste treatment and toxicity. The utilization of heterogeneous catalysts can overcome these drawbacks, but creating a reaction between solids, that is, the solid catalysts and the solid cellulose, is difficult (Wiredu and Amarasekara 2014). In fact, despite the fact that cellulose can be sequentially degraded into soluble sugars, lignin and humins are barely separated from the heterogeneous catalysts.

Furthermore, the activity of the catalysts decreases sensibly if the surface is covered by these sugars residues.

The use of magnetically recoverable heterogeneous catalysts on one hand facilitate the recycling of the catalyst and, on the other, help in the separation of the catalyst from the solid residues formed during the hydrolysis of cellulose. Furthermore, the acid hydrolysis of cellulose or lignocellulose with magnetic nanocatalysts can be accomplished both in water or in ionic liquids (Verma et al. 2013; Liu et al. 2014; Xiong et al. 2014).

Most approaches studied for the stabilization of MNPs for the hydrolysis of cellulose include the utilization of a porous supporting material, such as SBA, or the building of a functionalized magnetic core shell structure. For example, magnetic sulfonated mesoporous silica (Fe_3O_4-SBA-SO_3H) has been proved to be a good catalyst in water (Lai et al. 2011a; b). As illustrated in Fig. 6, the catalyst was prepared by a surfactant templated sol-gel method. Initially, Fe_3O_4 magnetic nanoparticles were dispersed in the block copolymer P123 for the co-condensation of tetraethoxylane (TEOS). The mercapto groups were introduced to the surface of the mesoporous silics (SBA-15) by the addition of 3-(mercaptopropyl)trimethoxysilane (MPTMS). The mercapto groups were finally oxidized into sulfonic groups inside the pores of the mesoporous silica by hydrogen peroxide. Compared with sulfuric acid, the so-produced catalyst showed higher catalytic activity for the hydrolysis of cellobiose (disaccharide of glucose, used as model compound). In 1 hr reaction at 120°C in water, 0.05 M sulfuric acid converted 54% of 1g of cellobiose into glucose while 1.5 g of Fe_3O_4-SBASO$_3$H catalyst converted 98%. A possible explanation of this behaviour was found in the channels of the magnetic catalysts, where concentrated acid sites were placed: the uniformity of the channels facilitated the easy entrance of the reactant and then the reaction with these active sites. Due to its high surface area, Fe_3O_4-SBA-SO_3H also showed higher activity than those of other strong acid catalysts, such as AC-SO_3H and commercial Amberlyst-15. The hydrolysis of a biomass compound, specifically corn cob, showed a total reducing (TRS) yield of 45% in 3 hr at 150°C. The catalysts

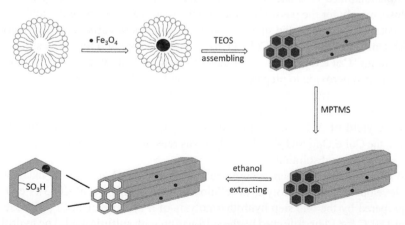

Fig. 6. Schematic illustration of the preparation of Fe_3O_4-SBA-SO_3H, by Lai et al. (2011). Reproduced by permission of The Royal Society of Chemistry.

was easily recovered by an external magnet, and showed no deactivation during three recycling tests after reactivation with 1 M sulfuric acid.

Following a similar approach, reduced graphene oxide (RGO) has also been used as supporting material instead of mesoporous SBA in the preparation of Fe_3O_4-RGO-SO_3H (Yang et al. 2015). Graphene oxide (GO) was prepared via the Hummer method from graphite powder. Sequentially, the GO was dispersed in water with $FeCl_3$ and $FeCl_2$. Later, ammonia was used to reduce the iron ions to produce the Fe_3O_4-RGO hybrid powder. Finally, the powder was sulfonated with 4-Benzenediazoniumsulfonate. The catalyst exhibited outstanding catalytic performance for the hydrolysis of cellulosic materials, with a yield of 98% of conversion of cellobiose in 3 hr at 150°C. The high activity of the catalysts was explained in terms of its crumpling feature with clear layers and with the coexistence of –COOH and –OH groups. The unique structure of Fe_3O_4-RGO-SO_3H favours the accessibility of cellulose to the active sites of the material and facilitates the diffusion of the product molecules. –COOH groups on the surface of the material provide a large concentration of acidic functionality for hydrolyzing cellulose, while –OH groups adsorb cellulose through forming strong hydrogen bonds between –OH groups and the oxygen atoms in beta-1,4 glycosidic bonds. Furthermore, the catalysts was separated from the reaction residue with an external magnetic force. The catalysts was reused at least five times without the loss of activity.

In addition, the sulfuric acid functionalized magnetic carbon nanotube catalyst (sulfonated MCNAs) was also prepared and used for the hydrolysis of crop stalks (Liu et al. 2014). The magnetic carbon nanotube arrays were synthesized by the chemical vapour deposition, using xylene and ferrocene as C and Fe precursors. Sulfonated magnetic carbon nanotube arrays were sequentially prepared by the treatment of magnetic carbon nanotube arrays with sulfuric acid at 250°C under nitrogen flow. The hydrolysis of corn stalk with MCNAs showed a TRS yield of 31% after 2 hr at 150°C. The TRS yield was also recorded after the four cycles reuse.

Core shell structured magnetic acid catalysts, where a magnetic core is covered with functionalized acid shells, have received more interest for the hydrolysis of cellulose, mainly for the possibility to avoid the reactivation step for the reutilization of the catalyst. Normally, the shell can be made of silica or carbonaceous materials.

Ebitani et al. (2011) used $CoFe_2O_4$ as a magnetic core and SiO_2 as a silica shell. The resulting $CoFe_2O_4@SiO_2$ was functionalized with thiol groups and oxidized with hydrogen peroxide to prepare $CoFe_2O_4@SiO_2$-SO_3H. The catalyst showed high catalytic activity, affording a glucose yield of 93% in the hydrolysis of cellobiose in water at 100°C in 20 min. In the same condition, commercial Amberlyst-15 showed a glucose yield of 22%. In contrast to the Fe_3O_4-mesoporous functionalized acid catalyst, the $CoFe_2O_4@SiO_2$-SO_3H catalyst was reused without standard regeneration in the sulphuric acid solution.

With a similar approach, a core shell structured $Fe_3O_4@C$–SO_3H catalyst has been developed for the hydrolysis of cellulose in water (Zhang et al. 2013). $Fe_3O_4@C$ was prepared by the one step hydrothermolysis of a mixture of $FeCl_3$, glucose and urea at 180°C for 14 hr, followed by the sulfonation with sulfuric acid. The hydrolysis of cellulose over the $Fe_3O_4@C$–SO_3H catalyst gave 48.6% cellulose conversion with 52.1% glucose selectivity, at140°C for 12 hr in water.

In the preparation of core-shell structured nanoparticles, it has been demonstrated that the carbonization and sulfonation processes have great influence on the quantity of the acidic groups (sulfonic, carboxy, and hydroxyl groups) and the stability of magnetic catalysts (Zhang et al. 2016). Therefore, a stable core-shell structured magnetic solid acid catalyst (MSAC), $Fe_3O_4/C-SO_3H$, has been prepared in a way to control these characteristics. The catalyst was prepared from glucose, sulfuric acid and modified magnetic nanoparticles of Fe_3O_4, which were used as the core. The best preparation conditions for MSAC were 3 hr of carbonization time, 450°C as the carbonization temperature, 9 hr of sulfonation time and 90°C as the sulfonation temperature. Corncobs was treated with the catalyst achieving 44.3% of xylose yield at 160°C for 16 hr. The catalyst has been proved to be stable and reusable up to three times. The catalyst was easily collected after evry cycle with the utilizaton of an external magnetic field.

Hydrolysis of cellulose and lignocelluloses by magnetic biocatalysts

As in the case of biodiesel, the hydrolysis of cellulose nanomagnetic biocatalysts have also been demonstrated to be efficient catalysts. The reason for the industrial applicability of these catalyst, still rise from the general peculiarities of biocatalysts described above. As for biodiesel, the nanoparticles are used as vehicles to support the enzymes and easily remove them from the reaction mixture after the reaction. The biocatalysts used for the hydrolysis of cellulose are cellulase. There are five types of cellulase: endocellulases, exocellulases or cellobiohydrolases, exocellulases, cellobiases or beta-glucosidases, oxidative cellulases and cellulose phosphorylases. Each type of enzyme decomposes cellulose via different mechanisms and with different outputs. However, mixtures of enzymes can also be used to optimize the yield of the reaction with a synergetic effect of the different biocatalysts.

For example, nonporous super paramagnetic nanoparticles (cyanuric chloride-activated and polyglutaraldehyde-activated) have been used to immobilize whole cellulase mixtures (endoglucanase, cellobiohydrolase and β-glucosidase) for the hydrolysis of cellulose (Alftren and Hobley 2014). The cellulase mixture was immobilized on silica coated magnetic nanoparticles. The magnetic biocatalyst showed a high mass of reducing sugar produced per mass of particles (2.8 g kg^{-1} min^{-1}). The magnetic was also tested in the hydrolysis of pretreated wheat straw biomass, confirming the potential of hydrolysing the real lignocellulosic substrate.

Recombinant cellulases has been supported on magnetic nanoparticles for the hydrolysis of carboxymethyl cellulose (CMC) and hemp hurd from the industrial residual (Abraham et al. 2014). With the best conditions, hydrolysis of CMC achieved 81%. With pretreated hemp hurd biomass, the maximum hydrolysis reached 93% in 48 hr. In addition, the immobilized enzyme retained 50% enzyme activity up to five cycles.

Beta-Glucosidase (BGL) from *Aspergillus niger* has been immobilized on functionalized iron oxide magnetic nanoparticles by covalent binding (Verma et al. 2013). Immobilized nanoparticles showed 93% immobilization binding. The immobilization of the enzymes was proved by Transmission electron microscopy (TEM) and Fourier transform infrared spectroscopy (FTIR) techniques. The thermal stability of the immobilized enzyme was enhanced at 70°C. The immobilized nanoparticle enzyme conjugate retained more than 50% of the enzyme activity up to 16 cycles. The maximum glucose yield of 90% from cellobiose hydrolysis by immobilized BGL was achieved in 16 hr.

Trichoderma reesei cellulase has also been employed as a nanomagnetic biocatalyst (Verma et al. 2013). The cellulase was anchored by covalent immobilization on chitosan-coated magnetic nanoparticles using glutaraldehyde as a coupling agent. Immobilized cellulase retained about 80% of its activity after 15 cycles of carboxymethylcellulose hydrolysis. The immobilized enzyme was able to hydrolyse lignocellulosic material from *Agave Atrovirens* leaves with a yield close to the amount detected with free enzyme and it was re-used in vegetal material conversion up to four cycles with 50% activity decrease.

Although the magnetic biocatalysts can promote the cellulose (or lignocelluloses) hydrolysis under mild conditions, these processes generally require a long reaction time and suffer the loss of the activity of the enzymes.

Dehydration of fructose into 5-HMF

When considering the possible downstream chemical processing technologies, the conversion of sugars to value-added chemicals is one of the most important (Lichtenthaler and Peters 2004). In fact, sugars are natural sources of hydrocarbons that can be converted in chemicals as an alternative to petroleum derivatives. Among the sugars, hexoses, that is, the six-carboned carbohydrates, are the most abundant monosaccharide existing in nature. The catalytic transformation of hexoses into furans mainly involves several steps such as dehydration, hydrolysis, isomerization, reforming, aldol condensation, hydrogenation and oxidation, which are of general interest. Among the hexoses, D-fructose and glucose (Fig. 7a and Fig. 7b) are economical and suitable to be used as the chemical feedstocks (Huber et al. 2005; Chheda et al. 2007; Corma et al. 2007).

The main industrial-captivating furanic products derivable from fructose and glucose include 5-hydroxymethylfurfural (5-HMF), 2,5-diformylfuran (2,5-DFF), 2,5-furandicarboxylic acid (2,5-FDCA), 2,5-bis(hydroxymethyl)-furan (2,5-BHF) and 2,5-dimethylfuran (2,5-DMF) (structures shown in Fig. 8).

As a dehydration product of hexoses, and therefore involving simple reaction steps, 5-HMF has been considered to be an important and renewable platform chemical for the green industry.

Actually, 5-HMF had been of great interest since the last decade of the 19th century. It was first separated with 20% yield from the reaction mixture of

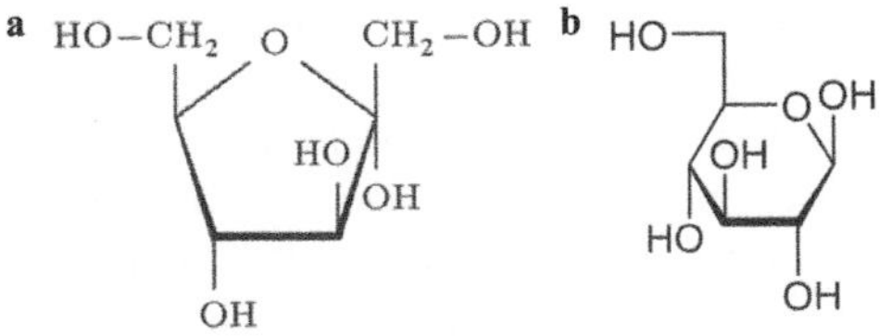

Fig. 7. (a) Alpha-D-Fructose. (b) Beta-D-glucose.

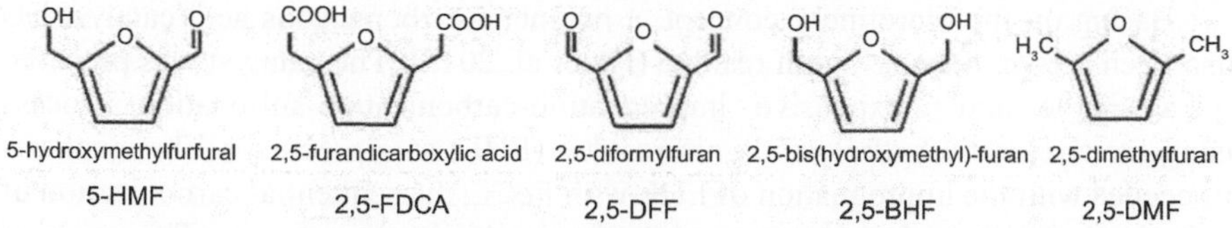

Fig. 8. Main industrial-captivating furanic products derivable from fructose and glucose.

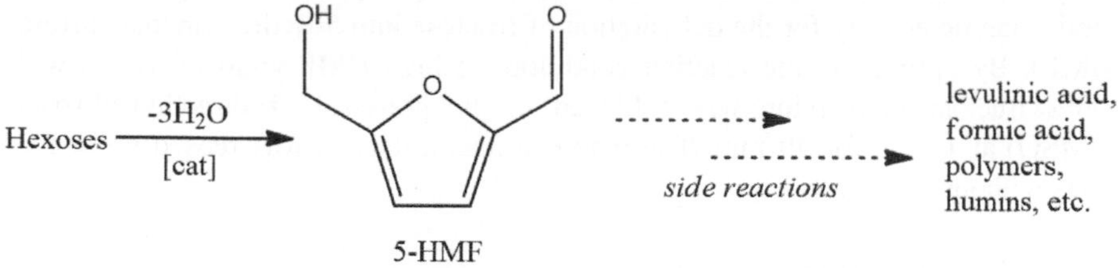

Fig. 9. Reaction scheme of the dehydration reaction of hexoses.

fructose and sucrose in 1895. Then, Fenton and his coworkers performed extensive investigations on 5-HMF (Fenton and Gostling 1901). After many year of research, a detailed characterization and explanation of the synthesis of 5-HMF have been provided. It has been proved that 5-HMF is formed from hexoses through removing three water molecules in the acid-catalyzed dehydration reaction (Antal et al. 1990). However, as illustrated in Fig. 9, the dehydration route of hexoses can produce several by-products.

Specifically, in the aqueous systems, 5-HMF enters into a consecutive reaction sequence, taking up two molecules of water and forming levulinic and formic acid as semi-final products. In the non-aqueous system, the hydrolysis of 5-HMF can be suppressed. However, the polymerization reactions are observed in all the systems, and lead to the production of coloured soluble polymers and insoluble brown humins (Lewkowski 2001). In order to prevent the side reactions and obtain a high yield of 5-HMF, one solution involves the designing and employment of a suitable catalyst which allows the formation of 5-HMF, while not promoting the consecutive reactions. The catalysts used are generally classified as mineral acid, organic acid and solid acid catalysts, and metal-containing catalysts (Tong et al. 2010).

Most of the magnetic nanocatalysts for the dehydration of fructose are of acidic nature. For example, a Fe_3O_4 core and sulfonic acid functionalized silica shell has been synthesized using the reverse microemulsion method (Zhang et al. 2014). For the preparation of the catalyst, the Fe_3O_4 nanoparticles were coated with phenyl modified silica shell nanolayers, and the phenyl groups were subsequently sulfonated to generate a solid sulfonic acid catalyst. The obtained acid catalyst showed higher activity than the conventional Amberlyst-15 catalyst and comparable activity to several homogeneous sulfonic acid catalysts with respect to the dehydration of fructose to 5-HMF. A fructose conversion of 99% with an HMF yield of 82% following 3 hr in dimethylsulfoxide at 110°C has been reported. After three recycles, no difference between the recycled catalyst and the fresh one were noticed.

Basing on the biorefinery concept, a magnetic carbonaceous acid catalyst has also been prepared using lignin residue (Hu et al. 2015). The catalyst was prepared by a simple and inexpensive impregnation-carbonization-sulfonation process using the enzymatic hydrolysis lignin residue (EHL) as a precursor. The synthesis proceeded with the impregnation of EHL with $FeCl_3$, the sequential carbonization of $FeCl_3$-loaded EHL under nitrogen and the final sulfurization process. The resulting catalyst possessed a porous structure with a high surface area and contained Fe_3O_4 components and $-SO_3H$, COOH and phenolic OH groups. The catalyst showed a good catalytic activity for the dehydration of fructose into 5-hydroxymethylfurfural (HMF). By optimizing the reaction conditions, a high HMF yield of 81.1% with 100% fructose conversion was achieved in the presence of dimethylsulfoxide (DMSO) at 130°C for 40 min. The magnetic nanocatalysts was reused up to five times without losing activity.

Derivatives of 5-HMF

5-HMF derivatives such as 2,5-diformylfuran (DFF), 2,5-furfuryldiisocyanate and 5-hydroxymethyl furfurylidenester, are particularly suitable starting materials for the preparation of polymeric materials such as polyesters, polyamides and polyurethane (Gandini and Belgacem 2002; Saha et al. 2012). In addition, these kinds of polymer, properly called furan-based polymers, display very good properties. For example, the polyurethane from furan shows very high resistance to thermal treatment and the kevlar-like polyamides produced from furan diamines and diacids exhibit liquid crystal behaviour. Also, the photo reactive polyesters have been used for printing ink formulation. Furthermore, the furan-based polyconjugated polymers possess good electrical conductivity (Saha et al. 2012). For all these reasons, 5-HMF has been called a well-known 'sleeping giant' in the field of industrial intermediate chemicals, just waiting for the opportunity to awake (Bicker et al. 2003). During the last years, some novel magnetic nanocatalysts have been designed for the conversion of 5-HMF into different derivatives.

Magnetic iron oxide-manganese oxide supported nanoparticles (Fe_3O_4/Mn_3O_4) were reported as efficient magnetically recoverable catalysts for the aerobic oxidation of the biomass-derived HMF into 2,5-diformylfuran (DFF) under benign reaction conditions (Liu et al. 2014). The high DFF yield of 82.1% has been obtained with 5-HMF conversion of 100% in 4 hr at 110°C.

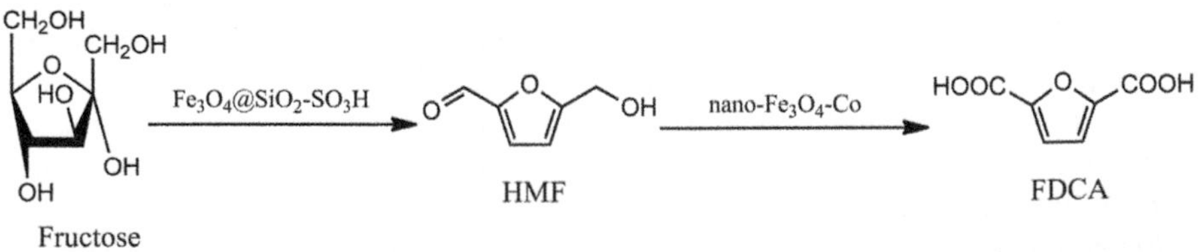

Fig. 10. Schematic illustration of the reaction route for the production of decaline from cyclopentanol. Taken from Wang et al. 2015. Copyright (2015), American Chemical Society.

More interesting, Wang et al. (2015) prepared two magnetic nanocatalysts, $Fe_3O_4@SiO_2-SO_3H$ and $Fe_3O_4-CoO_x$, for the one pot conversion of fructose to 2,5-Furandicarboxylic acid (FDCA) through the *in situ* formation of HMF, as shown in Fig. 10.

Both catalysts were produced via simple and easily reproducible wet impregnation methods. Firstly, the $Fe_3O_4-CoO_x$ catalyst was demonstrated to be active for the conversion of HMF to FDCA. In that way, several important reaction parameters were explored, with the highest FDCA yield of 68.6% obtained from HMF after 15 hr at a reaction temperature of 80°C. Sequentially, the one pot conversion of fructose into FDCA was performed. The catalytic conversion of fructose over $Fe_3O_4@SiO_2-SO_3H$ yielded 93.1% HMF, which was oxidized *in situ* into FDCA with a yield of 59.8% with $Fe_3O_4-CoO_x$.

Similarly, the co-utilization of $Fe_3O_4-RGO-SO_3H$ and a novel magnetic catalyst, $ZnFe_{1.65}Ru_{0.35}O_4$, was tested for the production of two different derivatives of HMF (Yang et al. 2017). In detail, the catalyst $ZnFe_{1.65}Ru_{0.35}O_4$ was prepared via a simple alkali-coprecipitation method, while $Fe_3O_4-RGO-SO_3H$ was prepared by treating GO with a solution of Fe^{2+}/Fe^{3+} in ammonia and sequentially functionalized. $ZnFe_{1.65}Ru_{0.35}O_4$'s catalytic performance was investigated for aerobic oxidation of 5-hydroxymethylfurfural (HMF) to 2,5-diformylfuran (DFF) and 2,5-furatidicarboxylic acid (FDCA) under different reaction conditions. A high DFF yield of 93.5% was obtained for the reaction at 110°C for 4 hr in DMSO, whereas the maximum FDCA yield of 91.2% was achieved for the reaction at 130°C for 16 hr in $H_2O/DMSO$. More importantly, a plausible reaction pathway was proposed for the aerobic oxidation of HMF into FDCA, with the rate-determining step determined through kinetic study. The sequential utilization of $Fe_3O_4-RGO-SO_3H$ and $ZnFe_{1.65}Ru_{0.35}O_4$ led to the production of DFF or FDCA directly from fructose, as shown in Fig. 11. After the removal of $Fe_3O_4-RGO-SO_3H$ from the reaction solution with a magnet, HMF was further oxidized in the catalysis of $ZnFe_{1.65}Ru_{0.35}O_4$. These two consecutive steps produced DFF in 73.3% yield and FDCA in 70.5% yield, under the respective optimum reaction conditions. The direct synthesis of DFF or FDCA from fructose avoids tedious separation of intermediate HMF, saving both time and cost.

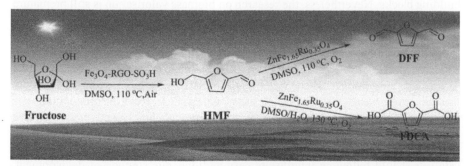

Fig. 11. Schematic illustration of the one pot reaction route for the production of DFF and FDCA from Fructose. Taken from Yang et al. 2017. Copyright (2017), American Chemical Society.

Aiming to a similar objective of a one pot production of HMF derivatives directly from fructose, but using only one catalyst, a magnetic bi-functional WO_3HO-VO(salten)-$SiO_2@Fe_3O_4$ nanocatalyst has been recently prepared (Mittal et al. 2016). The catalyst was used to synthesize 2,5-diformylfuran (2,5-DFF) from fructose. The chlorosilylated $SiO_2@Fe_3O_4$ (Cl-$SiO_2@Fe_3O_4$) nanoparticles served as the nanomagnetic platform for the two functionalities. Tungstic acid was generated via the protonation of sodium tungstate. Oxovanadium was complexed with a salten ligand. Under the optimal one-pot system, tungstic acid-mediated fructose dehydration afforded 82% conversion into 5-hydroxymethylfurfural (5-HMF) in 1 hr. The nanocatalyst was retrieved magnetically and re-used up to five times with marginal losses in the activity.

Conclusion

As described in the 2016 European Roadmap on Catalysis developed by the *European Cluster on Catalysis*, the utilization of catalysts is an indispensable key for a sustainable future for our society. The successful realisation of new catalytic materials requires researchers to integrate knowledge from fundamental research areas, including chemistry, physics, biology and mathematics, with engineering and applied sciences such as industrial chemistry. Current trends in catalysis include the integration of theoretical modelling with *in situ* studies to understand the reaction mechanism, the science of the catalyst at nanoscale dimension, the micro-kinetics and the reactor modelling. Another challenge is to describe a unified approach for homogeneous-, heterogeneous- and bio-catalysis. All these are examples of the generic challenge of "catalyst by design". Today, the area of catalysis is moving from description to prediction. In fact, every day the research in this field is increasingly pushing towards a more rational approach for the development of new catalytic materials for chemical processes. The fundamental elements of such an approach are computational modelling and advanced synthetic approaches aimed at delivering materials with improved catalytic performance, preferably starting from cheap, earth-abundant and easily accessible raw building blocks. As described above, the utilization of biomass is the only sustainable and green possible source of these building blocks. Nevertheless, the research on biomass valorisation is still premature compared to the research on other sources such as the petroleum ones. On the one hand, this opens the possibility of developing new reactions and pathways by also using the technologies developed for the other chemicals/fuel sources. On the other hand, the market of petroleum fuel/chemical is so mature that it is hardly attackable, so that, for the present, the economy might translate to a biomass-based one only through governments' directives. Unless new catalysts decrease the production costs to highly competitive goods. This is possible through highly active, recoverable catalysts. Therefore, the utilization of magnetically recoverable nanoparticles as innovative catalysts is an open challenge for the future. The research of MNPs catalysts should follow an interdisciplinary and dynamic strategy of development. This is possible though the modelling of new catalysts and reactions, and the focus on the end-life of the materials. Specifically, as the growth of the world standard of life is increasing, and the utilization of natural resources is growing too, the reutilization of metals is an inevitable passage. In addition, all the other components of the MNPs should be recoverable or at least nontoxic. Therefore, the MNPs catalysts should be designed in order to be easily recovered after every cycle, as to avoid the loss of metals in the reaction mixture and in the environment and to amortize the catalysts' costs. In addition, the MNPs should be easily recycled after the end-life of the catalysts. In that way, the utilization of biomass-derived supporting and stabilizing material is extremely important as they could track the circular life of the MNPs.

Besides reusability, one other fundamental characteristic that MNPs should possess is the robustness: in fact, raw biomass materials may contain impurities which could alter catalysts' selectivity and decrease activities. In addition, the robustness of the MNPs is important for the utilization of the newest reaction media such as ionic liquids, molten salt hydrates, supercritical fluids and energy sources such as microwaves and ultrasounds.

Lastly, the development of the MNPs should follow the growth of biorefineries. Therefore, as biorefineries aim to the same production rate of nowadays refineries, that is, of 5% of the total output to chemical products and the rest to transportation fuels and energy, the research on MNPs and biomass valorisation should always be correctly balanced in its targets.

Acknowledgements

Grants from the European Union's Horizon 2020 research and innovation Programme under the Marie Skłodowska-Curie grant agreement No. 721290. This publication reflects only the author's view, exempting the Community from any liability. Project website: http://cosmic-etn.eu/.

Reference cited

Abdelsalam, E., M. Samer, Y.A. Attia, M.A. Abdel-Hadi, H.E. Hassan and Y. Badr. 2017. Influence of zero valent iron nanoparticles and magnetic iron oxide nanoparticles on biogas and methane production from anaerobic digestion of manure. Energy 120: 842–853.

Abraham, R.E., M.L. Verma, C.J. Barrow and M. Puri. 2014. Suitability of magnetic nanoparticle immobilised cellulases in enhancing enzymatic saccharification of pretreated hemp biomass. Biotechnol Biofuels 7: 90.

Alftren, J. and T.J. Hobley. 2014. Immobilization of cellulase mixtures on magnetic particles for hydrolysis of lignocellulose and ease of recycling. Biomass Bioenerg 65: 72–78.

Aliramaji, S., A. Zamanian and Z. Sohrabijam. 2015. Characterization and synthesis of magnetite nanoparticles by innovative sonochemical method. 5th International Biennial Conference on Ultrafine Grained and Nanostructured Materials, UFGNSM-15 11: 265–269.

Anastas, P.T. and J.C. Warner. 1998. Green Chemistry: Theory And Practice. Oxford England; New York, Oxford University Press.

Anastas, P.T. and M.M. Kirchhoff. 2002. Origins, current status, and future challenges of green chemistry. Accounts Chem. Res. 35(9): 686–694.

Anr, R., A.A. Saleh, M.S. Islam and S. Hamdan. 2016. Active heterogeneous CaO catalyst synthesis from Anadara granosa (Kerang) seashells for Jatropha biodiesel production. 9th International Unimas STEM Engineering Conference—Innovative Solutions for Engineering and Technology Challenges (ENCON), Kuching, Malaysia, E D P Sciences.

Antal, M.J., W.S.L. Mok and G.N. Richards. 1990. Kinetic-studies of the reactions of ketoses and aldoses in water at high-temperature. 1—Mechanism of formation of 5-(hydroxymethyl)-2-furaldehyde from d-fructose and sucrose. Carbohydrate Research 199(1): 91–109.

Astruc, D., F. Lu and J.R. Aranzaes. 2005. Nanoparticles as recyclable catalysts: The frontier between homogeneous and heterogeneous catalysis. Angew. Chem. Int. Edit. 44(48): 7852–7872.

Bachmann, G. 2005. Winning the oil endgame. Innovation for profits, jobs, and security. Gaia 14(2): 186–187.

Baig, R.B.N. and R.S. Varma. 2014a. Magnetic carbon-supported palladium nanoparticles: An efficient and sustainable catalyst for hydrogenation reactions. ACS Sustain. Chem. Eng. 2(9): 2155–2158.

Baig, R.B.N. and R.S. Varma. 2014b. Magnetic silica-supported palladium catalyst: synthesis of allyl aryl ethers in water. Ind. Eng. Chem. Res. 53(49): 18625–18629.

Baig, R.B.N., M.N. Nadagouda and R.S. Varma. 2015. Magnetically retrievable catalysts for asymmetric synthesis. Coordin. Chem. Rev. 287: 137–156.

Baig, R.B.N., B.R. Vaddula, M.N. Nadagouda and R.S. Varma. 2015. The copper-nicotinamide complex: sustainable applications in coupling and cycloaddition reactions. Green Chem. 17(2): 1243–1248.

Baudoux, J., K. Perrigaud, P.J. Madec, A.C. Gaumont and I. Dez. 2007. Development of new SILP catalysts using chitosan as support. Green Chem. 9(12): 1346–1351.

Bentham, J. 2014. The scenario approach to possible futures for oil and natural gas. Energ Policy 64: 87–92.

Bicker, M., J. Hirth and H. Vogel. 2003. Dehydration of fructose to 5-hydroxymethylfurfural in sub- and supercritical acetone. Green Chem. 5(2): 280–284.

Boon-anuwat, N.N., W. Kiatkittipong, F. Aiouache and S. Assabumrungrat. 2015. Process design of continuous biodiesel production by reactive distillation: Comparison between homogeneous and heterogeneous catalysts. Chem. Eng. Process 92: 33–44.

Bozell, J.J. 2008. Feedstocks for the future—Biorefinery production of chemicals from renewable carbon. Clean-Soil Air Water 36(8): 641–647.

Bozell, J.J. 2010. Connecting biomass and petroleum processing with a chemical bridge. Science 329(5991): 522–523.

Bozell, J.J. and G.R. Petersen. 2010. Technology development for the production of biobased products from biorefinery carbohydrates-the US Department of Energy's "Top 10" revisited. Green Chem. 12(4): 539–554.

Cai, Z.Z., Y. Wang, Y.L. Teng, K.M. Chong, J.W. Wang, J.W. Zhang and D.P. Yang. 2015. A two-step biodiesel production process from waste cooking oil via recycling crude glycerol esterification catalyzed by alkali catalyst. Fuel Process Technol. 137: 186–193.

Canakci, M. and J. Van Gerpen. 2001. Biodiesel production from oils and fats with high free fatty acids. T ASAE 44(6): 1429–1436.

Carraretto, C., A. Macor, A. Mirandola, A. Stoppato and S. Tonon. 2004. Biodiesel as alternative fuel: Experimental analysis and energetic evaluations. Energy 29(12-15): 2195–2211.

Central Intelligence Agency. 2016. The World factbook. Washington D.C., National Government Publication.

Chen, Y., J.J. Cheng and K.S. Creamer. 2008. Inhibition of anaerobic digestion process: A review. Bioresource Technol. 99(10): 4044–4064.

Chheda, J.N., G.W. Huber and J.A. Dumesic. 2007. Liquid-phase catalytic processing of biomass-derived oxygenated hydrocarbons to fuels and chemicals. Angew Chem. Int. Edit. 46(38): 7164–7183.

Clark, J.H. and D.J. Macquarrie. 2002. Handbook of Green Chemistry and Technology. Abingdon, Blackwell Publishing.

Corma, A. and H. Garcia. 2003. Lewis acids: From conventional homogeneous to green homogeneous and heterogeneous catalysis. Chem. Rev. 103(11): 4307–4365.

Corma, A., S. Iborra and A. Velty. 2007. Chemical routes for the transformation of biomass into chemicals. Chem. Rev. 107(6): 2411–2502.

Cuenya, B.R. 2010. Synthesis and catalytic properties of metal nanoparticles: Size, shape, support, composition, and oxidation state effects. Thin Solid Films 518(12): 3127–3150.

Dalpozzo, R. 2015. Magnetic nanoparticle supports for asymmetric catalysts. Green Chem. 17(7): 3671–3686.

De Baere, L. 2006. Will anaerobic digestion of solid waste survive in the future? Water Sci. and Technol. 53(8): 187–194.

Dethlefsen, J.R., D. Lupp, A. Teshome, L.B. Nielsen and P. Fristrup. 2015. Molybdenum-catalyzed conversion of diols and biomass-derived polyols to alkenes using isopropyl alcohol as reductant and olvent. ACS Catal 5(6): 3638–3647.

Directive 2009/28/EC of the European Parliament and of the Council of 23 April 2009 on the promotion of the use of energy from renewable sources and amending and subsequently repealing Directives 2001/77/EC and 2003/30/EC. (Available at http://data.europa.eu/eli/dir/2009/28/oj).

Dong, N., F.Z. He, J.L. Xin, Q.Z. Wang, Z.Q. Lei and B.T. Su. 2015. Preparation of $CoFe_2O_4$ magnetic fiber nanomaterial via a template-assisted solvothermal method. Mater Lett. 141: 238–241.

Du, W., Y.Y. Xu, D.H. Liu and J. Zeng. 2004. Comparative study on lipase-catalyzed transformation of soybean oil for biodiesel production with different acyl acceptors. J. Mol. Catal B-Enzym 30(3-4): 125–129.

Fan, Y., F. Su, K. Li, C. Ke and Y. Yan. 2017. Carbon nanotube filled with magnetic iron oxide and modified with polyamidoamine dendrimers for immobilizing lipase toward application in biodiesel production. Sci. Rep.-UK 7: 45643.

Fan, X.Y., S.H. Li, H.H. Zhou and L. Lu. 2015. One-pot high temperature hydrothermal synthesis of Fe3O4@C/graphene nanocomposite as anode for high rate lithium ion battery. Electrochim. Acta 180: 1041–1049.

Fenton, H.J.H. and M. Gostling. 1901. Derivatives of methylfurfural. J. Chem. Soc. 79: 807–816.

Fjerbaek, L., K.V. Christensen and B. Norddahl. 2009. A review of the current state of biodiesel production using enzymatic transesterification. Biotechnol. Bioeng. 102(5): 1298–1315.

Fu, J.Y., L.G. Chen, P.M. Lv, L.M. Yang and Z.H. Yuan. 2015. Free fatty acids esterification for biodiesel production using self-synthesized macroporous cation exchange resin as solid acid catalyst. Fuel 154: 1–8.

Gandini, A. and M.N. Belgacem. 2002. Recent contributions to the preparation of polymers derived from renewable resources. J. Polym. Environ. 10(3): 105–114.

Gawande, M.B., R. Luque and R. Zboril. 2014. The rise of magnetically recyclable nanocatalysts. Chemcatchem. 6(12): 3312–3313.

Georgakilas, V., V. Tzitzios, D. Gournis and D. Petridis. 2005. Attachment of magnetic nanoparticles on carbon nanotubes and their soluble derivatives. Chem. Mater 17(7): 1613–1617.

Govan, J. and Y.K. Gun'ko. 2014. Recent advances in the application of magnetic nanoparticles as a support for homogeneous catalysts. Nanomaterials-Basel 4(2): 222–241.

Griffin Shay, E. 1993. Diesel fuel from vegetable oils: Status and opportunities. Biomass Bioenerg. 4(4): 227–242.

Gu, L., W. Huang, S.K. Tang, S.J. Tian and X.W. Zhang. 2015. A novel deep eutectic solvent for biodiesel preparation using a homogeneous base catalyst. Chem. Eng. J. 259: 647–652.

Guibal, E. 2004. Interactions of metal ions with chitosan-based sorbents: a review. Sep. Purif Technol. 38(1): 43–74.

Guignard, F. and M. Lattuada. 2015. Template-assisted synthesis of Janus silica nanobowls. Langmuir 31(16): 4635–4643.

Guldhe, A., B. Singh, T. Mutanda, K. Perrnaul and F. Bux. 2015. Advances in synthesis of biodiesel via enzyme catalysis: Novel and sustainable approaches. Renew. Sust. Energ. Rev. 41: 1447–1464.

Guo, P.M., F.H. Huang, M.M. Zheng, W.L. Li and Q.D. Huang. 2012. Magnetic solid base catalysts for the production of biodiesel. J. Am. Oil. Chem. Soc. 89(5): 925–933.

Hardy, J.J.E., S. Hubert, D.J. Macquarrie and A.J. Wilson. 2004. Chitosan-based heterogeneous catalysts for Suzuki and Heck reactions. Green Chem. 6(1): 53–56.

Hsu, A.F., K. Jones, T.A. Foglia and W.N. Marmer. 2002. Immobilized lipase-catalysed production of alkyl esters of restaurant grease as biodiesel. Biotec. Appl. Bioc. 36: 181–186.

Hu, L., X. Tang, Z. Wu, L. Lin, J.X. Xu, N. Xu and B.L. Dai. 2015. Magnetic lignin-derived carbonaceous catalyst for the dehydration of fructose into 5-hydroxymethylfurfural in dimethylsulfoxide. Chem. Eng. J. 263: 299–308.

Huber, G.W., J.N. Chheda, C.J. Barrett and J.A. Dumesic. 2005. Production of liquid alkanes by aqueousphase processing of biomass-derived carbohydrates. Science 308(5727): 1446–1450.

Iso, M., B.X. Chen, M. Eguchi, T. Kudo and S. Shrestha. 2001. Production of biodiesel fuel from triglycerides and alcohol using immobilized lipase. J. Mol. Catal. B-Enzym. 16(1): 53–58.

Jiang, X., X.L. Yang, Y.H. Zhu, Y.F. Yao, P. Zhao and C.Z. Li. 2015. Graphene/carbon-coated Fe3O4 nanoparticle hybrids for enhanced lithium storage. J. Mat. Chem. A 3(5): 2361–2369.

Jungmeier, G., R. Van Ree, E. de Jong, H. Stichnothe, I. de Bari, H. Jorgensen, M. Wellisch, P. Walsh, G. Garnier, J. Spaeth, K. Torr and K. Habu. 2014. Facts, figures and integration of biorefineries in a future bioeconomy—findings in IEA bioenergy task 42 "biorefining". Papers of the 22nd European Biomass Conference: Setting the Course for a Biobased Economy: 1480–1485.

Kabashima, H. and H. Hattori. 1998. Cyanoethylation of alcohols over solid base catalysts. Catal Today 44(1-4): 277–283.

Kamm, B. and M. Kamm. 2004. Principles of biorefineries. Appl. Microbiol. Biot. 64(2): 137–145.

Kappe, C.O. 2004. Controlled microwave heating in modern organic synthesis. Angew Chem. Int. Edit. 43(46): 6250–6284.

Kjarstad, J. and F. Johnsson. 2009. Resources and future supply of oil. Energ Policy 37(2): 441–464.

Kouzu, M. and J. Hidaka. 2012. Transesterification of vegetable oil into biodiesel catalyzed by CaO: A review. Fuel 93(1): 1–12.

Kramareva, N.V., A.Y. Stakheev, O.P. Tkachenko, K.V. Klementiev, W. Grunert, E.D. Finashina and L.M. Kustov. 2004. Heterogenized palladium chitosan complexes as potential catalysts in oxidation reactions: study of the structure. J. Mol. Catal. A-Chem. 209(1-2): 97–106.

Krausmann, F., D. Wiedenhofer, C. Lauk, W. Haas, H. Tanikawa, T. Fishman, A. Miatto, H. Schandl and H. Haberl. 2017. Global socioeconomic material stocks rise 23-fold over the 20th century and require half of annual resource use. Proceedings of the National Academy of Sciences of the United States of America 114(8): 1880–1885.

Kumari, A., P. Mahapatra, V.K. Garlapati and R. Banerjee. 2009. Enzymatic transesterification of Jatropha oil. Biotechnol. Biofuels 2: 1.

Lai, D.M., L. Deng, Q.X. Guo and Y. Fu. 2011a. Hydrolysis of biomass by magnetic solid acid. Energ. Environ. Sci. 4(9): 3552–3557.

Lai, D.M., L. Deng, J.A. Li, B. Liao, Q.X. Guo and Y. Fu. 2011b. Hydrolysis of cellulose into glucose by magnetic solid acid. Chemsuschem. 4(1): 55–58.

Lamer, V.K. and R.H. Dinegar. 1950. Theory, production and mechanism of formation of monodispersed hydrosols. J. Am. Chem. Soc. 72(11): 4847–4854.

Lancaster, M. 2002. Green Chemistry: An Introductory Text. Cambridge, RCS Editions.

Laurent, S., D. Forge, M. Port, A. Roch, C. Robic, L.V. Elst and R.N. Muller. 2008. Magnetic iron oxide nanoparticles: Synthesis, stabilization, vectorization, physicochemical characterizations, and biological applications. Chem. Rev. 108(6): 2064–2110.

Lenders, J.J.M., L.A. Bawazer, D.C. Green, H.R. Zope, P.H.H. Bomans, G. de With, A. Kros, F.C. Meldrum and N. Sommerdijk. 2017. Combinatorial evolution of biomimetic magnetite nanoparticles. Advanced Functional Materials 27(10): 9.

Lewkowski, J. 2001. Synthesis, chemistry and applications of 5-hydroxymethylfurfural and its derivatives. Arkivoc. 1: 17–54.

Li, S.-F., Y.-H. Fan, R.-F. Hu and W.-T. Wu. 2011. Pseudomonas cepacia lipase immobilized onto the electrospun PAN nanofibrous membranes for biodiesel production from soybean oil. J. Mol. Cat. B-Enzym. 72(1-2): 40–45.

Lichtenthaler, F.W. and S. Peters. 2004. Carbohydrates as green raw materials for the chemical industry. CR Chim. 7(2): 65–90.

Lin, L., S. Vittayapadung, X.Q. Li, W.W. Jiang and X.Q. Shen. 2013. Synthesis of magnetic calcium oxide hollow fiber catalyst for the production of biodiesel. Environ. Prog. Sustain. 32(4): 1255–1261.

Liu, B., Z.H. Zhang, K.L. Lv, K.J. Deng and H.M. Duan. 2014. Efficient aerobic oxidation of biomass derived 5-hydroxymethylfurfural to 2,5-diformylfuran catalyzed by magnetic nanoparticle supported manganese oxide. Appl. Catal. A-Gen. 472: 64–71.

Liu, B. and Z.H. Zhang. 2016. Catalytic conversion of biomass into chemicals and fuels over magnetic catalysts. ACS Catal. 6(1): 326–338.

Liu, C., P.M. Lv, Z.H. Yuan, F. Yan and W. Luo. 2010. The nanometer magnetic solid base catalyst for production of biodiesel. Renewable Energy 35(7): 1531–1536.

Liu, X.J., H.Y. He, Y.J. Wang, S.L. Zhu and X.L. Piao. 2008. Transesterification of soybean oil to biodiesel using CaO as a solid base catalyst. Fuel 87(2): 216–221.

Liu, Y.L., P.B. Zhang, M.M. Fan and P.P. Jiang. 2016. Biodiesel production from soybean oil catalyzed by magnetic nanoparticle MgFe2O4@CaO. Fuel 164: 314–321.

Liu, Z.W., X. Fu, S.R. Tang, Y. Cheng, L. Zhu, L.L. Xing, J. Wang and L.L. Xue. 2014. Sulfonated magnetic carbon nanotube arrays as effective solid acid catalysts for the hydrolyses of polysaccharides in crop stalks. Catal. Commun. 56: 1–4.

Lu, J., Y. Chen, F. Wang and T. Tan. 2009. Effect of water on methanolysis of glycerol trioleate catalyzed by immobilized lipase Candida sp. 99–125 in organic solvent system. J. Mol. Catal. B-Enzym. 56(2-3): 122–125.

Luna-del Risco, M., K. Orupold and H.C. Dubourguier. 2011. Particle-size effect of CuO and ZnO on biogas and methane production during anaerobic digestion. J. Hazard Mater 189(1-2): 603–608.

Luo, C.H., V. Shanmugam and C.S. Yeh. 2015. Nanoparticle biosynthesis using unicellular and subcellular supports. NPG Asia Mater 7: 11.

Ma, F.R. and M.A. Hanna. 1999. Biodiesel production: a review. Bioresource Technol. 70(1): 1–15.

Mahdavi, M., F. Namvar, M. Bin Ahmad and R. Mohamad. 2013. Green biosynthesis and characterization of magnetic iron oxide (Fe$_3$O$_4$) nanoparticles using seaweed (Sargassum muticum) aqueous extract. Molecules 18(5): 5954–5964.

Makridis, A., I. Chatzitheodorou, K. Topouridou, M.P. Yavropoulou, M. Angelakeris and C. Dendrinou-Samara. 2016. A facile microwave synthetic route for ferrite nanoparticles with direct impact in magnetic particle hyperthermia. Mat. Sci. Eng. C-Mater 63: 663–670.

Manzer, L.E. 2010. Recent developments in the conversion of biomass to renewable fuels and chemicals. Top Catal 53(15-18): 1193–1196.

Mehrasbi, M.R., J. Mohammadi, M. Peyda and M. Mohammadi. 2017. Covalent immobilization of Candida antarctica lipase on core-shell magnetic nanoparticles for production of biodiesel from waste cooking oil. Renewable Energy 101: 593–602.

Meunier, S.M. and R.L. Legge. 2012. Evaluation of diatomaceous earth supported lipase sol-gels as a medium for enzymatic transesterification of biodiesel. J. Mol. Catal. B-Enzym. 77: 92–97.

Mittal, N., G.M. Nisola, L.B. Malihan, J.G. Seo, H. Kim, S.P. Lee and W.J. Chung. 2016. One-pot synthesis of 2,5-diformylfuran from fructose using a magnetic bi-functional catalyst. RSC Adv. 6(31): 25678–25688.

Modi, M.K., J.R.C. Reddy, B.V.S.K. Rao and R.B.N. Prasad. 2007. Lipase-mediated conversion of vegetable oils into biodiesel using ethyl acetate as acyl acceptor. Bioresource Technol. 98(6): 1260–1264.

Okkerse, C. and H. van Bekkum. 1999. From fossil to green. Green Chem. 1(2): 107–114.

Ono, Y. 2003. Solid base catalysts for the synthesis of fine chemicals. J. Catal. 216(1-2): 406–415.

Osatiashtiani, A., A.F. Lee, M. Granollers, D.R. Brown, L. Olivi, G. Morales, J.A. Melero and K. Wilson. 2015. Hydrothermally stable, conformal, sulfated zirconia mono layer catalysts for glucose conversion to 5-HMF. ACS Catal. 5(7): 4345–4352.

Pessoa, F.L.P., S.P. Magalhaes and P.W.C. Falcao. 2009. Kinetic study of biodiesel production by enzymatic transesterification of vegetable oils. 10th International Symposium on Process Systems Engineering 27: 1809–1814.

Peterson, G.R. and W.P. Scarrah. 1984. Rapeseed oil trans-esterification by heterogeneous catalysis. J. Am. Oil Chem. Soc. 61(10): 1593–1597.

Poliakoff, M., J. Fitzpatrick, T. Farren and P. Anastas. 2002. Chemical emissions: An ongoing issue—Response. Science 298(5593): 542–542.

Polshettiwar, V. and R. Varma. 2008. Aqueous microwave chemistry: a clean and green synthetic tool for rapid drug discovery. Chem. Soc. Rev. 37(8): 1546–1557.

Polshettiwar, V., R. Luque, A. Fihri, H.B. Zhu, M. Bouhrara and J.M. Bassett. 2011. Magnetically recoverable nanocatalysts. Chem. Rev. 111(5): 3036–3075.

Prasad, C., S. Gangadhara and P. Venkateswarlu. 2016. Bio-inspired green synthesis of Fe$_3$O$_4$ magnetic nanoparticles using watermelon rinds and their catalytic activity. Applied Nanoscience 6(6): 797–802.

Prasad, C., G. Yuvaraja and P. Venkateswarlu. 2017. Biogenic synthesis of Fe3O4 magnetic nanoparticles using Pisum sativum peels extract and its effect on magnetic and Methyl orange dye degradation studies. J. Magn. Magn. Mater 424: 376–381.

Qiu, X.L., Q.L. Li, Y. Zhou, X.Y. Jin, A.D. Qi and Y.W. Yang. 2015. Sugar and pH dual-responsive snaptop nanocarriers based on mesoporous silica-coated Fe3O4 magnetic nanoparticles for cargo delivery. Chem. Commun. 51(20): 4237–4240.

Ragauskas, A.J., C.K. Williams, B.H. Davison, G. Britovsek, J. Cairney, C.A. Eckert, W.J. Frederick, J.P. Hallett, D.J. Leak, C.L. Liotta, J.R. Mielenz, R. Murphy, R. Templer and T. Tschaplinski. 2006. The path forward for biofuels and biomaterials. Science 311(5760): 484–489.

Robinson, I., L.D. Tung, S. Maenosono, C. Walti and N.T.K. Thanh. 2010. Synthesis of core-shell gold coated magnetic nanoparticles and their interaction with thiolated DNA. Nanoscale 2(12): 2624–2630.

Saha, B., S. Dutta and M.M. Abu-Omar. 2012. Aerobic oxidation of 5-hydroxylmethylfurfural with homogeneous and nanoparticulate catalysts. Catal. Sci. Technol. 2(1): 79–81.

Schachter, Y. and H. Pines. 1968. Calcium-oxide-catalyzed reactions of hydrocarbons and of alcohols. J. Catal. 11(2): 147–158.

Sharma, R.K., S. Dutta, S. Sharma, R. Zboril, R.S. Varma and M.B. Gawande. 2016. Fe_3O_4 (iron oxide)-supported nanocatalysts: synthesis, characterization and applications in coupling reactions. Green Chem 18(11): 3184–3209.

Sheldon, R.A. 2005. Green solvents for sustainable organic synthesis: state of the art. Green Chem. 7(5): 267–278.

Shi, D.M. and J.M. Vohs. 2016. Lignin-derived oxygenate reforming on a bimetallic surface: The reaction of benzaldehyde on Zn/Pt (111). Surf Sci. 650: 161–166.

Shimada, Y., Y. Watanabe, T. Samukawa, A. Sugihara, H. Noda, H. Fukuda and Y. Tominaga. 1999. Conversion of vegetable oil to biodiesel using immobilized Candida antarctica lipase. J. Am. Oil Chem. Soc. 76(7): 789–793.

Shylesh, S., V. Schunemann and W.R. Thiel. 2010. Magnetically separable nanocatalysts: Bridges between homogeneous and heterogeneous catalysis. Angew Chem. Int. Edit. 49(20): 3428–3459.

Shylesh, S. and W.R. Thiel. 2011. Bifunctional acid-base cooperativity in heterogeneous catalytic reactions: Advances in silica supported organic functional groups. Chemcatchem. 3(2): 278–287.

Sibikina, O.V., A.A. Iozep and A.V. Moskvin. 2009. Polysaccharide complexes with metal cations: structure and application (a review). Pharm. Chem. J. 43(6): 341–345.

Soltani, T. and B.K. Lee. 2017. Comparison of benzene and toluene photodegradation under visible light irradiation by Ba-doped BiFeO3 magnetic nanoparticles with fast sonochemical synthesis. Photoch. Photobio. Sci. 16(1): 86–95.

Sun, Y.L., Y.W. Yang, D.X. Chen, G. Wang, Y. Zhou, C.Y. Wang and J.F. Stoddart. 2013. Mechanized silica nanoparticles based on pillar 5 arenes for on-command cargo release. Small 9(19): 3224–3229.

Takagaki, A., M. Nishimura, S. Nishimura and K. Ebitani. 2011. Hydrolysis of sugars using magnetic silica nanoparticles with sulfonic acid groups. Chem. Lett. 40(10): 1195–1197.

The 2016 Cefic European Facts & Figures (Available at http://www.cefic.org/Facts-and-Figures/). The chemical industry (Available at http://www.essentialchemicalindustry.org/the-chemical-industry/thechemical-industry.html).

Thomas, G., F. Demoisson, R. Chassagnon, E. Popova and N. Millot. 2016. One-step continuous synthesis of functionalized magnetite nanoflowers. Nanotechnology 27(13): 15.

Tong, X.L., Y. Ma and Y.D. Li. 2010. Biomass into chemicals: Conversion of sugars to furan derivatives by catalytic processes. Appl. Catal. A-Gen. 385(1-2): 1–13.

Twidell, J. and R. Brice. 1992. Strategies for implementing renewable energy—lessons from europe. Energ. Policy 20(5): 464–479.

Varma, R.S. 2014. Journey on greener pathways: from the use of alternate energy inputs and benign reaction media to sustainable applications of nano-catalysts in synthesis and environmental remediation. Green Chem. 16(4): 2027–2041.

Venkateswarlu, S., Y.S. Rao, T. Balaji, B. Prathima and N.V.V. Jyothi. 2013. Biogenic synthesis of Fe3O4 magnetic nanoparticles using plantain peel extract. Mater Lett. 100: 241–244.

Verma, D., R. Tiwari and A.K. Sinha. 2013. Depolymerization of cellulosic feedstocks using magnetically separable functionalized graphene oxide. RSC Adv. 3(32): 13265–13272.

Verma, M.L., R. Chaudhary, T. Tsuzuki, C.J. Barrow and M. Puri. 2013. Immobilization of betaglucosidase on a magnetic nanoparticle improves thermostability: Application in cellobiose hydrolysis. Bioresource Technol. 135: 2–6.

Villeneuve, P., J.M. Muderhwa, J. Graille and M.J. Haas. 2000. Customizing lipases for biocatalysis: a survey of chemical, physical and molecular biological approaches. J. Mol. Catal B-Enzymatic 9(4-6): 113–148.

Virkutyte, J. and R.S. Varma. 2013. Green synthesis of nanomaterials: environmental aspects. ACS Sym. Ser. 1124: 11–39.

Vreeland, E.C., J. Watt, G.B. Schober, B.G. Hance, M.J. Austin, A.D. Price, B.D. Fellows, T.C. Monson, N.S. Hudak, L. Maldonado-Camargo, A.C. Bohorquez, C. Rinaldi and D.L. Huber. 2015. Enhanced nanoparticle size control by extending LaMer's mechanism. Chem. Mater 27(17): 6059–6066.

Wang, H.W., J. Covarrubias, H. Prock, X.R. Wu, D.H. Wang and S.H. Bossmann. 2015. Acidfunctionalized magnetic nanoparticle as heterogeneous catalysts for biodiesel synthesis. J. Phys. Chem. C 119(46): 26020–26028.

Wang, S.G., Z.H. Zhang and B. Liu. 2015. Catalytic conversion of fructose and 5-hydroxymethylfurfural into 2,5-furandicarboxylic acid over a recyclable Fe3O4-CoOX magnetite nanocatalyst. ACS Sustain. Chem. Eng. 3(3): 406–412.

Wang, X., F.L. Yuan, P. Hu, L.J. Yu and L.Y. Bai. 2008. Self-assembled growth of hollow spheres with octahedron-like Co nanocrystals via one-pot solution fabrication. J. Phys. Chem. C 112(24): 8773–8778.

Weir, A., P. Westerhoff, L. Fabricius, K. Hristovski and N. von Goetz. 2012. Titanium dioxide nanoparticles in food and personal care products. Environ. Sci. Technol. 46(4): 2242–2250.

Wiredu, B. and A.S. Amarasekara. 2014. Synthesis of a silica-immobilized Bronsted acidic ionic liquid catalyst and hydrolysis of cellulose in water under mild conditions. Catal Commun. 48: 41–44.

Wyk, J.P.V. 2001. Biotechnology and the utilization of biowaste as a resource for bioproduct development. Trends Biotechnol. 19 (5): 172–177.

Xie, W.L., X.L. Yang and M.L. Fan. 2015. Novel solid base catalyst for biodiesel production: Mesoporous SBA-15 silica immobilized with 1,3-dicyclohexyl-2-octylguanidine. Renewable Energy 80: 230–237.

Xiong, Y., Z.H. Zhang, X. Wang, B. Liu and J.T. Lin. 2014. Hydrolysis of cellulose in ionic liquids catalyzed by a magnetically-recoverable solid acid catalyst. Chem. Eng. J. 235: 349–355.

Yadvika, Santosh, T.R. Sreekrishnan, S. Kohli and V. Rana. 2004. Enhancement of biogas production from solid substrates using different techniques—a review. Bioresource Technol. 95(1): 1–10.

Yang, K.S., J.-H. Sohn and H.K. Kim. 2009. Catalytic properties of a lipase from photobacterium lipolyticum for biodiesel production containing a high methanol concentration. J. Biosci. Bioeng. 107(6): 599–604.

Yang, Z.Z., R.L. Huang, W. Qi, L.P. Tong, R.X. Su and Z.M. He. 2015. Hydrolysis of cellulose by sulfonated magnetic reduced graphene oxide. Chem. Eng. J. 280: 90–98.

Yang, Z.Z., W. Qi, R.X. Su and Z.M. He. 2017. Selective synthesis of 2,5-diformylfuran and 2,5-furandicarboxylic acid from 5-hydroxymethylfurfural and fructose catalyzed by magnetically separable catalysts. Energ. Fuels 31(1): 533–541.

Yin, M.Y., G.L. Yuan, Y.Q. Wu, M.Y. Huang and Y.Y. Jiang. 1999. Asymmetric hydrogenation of ketones catalyzed by a silica-supported chitosan-palladium complex. J. Mol. Catal. A-Chem. 147(1-2): 93–98.

Zhang, C.B., H.Y. Wang, F.D. Liu, L. Wang and H. He. 2013. Magnetic core-shell Fe3O4@C-SO3H nanoparticle catalyst for hydrolysis of cellulose. Cellulose 20(1): 127–134.

Zhang, D.H., C. Zhou, Z.H. Sun, L.Z. Wu, C.H. Tung and T.R. Zhang. 2012. Magnetically recyclable nanocatalysts (MRNCs): a versatile integration of high catalytic activity and facile recovery. Nanoscale 4(20): 6244–6255.

Zhang, P.B., Q.J. Han, M.M. Fan and P.P. Jiang. 2014. Magnetic solid base catalyst CaO/CoFe2O4 for biodiesel production: Influence of basicity and wettability of the catalyst in catalytic performance. Appl. Surf. Sci. 317: 1125–1130.

Zhang, P.B., M. Shi, Y.L. Liu, M.M. Fan, P.P. Jiang and Y.M. Dong. 2016. Sr doping magnetic CaO parcel ferrite improving catalytic activity on the synthesis of biodiesel by transesterification. Fuel 186: 787–791.

Zhang, X.C., M. Wang, Y.H. Wang, C.F. Zhang, Z. Zhang, F. Wang and J. Xu. 2014. Nanocoating of magnetic cores with sulfonic acid functionalized shells for the catalytic dehydration of fructose to 5-hydroxymethylfurfural. Chinese J. Catal. 35(5): 703–708.

Zhang, X.C., X.S. Tan, Y. Xu, W. Wang, L.L. Ma and W. Qi. 2016. Preparation of core-shell structure magnetic carbon-based solid acid and its catalytic performance on hemicellulose in corncobs. Bioresources 11(4): 10014–10029.

Zhang, Y., M.A. Dube, D.D. McLean and M. Kates. 2003a. Biodiesel production from waste cooking oil: 1. Process design and technological assessment. Bioresource Technol. 89(1): 1–16.

Zhang, Y., M.A. Dube, D.D. McLean and M. Kates. 2003b. Biodiesel production from waste cooking oil: 2. Economic assessment and sensitivity analysis. Bioresource Technol. 90(3): 229–240.

Zhang, Y.B., Y.W. Jing, J.X. Zhang, L.F. Sun and X. Quan. 2011. Performance of a ZVI-UASB reactor for azo dye wastewater treatment. J. Chem. Technol. Biot. 86(2): 199–204.

Zhao, X.B., F. Qi, C.L. Yuan, W. Du and D.H. Liu. 2015. Lipase-catalyzed process for biodiesel production: Enzyme immobilization, process simulation and optimization. Renew. Sust. Ener. Rev. 44: 182–197.

Ziderman, I.I. 1986. Purple dyes made from shellfish in antiquity. Review of Progress in Coloration and Related Topics 16(1): 46–52.

Chapter 5

Mixed-Oxide Nanocatalysts for Light Alkane Activation

M. Olga Guerrero-Pérez

Introduction: light alkanes as starting materials

Why should light alkanes be used?

The use of light alkanes (methane, ethane, propane and butane), originating from oil, natural and/or shale gas, or renewable resources, is increasingly attracting scientific interest since their direct utilization will offer alternative pathways to the preparation of high valuable chemicals, such as olefins, oxygenates and nitriles, that are nowadays produced mainly from olefins. In fact, Brazdil (Brazdil 2006; Cavani 2007) pointed out that the direct conversion of ethane and propane to commodity chemical intermediates has the potential to radically transform the chemical industry within the incoming years. However, the activation of these molecules is still a challenge, due to their high stability, since the conditions of pressure and temperature to activate C-H saturated bonds that are required are quite high, and subsequently, under such conditions, is difficult to control the selectivity of the desired products. In fact, the petrochemistry industry has evolved in the development of processes from acetylene, after olefins, and is currently involved in the development of the production of various chemical intermediates from alkanes directly. At the moment, the only well-established commercial alkane-based chemical process is the production of maleic anhydride from n-butane using catalysts type vanadium oxides and phosphorus, called VPO.

One of the main applications of alkanes is the production of olefins, such as propylene and ethylene, since their demand is expected to increase in the near future due to the wide range of applications that they have as starting materials for many

Departamento de Ingeniería Química, Escuela de Ingenierías D2111, Universidad de Málaga, E29017 Málaga, Spain.
Email: oguerrero@uma.es

chemicals and polymeric materials. Actually, olefins are produced mainly thought steam cracking, fluid catalytic cracking and dehydrogenation processes (Cavani 2007). These processes present many disadvantages: in steam cracking, ethylene is the preferred product and the yields of propylene are low; in addition, it is a high energy demanding process. On the other hand, dehydrogenation thermodynamic constraints limit alkane conversion, and the catalysts deactivate by coking. Due to that, during the last decade, significant efforts have been made towards the development of propane and ethane oxidative dehydrogenation (ODH) reactions. In the case of ethane ODH, the values reported in terms of ethylene yield are comparable to those obtained by steam cracking. However, in the case of propane ODH, the best results reported are still far from commercial implementation; thus, great efforts are being done in order to develop new catalytic systems for propane ODH and reactor configurations for such processes.

Light alkanes from fossil sources

Propane, ethane, methane and butane are obtained from petroleum and natural gas and in the last decades, in addition, and thanks to the improvement in extraction techniques, from shale gas. Shale gas, unlike oil and natural gas, is distributed geographically evenly, and its extraction becomes profitable when the price of oil rises for light alkanes are both abundant and accessible raw materials. However, due to the growing concern about the greenhouse effect and CO_2 emissions, the use of these raw materials has been questioned and much has been invested in the development of processes that use biomass as a raw material. However, the use of biomass as a raw material for the production of energy and chemical compounds also presents problems. Biomass is a raw material that competes with food resources, which are still to reach a large part of the world's population. In addition, processes that use biomass as a raw material are often energetically demanding, consume more water, and produce more waste and/or by-products than those that use fossil resources. Therefore, and given that there are proven fossil resources available for the coming decades, and despite not being renewable, the use of light hydrocarbons, from oil and natural gas, is expected to continue to be essential in the future. In fact, in recent months there has been a strong discussion in the European Parliament regarding the use of biodiesel from palm oil (the most common). In any case, the EU's energy future (also that of alternative fuels) will be decided in the European Parliament within the coming months/years. The meetings have already begun and are expected to end in June 2018 with the approval of the EU's energy plan for the next decade, which should allow compliance with the Paris Agreement against climate change. One of the issues under discussion is biofuels from the first generation, that is, those made from food materials such as palm oil. The EU encouraged them in the past decade, but now it tries to limit them by concluding that, far from helping in the fight against climate change, they can make the situation worse. Almost half of the palm oil that the EU imports is destined to make biodiesel, according to the data of the Transport and Environment organization. In the specific case of Spain, which made heavy investments in palm oil diesel plants, it practically does not have its own production, so the raw material must come from Indonesia and Malaysia, countries with serious deforestation problems. However, the European Parliament has proposed—in the framework of this three-way negotiation on the EU's energy future—that in 2021, diesel made from palm oil is no longer considered a biofuel. This would suppose its elimination of the obligatory quotas of biofuel that the Government fixes for the transport, discouraging its use and production. This is an example that demonstrates that some processes from biomass are not as economically and environmentally profitable as it seemed and that, at least during the next decades, fossil resources will continue to be necessary for the production of chemical intermediates and fuels.

Oxygenates and nitriles from light alkanes

Catalytic partial oxidation processes involving alkanes that have great promise for the fast, highly selective production of chemicals from readily available feedstocks. However, partial oxidation processes are so complex that they are always poorly understood, and because of safety concerns they are inevitably operated far from optimally. In these groups of chemical processes with commercial applications, ammoxidation to nitriles and oxidation to oxygenates can be found. Ammoxidation reactions (oxidation using molecular oxygen in the presence of ammonia) are interesting for the transformation of light alkanes into chemical intermediates. Acrylonitrile is a valuable chemical that can be synthesized thought propane ammoxidation. It is widely used as an intermediate for the preparation of synthetic rubbers, synthetic resins and carbon and acrylic fibers. It is a very common monomer to prepare several polymers: such as polyacrylonitrile, SAN (styrene-acrylonitrile), ABS (acrylonitrile butadiene styrene), ASA (acrylonitrile styrene acrylate) or NBR (acrylonitrile butadiene) (Guerrero-Pérez 2015). It is expected that propane-based technology will replace the propylene-based technology within the next years. The ammoxidation of ethane produces acetonitrile (Rojas 2013). It is used as a solvent in many commercial processes, such as high performance liquid chromatography or for the butadiene extraction in hydrocarbon streams. It is also used in several organic and inorganic syntheses, such as the synthesis of flavones and flavonol pigments. Currently, acetonitrile is obtained as a by-product during propylene ammoxidation since there is no method for the direct commercial synthesis of acetonitrile. Thus both reactions, ammoxidation of propane or ethane in presence of oxygen and ammonia to obtain the corresponding nitrile, are of interest nowadays, as has been already discussed. These reactions share similarities that result in different catalyst requirements (Guerrero-Pérez 2016). They also have great differences, since ethane only has methyl-CH_3 groups, while propane contains groups -CH_2-. This causes the reaction to start in different types of carbon and therefore, they need different types of catalysts. Hence, coupled with the complexity of having to activate alkanes, it joins the fact of having to develop very specific catalyst for each molecule, methane, ethane, propane or butane.

Oxygenates such as acrolein, propylene oxide, acrylic acid or propylene oxide are also of interest in the petroleum chemistry industry. Most of the propylene oxide is used for the production of polyether polyols by the process called alkoxydation.

These polyols are the building blocks in the production of polyurethane plastics. In addition, propylene oxide is used to produce propylene glycol by hydrolysis. Other major products are polypropylene glycol, propylene glycol ethers and propylene carbonate. Propylene glycol is used for the production of unsaturated polyester resins; it is also used as a humectant, as a solvent, and as a preservative in food and for tobacco products. Like ethylene glycol, propylene glycol is able to lower the freezing point of water, and so it is used as an aircraft de-icing fluid. Polypropylene glycol has also many applications, such as in the production of polyurethanes and it is also employed as a surfactant, wetting agent, and dispersant, among others. Propylene carbonate is used as a polar, aprotic solvent, and, due to its high dielectric constant, it is commonly used as a high permitivity component of electrolytes in lithium batteries, since, together with a low viscosity solvent, its high polarity allows for the creation of an effective solvation shell around lithium ions, resulting in a conductive electrolyte. In addition, propylene carbonate is used for the production of adhesives, cosmetics and paint strippers, and it is used as plasticizer. In the other hand, acrolein is used directly as a herbicide and a biocide. In addition, acrolein is used for preparing a high number of useful compounds such as methylpyridines and quinolines. Acrylic acids are industrially combined with themselves (for the production of polyacrylic acid) or with other monomers (such as acrylamides, acrylonitrile, styrene, butadiene and vinyl compounds) to form copolymers that result in the production of several types of coatings, plastics, adhesives, elastomers, polishes and paints. All these important applications of chemical intermediates (oxygenates and nitriles) that are obtained in the industrial petrochemistry from olefins (such as ethane, propane and butane) demonstrate that partial oxidation reactions are of high interest for the high number of useful products that are produced today and is why the improvement of these products.

The production of such molecules from alkanes is of interest since they are actually produced from other more valuable starting materials such as olefins. Although there has been intensive research effort that has been done during the last decades in order to develop these processes, the alkane transformation into oxygenates that is taking place at a commercial level is the catalytic transformation of n-butane into maleic anhydride over VPO (Vanadium and Phosphorous Oxides) catalytic materials.

Nanostructures in catalysis

Nanostructures are of wide importance in heterogeneously catalytic processes, including all the light alkane transformation mentioned since most of the industrial catalysts contain an active phase (active component) that is a nanoparticle dispersed and stabilized on the surface of a high area support. In this manner, the catalytic performance is affected by the particle diameter, shape, chemical structure and espoused surface of such nanoparticles and by the (nano)porous structure of the supports, making this area of study an important part of nanoscience and catalysis (Shiju 2009). It is also important to control the micro/nano shape of the catalytic materials, in order to develop miniature chemical devices and to design new reactors with improved properties. In this sense, recent advances in the design and fabrication of miniature chemical devices (micromixers, microseparators or microreactors) have shown that these devices are able to provide new opportunities to design more efficient chemical processes (Wegeng 2000). Therefore, there is a high interest in the development of smart nanostructured materials capable of very specific interactions with molecules, ions and atoms, since such materials have many applications. For example, microreactors can provide better energy and material utilization leading to more efficient chemical production and less pollution since the large surface area-to-volume ratio (attained in a microreactor) enhances heat and mass transfer rates (Tonkovich 2000). In this case, almost all of the catalytic surface is available for reaction at a fraction of the pressure drop that would be required for the same feed flow and particle size in a conventional fixed bed reactor. In addition, the large surface to volume ratio of catalysts-coated microchannels provides: (i) an excellent contact between reactants and catalyst, thus minimizing bypass, (ii) a better approach to matching permeation and reaction rates and (iii) the possibility of integrating microsensors and actuators that enable rapid and precise control of the reactor operation (Wan 2001). For manufacturing these devices, it is necessary to use structured catalysts supports, that is, rigid, orderly arranged support materials such as carbon nanofibers as well as bulk oxide nano and microfibers. These materials are microreactors on their own, and, in addition, can be grown or supported in stainless steel membranes or tubular nanodevices (Aran 2011; Martínez-Latorre 2009). These materials can find many applications as catalysis for gas and liquid processes, photocatalysis and electrochemistry, including as catalysts for alkane activation processes.

During the last decades, several groups have investigated the role of dispersed metal nanoparticles based catalysts in different alkane based reactions, especially gold nanoparticles based catalysts. It was shown that these catalytic materials are active for the total oxidation reaction of several hydrocarbons at low temperatures (Gasior 2004). Total oxidation hydrocarbon reactions are of high importance for pollutant abatement, and when such processes are carried out at low temperature, the NOx emissions are decreased. Besides total oxidation reactions, dispersed nanoparticles catalysts have been studied for several partial oxidation reactions; for example, gold nanoparticles catalysts have shown to be promising for the propane oxidation into propylene oxide (Bravo-Suarez 2008). Propylene oxide is a major chemical intermediate produced on a large scale industrially, its major application being its use in the production of polyether polyols for use in making polyurethane plastics. Its current industrial production starts from propylene in a multi-step process; the possibility of its production directly from propane being quite convenient. Oyama and coworkers (Bravo-Suarez 2008) demonstrated that the direct propane epoxidation through sequential propane dehydrogenation-propylene epoxidation was possible over Au/Ti and Au/Ti-Si catalysts. Weckhuysen and coworkers (Nijhuis 2008) demonstrated the adsorption of propene molecule on gold nanoparticles. Such adsorption can be a key step in the epoxidation reaction mechanism, and support the that gold nanoparticles on titania can make propene reactively adsorb on titania to produce a bidentate propoxy species. This adsorption of propene on the gold nanoparticles reduces the direct water formation which is an undesirable competitive reaction during the hydro-eposication of propene. This finding was possible by applying a detailed analysis of *in situ* measured XANES spectra over gold/silica catalysts. In addition to gold nanoparticles catalysts for total oxidation and propane reaction into propyelene oxide, Cr nanoparticles catalysts have also been used for the oxidative dehydrogenation of ethane into ethylene in the presence of carbon dioxide (Rahmani 2015).

Although these few metal nanoparticles catalytic applications for light alkane activation, it is the main objective of present article to focus on the mixed oxide nanoparticles based catalysts. Mixed oxide catalytic materials have been extensively studied during the last decades for alkane activation reactions. The use of at least two transition metals to prepare the mixed oxide based catalysts confers to the material interesting catalytic properties, although mixed oxides are complex materials, and the elucidation of their structures is not easy. For alkane activation, those containing Vanadium have proven to be quite promising (Haber 2009). There are several well know mixed oxide vanadium containing catalytic systems that have been developed during the last years: for example, the VPO catalyst, that is commercially used for the transformation of n-butane into maleic anhydride; VSbO, used mainly for ammoxidation reactions; or Mo-V multioxide system, used for propane transformation into acrylic acid. A review about the knowledge of nanocatalyst structures based on these catalytic systems will be the objective of this study.

Mixed oxides nanoparticles as catalysts for light alkane activation

It has been shown that transition metal nanoparticles are very attractive to use as catalysts, due to their high surface-to-volume ration and their high surface energy, which makes their surface atoms very active. However, oxide nanoparticles are important in the catalytic processes of light alkane activation, for example, mixed oxide materials, and especially those containing Vanadium, which are well-known as promising catalysts for light alkane partial oxidation reactions. Supported vanadia catalysts have been extensively studied during the last decades for the propane ODH reaction (Guerrero-Pérez 2017). The nature of the VOx phases, the nature of the support, and the influence of the Vanadium precursor, have been extensively studied during the last years. Wachs et al. (Carrero 2013) studied these catalysts during the propane ODH reaction, with different oxide precursors (2-propanol/ vanadyl trisoproposide, oxalic acid/ammonium metavanadate and toluene/vanadyl acetylacetonate) and three different supports (SiO_2, TiO_2 and Al_2O_3). *In situ* Raman spectroscopy studies was used in this work to elucidate the structure of VOx species since such a technique is able to detect monomeric and polymeric VOx dispersed species as well as V_2O_5 crystals of quite small size, when such species are not able to be detected by XRD (Bañares 2002). It was shown that the choice of precursor does not affected the dispersion of the supported V oxide phases for low coverages (around four atoms of V per nm^2 of support). For such coverage, only isolated and oligomeric surface Vanadium oxide species are present, and, in this case, only crystalline V_2O_5 nanoparticles are detected for the ammonium metavanadate precursor, indicating that such a precursor favors crystalline vanadium oxide nanoparticle formation at low vanadium coverages. In this work, the turnover frequently (TOF) values were calculated during propane ODH and it was demonstrated that there is no significant difference in TOF for the isolated and oligomeric surface vanadium oxide sites. However, surprisingly, vanadium[1] oxide nanoparticles (in the one to two nm range) presented very high values, and demonstrated that these nanoparticles are quite active and selective, since crystalline V_2O_5 nanoparticles at low coverages present very high activity during propane ODH. Also, a careful examination of the literature for propane ODH over vanadium oxide catalysts (Carrero 2014) revealed that the activity trends that have been found are due to the use of different vanadium precursors and their ability to form V_2O_5 nanoparticles at low coverage. Researchers that used vanadium precursors that form vanadium oxide nanoparticles at low coverages found that the propane ODH TOF was dependent on the surface vanadium density. It seems that the origin of such activity is the energy of the formation of oxygen defects at the vanadia-covered surface that expresses the reducibility capability of the active site. However, the higher reactivity of propane that causes a rate constant for the deep oxidation reaction will always be higher than that of the dehydrogenation reaction, resulting in a maximum yield of propane below 100%. Thus, in order to obtain the maximum yield, is the operation at the maximum temperature, in order to avoid the non-selective gas phase reaction. The appropriate catalysts needs to have a very low amount of acid sites in order to avoid the propene deep oxidation.

Since several studies as those presented about V_2O_5 nanoparticles had shown the potential of oxide and metal nanoparticles in catalysis, our group designed some catalysts by the incorporation of bulk-mixed nanoparticles on the surface of a support for their stabilization, based on catalytic systems that were used as bulk materials and that were well-known as active and selective for light alkane partial oxidation reactions. Another advantage, when such nanoscaled materials are used at a basic research level, is that the surface-to-volume ratio is higher than when conventional bulk-mixed metal oxide catalysts are used. When characterization techniques are used with conventional bulk oxide catalytic materials, the signal from the bulk overwhelms the signal from the outermost layer. The outermost layer is the end of the tridimensional bulk pattern and is not like the bulk; in addition, the nature of the outermost layer can be shaped by the reaction conditions. Thus, one approach to study this issue is the use of nano-scaled mixed oxide catalytic materials, that possess a very high surface to volume ratio. Following this idea, our group has used alumina-stabilized nanoscaled V-Sb-O catalysts for the propane ammoxidation reaction. In the first study, it was the ideal conditions to form the Sb-V-O nanoparticles on Al_2O_3 (Guerrero-Pérez 2002) and Nb_2O_5 supports (Guerrero-Pérez 2003; Guerrero-Pérez 2003) were optimized, and the role of the support was also analyzed. It was determined that with catalytic formulations close to Sb/V molar ratio of 1, and with Sb+V coverages close to 8 atoms of Sb and V per nm^2 of support, $VSbO_4$ nanoparticles were detected; in some cases, the formation of such nanoparticles was detected under reaction conditions, such catalysts being the most active and selective to acrylonitrile formation, in line with the ODH studies with VOx catalysts (Carrero 2014) that detected very high activity in the catalysts that presented V_2O_5 nanoparticles.

An operando Raman study was performed (Guerrero-Pérez 2002; Guerrero-Pérez 2004; Guerrero-Pérez 2007) in order to determine the structure of the catalysts under reaction conditions. The results can be found in Fig. 1. The interaction between surface Sb (no Raman bands) and surface V oxide species (Raman bands at 1020 and

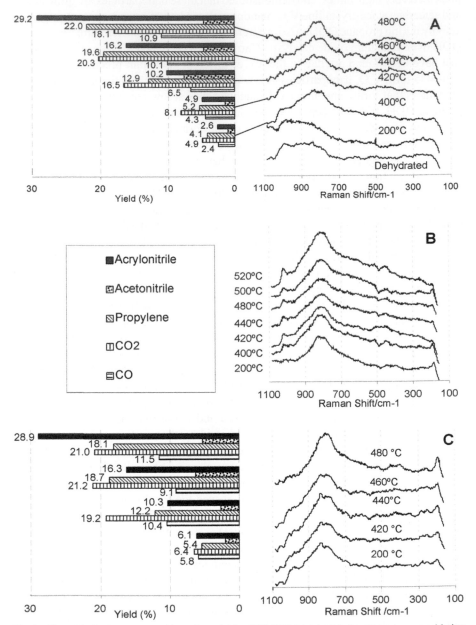

Fig. 1. Operando Raman spectra with product yields of 1Sb1V/Al catalyst during propane ammoxidation (A, first run), under oxidizing conditions (B, O_2+He) and during propane ammoxidation (C, second run). Reaction conditions: 200 mg of catalyst and total flow 20 ml/min. Reproduced with permission from M.O. Guerrero-Pérez, M.A. Bañares, Catal. Today 96 (2004) 265–272.

900 cm^{-1}) on alumina takes place under reducing or non net-oxidizing environments (e.g., ammoxidation reaction conditions) leading to the formation of VSbO$_4$ (broad Raman band at 840 cm^{-1}) and segregated α-Sb$_2$O$_4$. The oxidation states in VSbO$_4$, are Sb(V) and V(III). The reduction of surface vanadium(V) species under ammoxidation conditions can be the driving force to build VSbO$_4$ during the reaction. The interaction between Sb and V appears partially reversed upon reoxidation, where V(III) species in VSbO$_4$ segregate as surface V(V) species. This trend increases with the reoxidation temperature. Interestingly, a concomitant decrease of segregated α-Sb$_2$O$_4$ is evident, and the VSbO$_4$ phase does not appear significantly affected by these changes. Actually, the VSbO$_4$ phase is rather stable under different environments (oxidizing, and reducing) and accepts several stoichiometries close to V$_1$Sb$_1$O$_4$. Apparently, vanadia cations leave the VSbO$_4$ structure during reoxidation cycles and antimony enters it. A reverse trend is observed under ammoxidation reaction conditions. α-Sb$_2$O$_4$ is a mixed-valence oxide possessing Sb(III) and Sb(V) cations. Thus, it would make sense that such stability of the SbVO$_4$ phase must be due to the migration of the corresponding amount of Sb(V) cations from α-Sb$_2$O$_4$ into VSbO$_4$ to compensate for the removal of V(III) cations. It is interesting to underline that surface vanadium species are pentavalent while bulk vanadium species in VSbO$_4$ are trivalent. This cycle appears to be the redox cycle for vanadium sites. The migration cycle of Sb (V) species would facilitate the redox cycle of vanadium species stabilizing the VSbO$_4$ lattice. This migration cycle of Sb and V species appears important for the propane ammoxidation reaction, and is consistent with the structural transformations observed in bulk vanadium antimonate catalysts.

Thus, the Sb-V-O case has proved how the bulk-mixed oxide nanoparticles based catalysts presented several advantages. They have good catalytic behavior during propane ammoxidation, due to the fact that the active phase is isolated on a support. They also have better physical properties when these catalysts are compared with bulk oxide materials and are less expensive since a low amount of the active phase is required. From a fundamental research point of view, they are more convenient for performing structure-activity relationship studies, as have been shown in the Sb-V-O, where the mechanism and the active phase were determined during the propane ammoxidation reaction. A similar approach was used for the design of Mo-V-Nb-Te-O nanoscaled catalysts. This multioxide catalytic system is well known as active and selective for the selective oxidation of propane into acrylic acid and acrylonitrile. These catalytic materials present several active phases: M1, M2 and rutile. The M1 phase (TeM$_{20}$O$_{31}$ orthorhombic) crystallizes in the orthorhombic system and is able to undergo oxidation and reduction in a certain degree without changes in this structure whereas the M2 phase (Te$_{0.33}$MO$_{3.33}$) crystallizes in the hexagonal system. It seems that there is a synergistic effect between these two phases (Grasselli 2011; Grasselli 2008; Holmber 2004; Celaya Sanfiz 2010; Korovchenko 2008; Balcells 2003; López-Medina 2012). The catalysts' formulation and synthesis method in order to obtain the nanoscaled M1, M2 and rutile phases on an alumina support was optimized (López-Medina 2012), obtaining the best configuration for a Mo$_5$V$_4$Nb$_{0.5}$Te$_{0.5}$O nanoscaled catalysts supported on alumina. This is illustrated in Scheme 1. Such catalytic material presented very good catalytic behavior for the propane ammxoidation reaction, with almost 50% of acrylonitrile yield at 80% propane

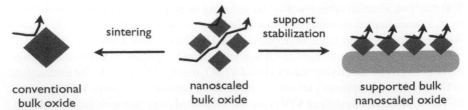

conventional bulk oxide   nanoscaled bulk oxide   supported bulk nanoscaled oxide

Scheme 1. Bulk catalyst nanoparticles are illustrated in the middle. Leftwards arrow illustrates sintering phenomena occurring during reaction. Rightwards arrow illustrates support-stabilized bulk nanoparticles. Reproduced with permission from R. López Medina et al. Catal. Today 192 (2012) 67–71.

conversion, in the same range than the results reported with bulk oxide mesoporous oxides (Albonetti 1997). The supported bulk nanoscaled catalyst, with a much lower amount of MoVNbTe multioxide phase, performs better than the corresponding unsupported bulk (i.e., conventional bulk) catalysts, which have a higher amount of active oxide phase; this is illustrated by the data shown in Fig. 2, that shows the formation of acrylonitrile per unit of mass of active phase (Mo+V+Te+Nb) and also per unit of time. In both cases, and for all the catalysts formulations studied in the referred work, the acrylonitrile produced per gram of active phase is much higher in the case of the supported bulk nanoscaled catalysts. Thus, when the active oxide phases are nanoscaled the performance of such phases is enhanced. As in the case of VOx and V-Sb-O catalytic systems, supporting nanoscaled bulk catalysts improves their cost and mechanical properties; in addition, they increase the exposure of the active site, triggering the activity per gram of active component.

Nano- and micro-shaped catalysts for alkane activation

It has been shown how important it is to nanoscale the active phases in order to maximize the catalytic activity and selectivity of these active phases, but another important issue is to control the shape of the catalytic particles. This issue is important in order to improve some physical properties of the catalytic materials since catalytic materials in powder form entail some practical constraints to potential industrial applications. This is also necessary in order to design new reactors and processes, including nanodevices such as microseparators or microreactors, since it has been shown that these devices are able to provide new opportunities to design more efficient chemical processes (Wegeng 2000). For example, microreactors can provide better energy and material utilization leading to more efficient chemical production and less pollution since the large surface area-to-volume ratio (attained in a microreactor) enhances heat and mass transfer rates (Tonkovich 2000). In this case, almost all of the catalytic surface is available for reaction at a fraction of the pressure drop that would be required for the same feed flow and particle size in a conventional fixed bed reactor. In addition, the large surface to volume ratio of catalysts-coated microchannels provides: (i) excellent contact between reactants and catalyst, thus minimizing bypass, (ii) a better approach to matching permeation and reaction rates and (iii) the possibility of integrating microsensors and actuators that enable rapid and precise control of the reactor operation (Wan 2001). In this sense, a great effort

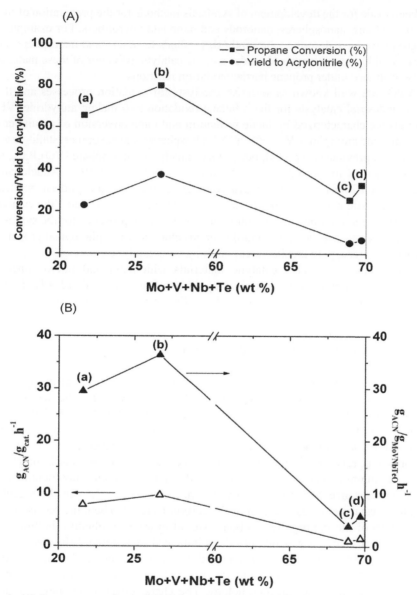

Fig. 2. (A) Propane conversion and yield to acrylonitrile versus Mo + V + Nb + Te content. (a) Mo6V3Nb0.5Te0.5O-N, (b) Mo5V4Nb0.5Te0.5O-N, (c) Mo6V3Nb0.5Te0.5ON- bulk and (d) Mo5V4Nb0.5Te0.5O-N-bulk and (B) formation of acrylonitrile per unit mass of catalyst per unit time. Reaction conditions: feed composition (vol.%) C3H8/O2/NH3/He 9.8/25/8.6/56.6, 0.2 g of catalyst with particle dimensions 0.25–0.125 mm, and total flow rate of 20 ml/min. T = 450°C. Bulk catalyst nanoparticles are illustrated in the middle. Leftwards arrow illustrates sintering phenomena occurring during reaction. Rightwards arrow illustrates support-stabilized bulk nanoparticles. Reproduced with permission from R. López Medina et al. Catal. Today 192 (2012) 67–71.

has been made for the development of synthesis methods for the preparation of thin-film oxide films, nanospheres, nanorods and nano and microfibers. For example, in our group, we have described the synthesis methods for micro and nanospheres and fibres based in Vanadium mixed oxides and the catalytic behavior of these materials has been studied under propane partial oxidation reactions.

VPOs are well known as selective catalysts for oxidation processes and they are the industrial catalysts for the n-butane oxidation into maleic anhydride. VPO materials are characterized by facile formation and interconversion of a number of crystalline and amorphous V^{3+}, V^{4+} and V^{5+} phosphates; and several crystallographic phases were previously reported, being bulk vanadyl pyrophosphate $(VO)_2P_2O_7$, the main component in the industrial catalyst (Centi 1993; Centi 1988; Abdelouahab 1992; Carreon 2004). Some of the disadvantages of the classical synthesis methods of these catalytic materials could be solved by the use of nanoscaled VPO catalytic materials. The main problems regarding these classical approaches for the synthesis of these materials are the poor control over structural and morphological properties (i.e., surface areas, pore architectures), limiting their catalytic behavior.

The synthesis of VPO catalytic materials with nano and micro spherical controlled morphology was described by Valero-Romero et al. (2014). In this study, carbon nanospheres were prepared in a first step, using cellulose as a carbon precursor. In this synthesis, phosphoric acid was used as precursors with two objectives, first to incorporate P to the carbon spheres (necessary for the preparation of the VPO catalytic material) and second, because it had been described that phosphoric groups on the surface of carbon materials are able to make them resistant in air at high temperatures. This is due to the presence of different phosphorous functional surface groups, such as $C-O-PO_3$, $C-PO_3$ and C-P groups, that act as a physical barrier blocking the active carbon sites (Labruquere 1998; Hall 1989; Wu 2006; Rosas 2009). This oxidation resistance makes the carbon materials suitable for partial oxidation reactions, since the materials can be used under the oxidant atmosphere at relatively high temperatures. The carbon spheres were impregnated with a vanadium precursor. It was described that after vanadium incorporation, the oxidation resistance decreased to a lower temperature, since the vanadium centres are able to catalyse the gasification of the carbon material. Then, the procedure for calcinations of the impregnated carbon material in order to obtain a hollow VPO microspherical material was optimized, as indicated in Scheme 2. In this manner, two different spherical VPO nanostructured catalysts with controlled morphology, were prepared, that is, a carbon-supported material and a bulk hollow mixed oxide material, both with spherical morphology. The characterization of these series of

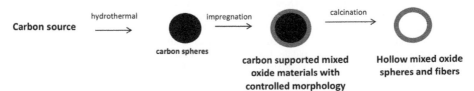

Scheme 2. Proposed procedure for the preparation of the mixed oxide spheres. Reproduced with permission from M.J. Valero et al. 2014. Catal. Today 227: 233–241.

nanostructured catalytic materials indicated that the VPO phases, that had been described as active for alkane activation, were identified in these nanostructured materials, such as $(VO)_2P_2O_7$ and $VOPO_4$ phases. In these structures, vanadium is present as V^{4+} and V^{5+}. It is important that Vanadium sites can be present with both oxidation states since under reaction conditions, the active sites have to change under reaction conditions in order to facilitate the redox cycle necessary to partially oxidize the corresponding alkane. Thus, the hollow mixed VPO spheres were tested under propane partial oxidation. The carbon supported VPO materials were not tested as oxidation catalysts since they were not resistant under oxidizing reaction conditions, although they can present other applications, catalytic and non-catalytic. The results showed that these materials presented a very good performance for propane ODH, being selective to propylene molecules. The characterization of used samples showed that the vanadium-phosphate mixed phases are stable under reaction conditions. It was also observed that the larger particles did not keep the spherical morphology under reaction conditions. This morphology, on the contrary, was kept by the smaller particles (below 1 μm). Thus, a simple and low cost method for preparing VPO catalysts with spherical morphology and with a high surface area was described in this example, underlining that this technology of preparing hollow oxide nano or micro spheres can be applied to other oxide bulk materials.

Due to the good results described in the last example that has been given, regarding the VPO nanostructured catalytic materials, in another contribution (Berenguer 2016), a similar procedure was performed for the preparation of VPO nano and micro fibres. In this case, a series of VPO catalytic materials with fibrous morphology was prepared by using phosphorous-functionalized carbon fibres (P-CFs) as support (Scheme 3). The preparation of this unique material by electrospinning of lignin/H_3PO_4 (1/0.3 wt. %) solutions was described in a previous contribution of the same group (Berenguer 2014). The introduction of phosphorous enables for greatly shortening the conventional stabilization process of lignin fibres and, at the same time, promote the chemical activation of CFs during carbonization. As in the case of the example of the spheres, the P-CFs were used as support for the preparation of carbon supported VPO fibrous catalysts. For the impregnation procedure, it was used such an amount of V precursor to reach 4 V atoms per nm^2 of carbon support. This amount was used because a previous study (Guerrero-Pérez 2012) showed that this is the maximum amount of vanadium oxide species that remain dispersed without the formation of V_2O_5 crystals on a phosphorous containing activated carbon. In addition, it was calculated that with this amount, the V/P atomic ratio in the supported catalysts would be close to 1, considering the atomic concentration of P calculated by XPS, and necessary for the formation of VPO mixed phases. As in the case of the fibres, by optimizing the calcinations procedure of these supported materials, bulk VPO catalysts with fibrous morphology were prepared. Thus, the synthesis of two quite different materials was optimized, a bulk mixed VPO material with fibrous structure, and a VPO carbon nanofibers supported material. Both materials find many applications in catalysis, and in the activation of alkanes.

A similar approach can be also be followed for the preparation of VOx nano and microfibers (Berenguer 2017). In the cited work, a new approach to produce VOx nanofibers, allowing variable +4/+5 oxidation state and crystallinity, was proposed.

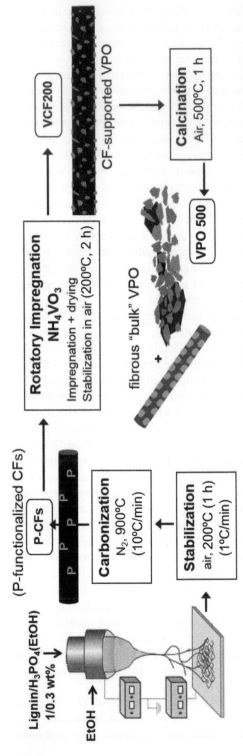

Scheme 3. Procedure for the preparation of VPO catalysts with fibrous morphology. Reproduced with permission from R. Berenguer et al. 2016. Catal. Today 277: 266–273.

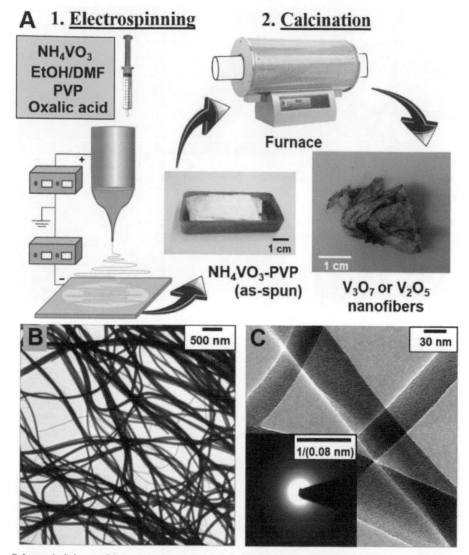

Scheme 4. Scheme of the preparation procedure of VOx nanofibers. Inset: photographs of the NH4VO3-PVP fiber mesh in the crucible before and after calcination (example, VF-350-0.2 sample). B and C. TEM images at different magnification (inset: SAED pattern) of the electrospun NH4VO3-PVP fibers. Reproduced with permission from R. Berenguer et al. 2017. ACS Omega 2(11): 7739–7745.

The proposed methodology is outlined in Scheme 4 and involves the use of ammonium metavanadate as Vanadium precursor and ethanol, since this alcohol is an excellent alternative to water as a solvent in electrospinning processes. Basically, it gathers a relatively high dielectric constant and volatility, promoting the yield of regular fibres (Fong 1999); it can be sustainably produced from biomass and is relatively cheap and not harmful. However, the solubility of NH_4VO_3 in ethanol is very low. This is solved by the addition of oxalic acid, which increases solubility, leading to homogeneity and sufficient consistency of the resulting fibres after calcination. At the

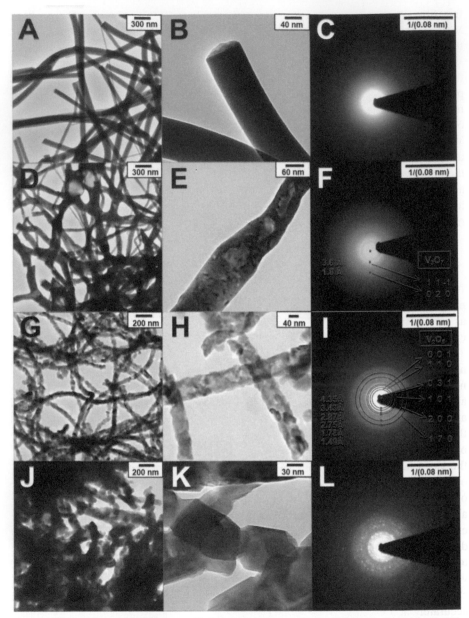

Fig. 3. TEM images and SAED patterns of VF-300-0.2 (A, B and C), VF-350-0.2 (D, E and F), VF-350-0.5 (G, H and I), and VF-400-0.2 (J, K and L). Reproduced with permission from R. Berenguer et al. 2017. ACS Omega 2(11): 7739–7745.

same time, oxalic acid plays a key role as a vanadium reductant. It was also necessary to incorporate polyvinylpyrrolidone (PVP) and another solvent, dimethylformamide (DMF), to improve the spinnability of the ethanolic solutions. TEM characterization (Fig. 3) revealed that these materials consisted of long-range continuous and defect-free fibres (Fig. 3B), with diameters ranging between 20 and 90 nm (Fig. 3C). V^{5+}/V^{4+} ratios in the final materials were detected by XPS. It was observed how the rise in the heating rate or calcination temperature reduces the proportion of V^{4+}, obtaining fibres with different V^{4+}/V^{5+} ratios. The further characterization of these materials can be found in the cited paper (Rosas 2009), but it showed were prepared with this method. The efficient incorporation of the mentioned ammonium metavanadate reductant results in VOx nanofibers with a considerable proportion of V^{4+}. Further thermal treatment of these fibres permitted the decrease of the V^{4+}/V^{5+} ratio. This opens the possibility to obtain long-aspect fibrous materials with different crystalline phases that are difficult to obtain by conventional methods. By the optimization of the parameters of this synthetic route, for example, the nature and concentration of the reductant, the calcination conditions, etc., it is expected to obtain other interesting mixed-valence VOx materials, which are useful catalytic materials for several alkane activation reactions.

Perspective

Although in recent years there has been much talk about the applications of raw materials of biomass origin, the reality shows that raw materials of fossil origin, such as oil and natural gas, will continue to play an important role in the production of chemical intermediates for commodities production and energy in the coming decades. The future of catalysis science during the next years will be in the development of nanostructured catalysts, since it has been shown that the synthesis of controlled nanostructures at the atomic scale, will lead to breakthroughs in catalysis, including alkane activation processes (which are a major challenge in energy and environmental applications), which shall be developed and commercialized in the next years. In this sense, light alkanes, such as methane, ethane, propane and butane, which are obtained directly from oil and natural gas, will play a decisive role. So far and for the vast majority of applications, it is necessary to transform these molecules into olefins to, produce other molecules such as nitriles and oxygenates. The development of the production processes of these molecules (oxygenates and nitriles) from alkanes implies without a doubt a great reduction in production costs and significant energy saving. The development of these processes is certainly linked to the development of more efficient catalysts, and this is where nanotechnology applied to the development of catalysts "A la carte" will play a decisive role in the coming years. Thus, it is expected that the advances in the synthesis of nanostructured catalysts will develop new applications and improve the existing ones since more accurately and efficiently control reaction pathways through a precise control of the active species present on the surface of such structures will be performed. There are still some issues that must be solved, such as the characterization techniques of the surface of these materials under reaction conditions, and the appropriate synthesis methods that will allow the required precise control of the structures at a molecular scale.

Acknowledgements

This article is dedicated to all those people who dedicate their daily work to the development of catalysis. This science is essential to achieve a sustainable development of humanity and to protect the environment.

Reference cited

Abdelouahab, F.B., R. Olier, N. Guihaume, F. Lefebvre and J.C. Volta. 1992. A study by *in situ* laser Raman spectroscopy of VPO catalysts for n-butane oxidation to maleic anhydride I. Preparation and characterization of pure reference phases. J. Catal. 134: 151–167.

Albonetti, S., G. Blanchard, P. Buratin, T.J. Cassidy, S. Masetti and F. Trifiro. 1997. Mechanism of ammoxidation of propane on a Sb/V/O system. Catal. Lett. 45: 119–123.

Aran, H.C., S. Pachecho Benito, M.W.J. Luiten-Olieman, S. Er, M. Wessling, L. Lefferts, N.E. Benes and R.G.H. Lammertink. 2011. Carbon nanofibers in catalytic membrane microreactors. J. Membrane Sci. 381: 244–250.

Balcells, E., F. Borgmeier, I. Gribtede and H.G. Lintz. 2003. Partial oxidation of propane and propene to acrylic acid over a Mo-V-Te-Nb oxide catalyst. Catal. Letters 87: 195–199.

Bañares, M.A. and I.E. Wacs. 2002. Molecular structures of supported metal oxide catalysts under different environments. J. Raman Spectros. 33: 359–380.

Berenguer, R., F.J. García-Mateos, J. Rodríguez-Mirasol and T. Cordero. 2014. Electrospinning of P-containing lignin solution for the preparation of carbon fibers with enhanced surface area and oxidation resistant. The World Conference on Carbon (Carbon 2014), Jeju Island (Korea).

Berenguer, R., J. Fornells, F.J. García-Mateos, M.O. Guerrero-Pérez, J. Rodríguez-Mirasol and T. Cordero. 2016. Novel synthesis method of porous VPO catalysts with fibrous structure by electrospinning. Catal. Today 277: 266–273.

Berenguer, R., M.O. Guerrero-Pérez, I. Guzmán, J. Rodríguez-Mirasol and T. Cordero. 2017. Synthesis of vanadium oxide nanofibers with variable crystallinity and V5+/V4+ ratios. ACS Omega 2(11): 7739–7745.

Bravo-Suarez, J.J., K.K. Bando, J. Lu, T. Fujitani and S.T. Oyama. 2008. Oxidation of propane to propylene oxide on gold Catalysts. J. Catal. 255: 114–126.

Bravo-Suarez, J.J., K.K. Bando, T. Akita, T. Fujitami, T.J. Fuhrer and S.T. Oyama. 2008. Propane reacts with O_2 and H_2 on gold supported TS-1 to form oxygenates with high selectivity. Chem. Commun. 3272–3274.

Brazdil, J.F. 2006. Strategies for the selective catalytic oxidation of alkanes. Top. Catal. 38: 289–294.

Carreon, M.A., V.V. Guliants, M.O. Guerrero-Pérez and M.A. Bañares. 2004. Phase transformations in mesostructured VPO/Surfactant composites. Micro. Meso. Mat. 71: 57–63.

Carrero, C.A., C.J. Keturakis, A. Orrego, R. Schomäcker and I.E. Wachs. 2013. Anomalous reactivity of supported V_2O_5 nanoparticles for propane oxidative dehydrogenation: influence of the vanadium oxide precursor. Dalton Trans. 42: 12644–112653.

Carrero, C.A., R. Schloegl, I.E. Wachs and R. Schomaecker. 2014. Critical literature review of the kinetics for the oxidative dehydrogenation of propane over well defined supported vanadium oxide catalysts. ACS Catal. 4: 3357–3380.

Cavani, F., N. Ballarini and A. Cericola. 2007. Oxidative dehydrogenation of ethane and propane: How far from commercial implementation? Catal. Today 127: 113–131.

Celaya Sanfiz, A., T.W. Hansen, D. Teschner, P. Schnorch, F. Girgsdies, A. Trunschke, R. Schlögl, M. Hong Looi and S. Bee Abd Hamid. 2010. Dynamics of MoVTeNb oxide M1 phase in propane oxidation. J. Phys. Chem. C 114: 1912–1921.

Centi, G., F. Trifiro, J.R. Ebner and V.M. Franchetti. 1988. Mechanistic aspects of maleic anhydride synthesis from C_4 hydrocarbons over phosphorus vanadium oxide. Chem. Rev. 88: 55–80.

Centi, G. 1993. Vanadyl pyrophosphate—A critical overview. Catal. Today 16: 5–26.

Factsheet palm oil biodiesel EU TE May 2018. European Federation for Transport and Environment AISBL (available here https://www.transportenvironment.org/sites/te/files/Factsheet%20palm%20 oil%20biofuels%20TE%20May%202018.pdf).

Fong, H., I. Chun and D.H. Reneker. 1999. Beaded nanofibers formed during electrospinning. Polymer 40: 4585–4592.

Gasior, M., B. Grzybowska, K. Samson, M. Ruszel and J. Haber. 2004. Oxidation of CO and C_3 hydrocarbons on gold dispersed on oxide supports. Catal. Today 91-92: 131–135.

Grasselli, R.K., C.G. Lugmair and A.F. Volpe. 2008. Doping of MoVNbTeO (M1) and MoVTeO (M2) phases for selective oxidation of propane and propylene to acrylic acid. Top. Catal. 50: 66–73.

Grasselli, R.K., C.G. Lugmair and A.F. Volpe. 2011. Towards an understanding of the reaction pathways in propane ammoxidation based on the distribution of elements at the active centers of the M1 phase of the MoV(Nb,Ta)TeO system. Top. Catal.

Guerrero-Pérez, M.O., J.L.G. Fierro, M.A. Vicente and M.A. Bañares. 2002. Effect of Sb/V ratio and of Sb+V coverage on the molecular structure and activity of alumina-supported Sb-V-O catalysts for the ammoxidation of propane to acrylonitrile. J. Catal. 206: 339–348.

Guerrero-Pérez, M.O. and M.A. Bañares. 2002. Operando Raman study of alumina-supported Sb-V-O catalyst during propane ammoxidation to acrylonitrile with on line activity measurement. Chem. Commun. 12: 1292–1293.

Guerrero-Pérez, M.O., J.L.G. Fierro and M.A. Bañares. 2003. Effect of the oxide support on the propane ammoxidation with Sb-V-O based catalysts. Catal. Today 78: 387–396.

Guerrero-Pérez, M.O., J.L.G. Fierro and M.A. Bañares. 2003. Niobia-supported Sb-V-O catalysts for propane ammoxidation: effect of catalyst composition on the selectivity to acrylonitrile. Phys. Chem. Chem. Phys. 5: 4032–4039.

Guerrero-Pérez, M.O. and M.A. Bañares. 2004. Operando Raman-GC studies of alumina-supported Sb-V-O catalysts and role of the preparation method. Catal. Today 96: 265–272.

Guerrero-Pérez, M.O. and M.A. Bañares. 2007. Operando Raman-GC study of supported alumina Sb and V based catalysts: effect of Sb/V molar ratio and total Sb+V coverage in the structure of catalysts during propane ammoxidation. J. Phys. Chem. C 111: 1315–1322.

Guerrero-Pérez, M.O., J.M. Rosas, R. López-Medina, M.A. Bañares, J. Rodríguez-Mirasol and T. Cordero. 2012. J. Phys. Chem. C 116: 20396–20403.

Guerrero-Pérez, M.O. and M.A. Bañares. 2015. Metrics of acrylonitrile: from biomass vs. petrochemical route. Catal. Today 239: 25–30.

Guerrero-Pérez, M.O., E. Rojas-García, R. López-Medina and M.A. Bañares. 2016. Propane versus ethane ammoxidation on mixed oxide catalytic systems: influence of the alkane structure. Catal. Lett. 146: 1838–1847.

Guerrero-Pérez, M.O. 2017. Supported, bulk and bulk-supported vanadium oxide catalysts: A short review with an historical perspective. Catal. Today 285: 226–233.

Haber, J. 2009. Fifty years of my romance with vanadium oxide catalysts. Catal. Today 142: 100–113.

Hall, P.J. and J.M. Calo. 1989. Secondary interactions upon termal desoption of surface oxides from coal chars. Energ. Fuel 3: 370–376.

Holmber, J., R.K. Grasselli and A. Andersson. 2004. Catalytic behaviour of M1, M2, and M1/M2 physical mixtures of the Mo-V-Nb-Te-oxide system in propane and proene ammoxidation. Appl. Catal. A: Gen. 270: 121–134.

Korovchenko, P., N.R. Shiju, A.K. Dozier, U.M. Graham, M.O. Guerrero-Pérez and V.V. Guliants. 2008. M1 to M2 phase transformation and phase cooperation in bulk mixed metal Mo-V-M-O (M=Te, Nb) catalysts for selective ammoxidation of propane. Top. Catal. 50: 43–51.

Labruquere, S., R. Pailler, R. Naslain and B. Desbat. 1998. Oxidation inhibition of carbon fibre preforms and C/C composites by H_3PO_4. J. Eur. Ceram. Soc. 18: 1953–1960.

López-Medina, R., E. Rojas, M.A. Bañares and M.O. Guerrero-Pérez. 2012. Highly active and selective supported-bulk nanostructured MoVNbTe catalysts for the propane ammoxidation process. Catal. Today 192: 67–71.

Martínez-Latorre, L., P. Ruiz-Cebollada, A. Monzón and E. García-Bordejé. 2009. Preparation of stainless steel microreactors coated with carbon nanofiber layer: impact of hydrocarbon and temperature. Catal. Today 147: S87–S93.

Nijhuis, T.A., E. Sacaliuc, A.M. Beale, A.M.J. van der Eerden, J.C. Schouten and B.M. Weckhuysen. 2008. Spectroscopic evidence for the adsorption of propene on gold nanoparticles during the hydro-eposication of propene. J. Catal. 258: 256–264.

Rahmani, F., M. Haghighi and M. Amini. 2015. The beneficial utilization of natural zeolite in preparation of Cr/clinoptilonite nanocatalyst used in Co_2-oxidative dehydrogenation of ethane to ethylene. J. Ind. Eng. Chem. 31: 142–155.

Rojas, E., J.J. Delgado, M.O. Guerrero-Pérez and M.A. Bañares. 2013. Performance of NiO and Ni-Nb-O active phases during the ethane ammoxidation into acetonitrile. Catal. Sci. & Tech. 3: 3173–3176.

Rosas, J.M., J. Bedia, J. Rodríguez-Mirasol and T. Cordero. 2009. HEMP-derived activated carbón fibers by chemical activation with phosphoric acid. Fuel 88: 19–26.

Shiju, N.R. and V.V. Guliants. 2009. Recent developments in catalysis using nanostructured materials. App. Catal. A: Gen. 356: 1–17.

Tonkovich, A.L.Y., S.P. Fitzgerald, J.L. Zilka, M.J. LaMont, Y. Wang, D.P. VanderWiel and R.S. Wengeng. 2000. Microreaction technology: industrial prospects. pp. 364. *In*: Ehrfeld, W. (ed.). Proceedings of the Third International Conference on Microreaction Technology, Springer, Berlin.

Valero-Romero, M.J., A. Cabrera-Molina, M.O. Guerrero-Perez, J. Rodriguez-Mirasol and T. Cordero. 2014. Carbon materials as template for the preparation of mixed oxides with controlled morphology and porous structure. Catal. Today 227: 233–241.

Wan, Y.S.S., J.L.H. Chau, A. Gavriilidis and K.L. Yeung. 2001. Design and fabrication of zeolite-based microreactors and membrane microseparators. Micro. Meso. Mat. 42: 157–175.

Wegeng, R.S., M.K. Drost and D.L. Brenchley. 2000. Microreaction technology: industrial prospects. pp. 2. *In*: Ehrfeld, W. (ed.). Proceedings of the Third International Conference on Microreaction Technology, Springer, Berlin.

Wu, X. and L.R. Radovic. 2006. Inhibition of catalytic oxidation of carbón/carbón composites by phosphorus. Carbon 44: 141–151.

Chapter 6

Nanocatalysts from Biomass and for the Transformation of Biomass

María Luisa Rojas Cervantes

INTRODUCTION

The catalysis and the use of catalytic systems in numerous industrial processes is a field in constant development and meets the ninth principle of Green Chemistry. Although the catalysts are used in small amounts, many times the cost of their production is quite high, increasing the final expense of the global catalytic process. Therefore, there is a necessity of looking for alternative cheaper ways for productions of catalysts. In this sense, in the last twenty years, the synthesis of heterogeneous catalysts from waste materials derived from both industrial and biological sources has aroused a great interest and numerous studies have been reported (Balakrishnan et al. 2011; Bennett et al. 2016). The consumption of waste materials is quite advantageous because it contributes to the reduction of the economic and environmental costs associated with their disposal. Among these waste materials, biomass, referred to renewable organic materials derived from plants and animals than can serve and sources of energy, stands out as a source for the production of nanocatalysts, mainly materials based on carbon (Hu et al. 2010; Matthiesen et al. 2014; Balu et al. 2015; Deng et al. 2016; Jain et al. 2016; Hill 2017) and metal nanoparticles (Kharissova et al. 2013; Kou et al. 2013; Varma 2013; Varma 2016).

The structural analysis of biomass is of particular importance in the development of processes to produce other fuels and chemicals and in the study of combustion phenomenon. The lignocellulosic materials, which are the most renewable organic resource on earth, are composed of cellulose (40–45%), hemicellulose (25–35%) and lignin (20%) covalently crosslinked or bonded by non-covalent forces (Fig. 1). They also contain organic extractives and small amounts of inorganic materials (Lam and Luong 2014; Bhanja and Bhaumik 2016).

Departamento de Química Inorgánica y Química Técnica, Facultad de Ciencias-UNED, Paseo Senda del Rey n° 9, Madrid 28.040.
Email: mrojas@ccia.uned.es

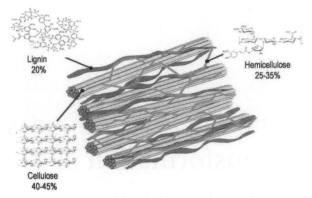

Fig. 1. Composition of lignocellulosic materials.

Although the main source of the biomass proceeds from lignocellulosic materials, other alternatives, such as the shellfish biomass and the non-edible oils can be also considered as renewable sources for the production of nanocatalysts and for the transformation into high value added products.

In this chapter, we study the synthesis of nanocatalysts starting from biomass and the transformation of biomass into high value added products by using nanocatalyts in the involved reactions. In the first section, we present the production of several types of nanocatalysts from biomass, classified in different categories, dedicating a more detailed study to the materials based on carbon due to their high production. We show the alternative strategies applied for the synthesis of porous carbons and their different applications. Another type of nanocatalysts which is revised is the group of metal-nanocomposites and metal nanoparticles. In some cases, the biomass is used not only to generate the nanocatalyst, but is also transformed into the target product. This alternative makes the system more cost effective and environmentally friendly. Some examples of this alternative are studied in the two groups of catalysts, based on carbon and nanometallic systems.

The second section of the chapter is devoted to summarize the reactions transforming the biomass in platform chemicals by using nanocatalysts different from the nanomagnetic particles, which are revised in another chapter of this book. The multiple reactions leading to the transformation of the biomass as well as the nanocatalysts more usually employed for each one of these reactions will be presented in different sub-sections.

Nanocatalysts from biomass

Materials based on carbon

The synthesis of biomass-derived porous carbonaceous materials has increased significantly in the last two decades, in view of their wide field of application in numerous areas, including catalysis. Moreover, the use of bio-wastes to produce carbons can solve some of the problems derived from the waste management of many industries (Abioye and Ani 2015; Yahya et al. 2015). In this section, we revise different approaches for synthesizing this kind of material, from the conventional thermal activation to the ionothermal carbonization.

Conventional thermal activation: activated carbons

Activated carbons are carbonaceous materials predominately amorphous in nature with high porosity, high physicochemical stability and high degree of surface reactivity. They differentiate from elemental carbon by the presence of oxygen functionalities at the outer and inner surfaces in the form of acidic (carboxylic, lactonic, phenolic) and non-acidic (carbonyl, ether, quinone) groups.

Activated carbons can be produced from naturally occurring and synthetic carbonaceous solid precursors. The occurring raw materials for activated carbons are from animal, mineral and vegetable origin (Mohammad-Khah and Ansari 2009) and the classification of activated carbons can be made based on the starting materials, whose type and structure play a significant role in influencing the properties and characteristics of the final product (Lua and Yang 2005; Cagnon et al. 2009; Mohammad-Khah and Ansari 2009).

The activated carbons are produced in a process consisting of two steps, carbonization and activation, which can be carried out subsequently or simultaneously. The carbonization of a precursor takes place at a low temperature usually between 400°C and 850°C and in the absence of oxygen. Carbonization is a process of converting carbonaceous materials into activated carbon by thermal decomposition in a furnace (convectional heating) or a microwave using a controlled atmosphere and heating. The activation can be physical (with some mild oxidizing gases such as carbon dioxide or steam), chemical (in the presence of a dehydrating agent, e.g., H_2SO_4, $ZnCl_2$, KOH, and H_3PO_4), combined (physicochemical) and induced by microwave. More details about the activation process and the variables involved in the activation of activated carbons can be found in Abioye and Ani (2015) and Yahya et al. (2015).

Although most of the commercial activated carbons are produced from fossil fuels (petroleum and coals), the use of agricultural waste biomass for their production has increased significantly in the last years (Abioye and Ani 2015; Yahya et al. 2015; Aguayo-Villarreal et al. 2017), leading to cheaper, renewable and environmental friendly porous carbons. Any inexpensive material that has a high carbon and a low inorganic content can be utilized as a precursor for activated carbon productions. Therefore, any lignocellulosic material can be used as a suitable, low-cost and abundant alternative to the conventional raw materials for an activated carbon purpose (Ioannidou and Zabaniotou 2007). Different examples of lignocellulosic materials used for the production of activated carbons are banana waste, bamboo, coconut shell, almond shell, walnut shell, olive stone, rice husk, waste tea, sunflower shell, etc., among many others (Abioye and Ani 2015; Yahya et al. 2015).

The main application of activated carbons derived from biomass is as adsorbents for the removal of different species, such as dyes (Demirbas 2009; Shin 2012; Hadi et al. 2015), pharmaceuticals (Rivera-Utrilla et al. 2013; Kyzas et al. 2015) and heavy metals (Sen et al. 2015; Tang et al. 2015; Hu and Cao 2016). However, there are also many studies reporting the use of these materials as catalytic supports and catalysts.

Some examples of the use of activated carbons in the catalytic field, mainly for fine chemical synthesis, can be found in Matos et al. (2017).

The activated carbons obtained by traditional pyrolysis are mainly microporous, with extremely high surface areas. Therefore, it is necessary to recur to other alternatives to produce carbons with hierarchical porous structure of micro-, meso- and macropores, required for different applications.

Hydrothermal carbonization

In the past few years, the use of hydrothermal carbonization (HTC) for the conversion of waste biomass into novel carbon-based materials with a wide variety of applications has received considerable attention (Titirici et al. 2008; Hu et al. 2010; De et al. 2015; Deng et al. 2016; Jain et al. 2016). Hydrothermal carbonization is a thermo-chemical conversion technique which uses subcritical water (over 100°C and 0.1 MPa) for the transformation of wet/dry biomass to carbonaceous products through fractionation of the feedstock. The raw materials for the preparation of carbon materials through HTC can be isolated carbohydrates (sucrose, starch), organic molecules and waste biomass. According to the temperature used in the treatment, the HTC process can be classified in high-temperature hydrothermal carbonization and low-temperature hydrothermal carbonization. A variety of carbon-based materials including carbon nanotubes, graphite, microspheres and activated carbons can be obtained by the first process (Hirose et al. 2002; Salvador et al. 2007). The chemical reactions involved in the hydrothermal process at low temperature are very complex. It is commonly accepted that the hydrothermal treatment of biomass consists of five procedures, that is, hydrolysis, dehydration, decarboxylation, polymerization and aromatization. During the initial stages of the hydrothermal treatment, the dehydration of hexose (i.e., glucose, sucrose, starch, etc.) and pentose (xylose, ribose) leads to the formation of 5-hydroxymethyl furfural (HMF) and furfural, respectively (Titirici et al. 2008), as well as some organic acids, which lower the pH and catalyse the *in situ* dehydration of carbohydrates. The intermediate products obtained condense to form substances denoted as *humins*, which are transformed upon dehydration or pyrolysis into the final carbonaceous material, sometimes classified as *biochars*, which is defined as "fine-grained charcoal, high in organic carbon and largely resistant to decomposition". The materials obtained by HTC at low temperature are C-H-O structures rich in functional surface groups, micro- or mesoporous and with a high reactivity.

Different parameters governing the hydrothermal carbonization can be found in Jain et al. (2016). Additionally, the initial structure and composition of the biomass material can significantly influence the size, shape and porous structure of the final biochar (Ioannidou and Zabaniotou 2007; Deng et al. 2016). In this sense, besides the different structures found in hexose-based carbons and pentose-based ones, the morphologies also vary from interconnected spherical particles in the first case to separated carbon spheres in the second one. The metal ion catalysts also sometimes play an effective role in enhancing the HTC process of carbohydrates and influence the shape of the final materials. In this regard, the HTC of starch and rice grains under mild conditions can be catalysed by iron ions and iron oxide nanoparticles (Cui et al. 2006). Fe^{2+} ions facilitate both hollow and massive carbon microspheres, whereas the Fe_2O_3 nanoparticles lead to the formation of very fine, rope like carbon nanostructures.

The HTC process can be assisted by microwave heating, resulting in reduced carbonization time and milder temperatures (Guiotoku et al. 2009; Jung et al. 2013).

The morphology and porosity of the HTC carbons can be tailored using a template and the surface chemistry can be finely adjusted by carbonizing the monosaccharide together with a precursor containing the desired functionality (Zhao et al. 2010; Wohlgemuth et al. 2012).

Although the carbon materials synthesized by HTC are mainly used for the absorption of heavy metals, CO_2 capture, and energy production and storage, they have also been applied in the field of catalysis, such as selective hydrogenation of phenolic derivatives to ketones (Makowski et al. 2008), Diels-Alder reaction, aldol condensation and transesterification (Demir-Cakan et al. 2010). Some authors (Xu et al. 2007) reported the synthesis of Pt-/Pd-loaded carbon microspheres (CMS) as catalysts for DMFCs, showing higher catalytic activity for methanol oxidation in alkaline media than commercial Pt (Pd)/carbon black. In a similar way, other carbon materials derived from HTC loaded with Pt nanoparticles also showed better performance for methanol oxidation (Sevilla et al. 2007).

Lastly, hybrid nanostructures of carbonaceous nanofibres embedded with noble-metal nanoparticles synthesized performed as efficient catalysts for the 100% conversion of CO to CO_2 at low temperature (Qian et al. 2007).

Starch carbonization (Starbon)

A wide range of mesoporous carbon materials with tuneable surface and bulk chemistry (Starbon) can be produced starting from starch as crude biomass (Budarin et al. 2006). The production process consist of three steps: (1) expansion: the formation of gel occurs by heating the starch in water, producing an opening and disordering of the polymer network; (2) exchange of water with a low surface tension solvent at low temperature ($\approx$ 5ºC) (in order to prevent the network structure from collapsing during the drying process) and drying of the gel by evaporation of the solvent; and (3) pyrolysis of the starch materials previously doped with a catalytic amount of organic acid in the 150–800ºC range under vacuum.

The final product exhibits a variable structure, stability and chemistry that ranges from starch-like amorphous carbon to commercial activated carbons (Budarin et al. 2006; White et al. 2009).

Apart from starch, the Starbon process has also been applied to prepare mesoporous carbons by starting from other polysaccharides with linear backbone, such as amylose, pectin (Yu et al. 2002) or alginic acid (derived from algal biomass) (Dodson et al. 2013).

As a result of the treatment in the Starbon process, the hydrophilic starch is converted to a highly hydrophobic carbon with high stability in acid or base media, which is a key property for application in the field of catalysis. The materials can be also functionalized during the synthesis procedure, mainly with sulfonic groups in order to be conferred with an acid character. The incorporation of $-SO_3H$ groups provides these sulfonated carbons with a hydrophilic character, which generally leads to high catalytic activity in the solution. In this context, sulfonated Starbon carbons have been used as heterogeneous catalysts in the esterification of different acids, such as succinic, fumaric, itaconic and levulinic (Budarin et al. 2007; Budarin 2007; Clark et al. 2008). Other reactions catalysed by Starbon-SO$_3$H carbons are the transformations of toluene into *p*-benzyltoluene, oleic acid into ethyl oleate or phenol into O-/C-alkylated phenol (De et al. 2015). In addition to their use as catalysts, Starbon carbons have also been used as catalytic supports for noble metals (Pd, Pt, Ru and Rh). These catalysts have shown a good performance in the hydrogenation of succinic acid, displaying higher turnover frequency and stability when compared to other commercial carbons (Luque and Clark 2010).

Template method

The template method or template–directed synthesis is a powerful approach to obtain carbon materials with controlled and ordered pore structure, ranging from 1 to 200 nm, which cannot be achieved by traditional activation. The template acts as a scaffold and a directing agent of the formation of pores during the carbonization step.

In the hard-templated method, a pre-formed porous solid is first prepared and then the carbon precursor is introduced into the pores of the template. The polymerization and carbonization of the precursor is produced at the confined zone and, finally, the inorganic template is removed yielding to a porous carbon which is a reverse replica of the inorganic precursor.

Since the pioneering work of Knox et al. (1986), many other hierarchical carbon materials with functional groups at the surface have been obtained by using the appropriate nanostructured silica materials, such as MCM-48 (Ryoo et al. 1999) and SBA-15 (Lu et al. 2003), as templates and a wide number of carbon precursors (e.g., sucrose, phenolic resin, furfuryl alcohol and mesophase pitch) as carbon sources (Ryoo et al. 2001). One of the most interesting works is that of Titirici et al. (2007), in which different hierarchical carbon materials (hydrophilic macroporous carbon casts, hollow carbon spheres or mesoporous carbon materials), rich in polar functional groups (such as COOH, OH and CO) at the surface, have been obtained carrying out the HTC of furfural in the presence of nanostructured silica templates by tuning the polarity of the surface of silica.

The main drawback of the hard template process is the necessity of removal of the silica template by using toxic reagents (for example, NaOH or HF). This drawback can be solved by using the soft-template method, based on the cooperative organic-organic assembly between amphiphilic molecules (block polymers and surfactants), which act as structures–directing agents and organic precursors solution. Pluronic® F127 is the typical triblock copolymer used as the soft template. When the concentration of F127 is higher than its critical micelle concentration, micelles are formed in the aqueous solution, acting as source of pores before the hydrothermal treatment. This triblock copolymer has been used for the synthesis of carbons using sugarcane bagasse (Huang and Doong 2011), fructose (Kubo et al. 2011) and lignin (Saha et al. 2014) as a carbon precursor. One disadvantage of the soft-template method is that the carbonization process cannot be carried out at high temperatures, because of the instability of surfactants and block polymers at elevated temperatures.

Dual-template synthesis combining both a hard template to generate macropores and a soft template to form the mesopores is a good alternative for the synthesis of carbonaceous materials. For example, nitrogen-doped hierarchically porous carbon materials were fabricated by using an Al-based composite and triblock copolymer Pluronic F127 as co-templates, and natural banana peel as the precursor (Liu et al. 2014).

Ionothermal carbonization

Ionic liquids (ILs), which are liquids below 100°C, formed entirely by cations and anions, play a significant role as benign solvents in the synthesis of carbon materials derived from biomass owing to their attractive properties, such as good biomass solubility, suppressed solvent volatility and excellent thermal stability. The ability to form hydrogen bonds with solutes contributes to enhancing the solubility of carbohydrates. The process in which ILs are involved in the synthesis of mesoporous carbons is called ionothermal carbonization (ITC), and the carbon materials prepared by this way possess higher surface areas and higher carbon content than those obtained under HTC (Zhang et al. 2015). Lee et al. (2009) reported the first ITC strategy to obtain nitrogen-rich mesoporous carbon, starting from imidazolium ILs bearing one or more nitrile side chains. Other N-doped carbons, with high nitrogen content and partially developed graphitic structure were obtained by the direct carbonization of dicyanamide (DCA)-anion based ILs (Paraknowitsch et al. 2010).

The use of ILs allows the doping of carbon with heteroatoms, that is, not only from nitrogen, but also from boron and sulphur. Some examples of co-doped carbons prepared from ILs are described in Zhang et al. (2015).

The poly(ionic liquid)s, Poly-ILs, which are the polymeric form of ILs and are prepared by polymerization of a monomer IL using an initiator, can be also used to synthesize carbon materials with well-defined morphologies, such as fibres and monoliths (Yuan et al. 2010). Thus, polysaccharide biomass and sugar-based molecules can be transformed into functional carbons in the presence of low fractions of ILs or poly-ILs at 400°C under nitrogen.

Some examples of the applications of carbons derived from ILs can be found in Zhang et al. (2015). Thus, metal-nanoparticle-doped carbons prepared from ILs in the presence of metal salts or metal-containing ILs can exhibit potential applications in electrocatalysis and chemical catalysis, such as biomass refining, Ullmann coupling reactions, hydrogenation, aerobic oxidation, carbonylation and phenylacetylene reduction. In addition, nitrogen-doped carbons were tested as a solid base in base-catalyzed reactions, such as Knoevenagel condensation and oxidative dehydrogenation.

Metal nanoparticles and metal nanocomposites

The preparation of noble metal nanoparticles (MNPs) has aroused a great interest in the field of nanomaterials research, owing to their multiple applications in different fields, as energy conversion, chemical reaction, environmental remediation, and so on. MNPs exhibit a dramatically increased surface area, chemical activity, specificity of interaction and unprecedented efficiency compared to their bulk counterparts and offer a potential opportunity as recyclable heterogeneous catalysts (Campelo et al. 2009).

However, the MNPs suffer from the tendency to aggregate, resulting in a loss of catalytic activity. In order to avoid it, it is necessary to functionalize the surface through post grafting with different capping agents, mainly polyvinylpyrrolidone (PVP), which generally implies the use of extreme conditions of pressure and temperature and toxic chemicals. Furthermore, the general bottom-up approach for the production of MNPs often entails the use of hazardous reducing agents, typically hydrides or hydrazines. Therefore, the choice of greener routes for the synthesis of MNPs involving an eco-friendly reducing agent, environmental friendly solvent and non-hazardous capping entity is imperative (Varma 2012).

One promising strategy to address the stability and reusability of MNPs is to immobilize them on a suitable substrate. In recent years, the production of biomass derived metal nanoparticles supported catalysts has intensified. Biomass contains numerous oxygenated functional groups (carboxyl, carbonyl, hydroxyl and ether) which can act as ligands for metal complexation in an aqueous solution. This ability can be exploited to obtain a high dispersion of MNPs into the biomass matrix. In this sense, metal nanoparticles have been successfully immobilized onto many biomass-derived materials, such as cellulose (Lin et al. 2011), chitosan (Xue et al. 2015), starch (Budarin et al. 2008), beech woods chips (Richardson et al. 2010) and sawdust (Sevilla et al. 2007) by directly adding the metal precursor in the biomass solid fuel before thermochemical conversion. In addition to the high sorption capacity and the affinity with metal ions, some biomaterials have also the advantage of physical and chemical versatility that allow easy surface functionalization and production of materials in different forms, including flakes, gel beads, membranes and fibres (Nadagouda et al. 2009).

Many of the biomass synthesized noble metal nanoparticles supported catalysts have excellent performance in photochemical catalysis and selective hydrogenation (Jia et al. 2009; Tian et al. 2013) for the *p*-nitrophenol reduction (Lin et al. 2011) and for C-C coupling reactions (Budarin et al. 2008).

In another method for the green synthesis of MNPs, the biomass is not used as a support for the MNPs, but as a reducing agent, capping agent, or both. An example is the preparation of AuPd/C and AgPd/C electrocatalysts by heating the metallic precursors in the presence of an aqueous pecan nutshell extract under microwave activation. Further, the colloidal dispersion of MNPs is supported in an appropriate amount of Vulcan XC72 and pyrolyzed in order to obtain the bimetallic catalysts (Casas Hidalgo et al. 2016). Another example is the synthesis of the Au-SrTiO$_3$ composite prepared by wet impregnation of the support (SrTiO$_3$) with the Au salt in the presence of the *syzygium* extract solution (Pan et al. 2015). The catalyst performed well in the direct hydrogen generation from formaldehyde at a low temperature, the oxygen vacancies and organic biomass groups on the support playing an important role in the activation of formaldehyde.

The antioxidants present in wine waste or tea polyphenols (Virkutyte and Varma 2011; Varma 2012) serve both as reducing and capping agents, to produce a great variety of extremely stable spherical-, rod- or flower-shaped MNPs under mild reaction conditions, that open up many opportunities for their utilization and potential mass production. Plant leaves or plant seed extracts contain reducing sugars (aldoses), terpenoids, amino acids and other organic compounds (Huang et al. 2007; Nadagouda et al. 2009; Yilmaz et al. 2011), which produce important changes in nanoparticles formation and stabilization.

Other bio-systems used for the synthesis of not supported MNPs are coffee- and tea-extracts, agricultural products (beet juice) and polyphenols from varied origins (Varma 2016). When these MNPs are borne in supports not derived from biomass, they find applications in the field of catalysis (Vilchis-Nestor et al. 2009; Huang et al. 2014).

Many of the MNPs are produced using polyalcohol solvents, mainly ethylene glycol, as reducing agents, owing to the simplicity of the process and its compatibility in open-air environments. However, the metabolites of ethylene glycol, glycolic acid and oxalic acid may cause different health problems, such as acute renal failure, cerebral damage and injury to other organs, which prohibits the widespread use of the polyol process.

One alternative green polyol is the glycerol, which is an abundant, inexpensive and nontoxic by-product from biofuel processing. Kou et al. (2013) have combined the use of glycerol and MW irradiation to produce nanoparticles of noble metals whose shapes can be changed by adding different surfactants.

Other catalysts from biomass

In addition to pure metal nanoparticles, nanoparticles of metal oxides have been obtained from biomass sources (Kharissova et al. 2013). In this regard, iron oxide and iron oxohydroxide, which turned out to be active Fenton-like catalysts for the discoloration of dyes in an aqueous solution, were synthesized by using the extracts of green tea leaves. Nanoparticles of TiO_2 (25–150 nm) were formed by treating titanium alkoxides with nyctanthes leaf extract or with *Jatropha curcas* L. Smaller spherical nanoparticles (25–40 nm), were produced by using *Aloe barba-densis Miller* leaf extract and zinc nitrate. Cu_2O and CuO nanoparticles were formed by using agriculture wastes of *Arachis hypogaea* L. (*Fabaceae*) leaf extracts and *Aloe vera* extract, as biomass feedstock, respectively. This last feedstock, *Aloe vera*, was also used for the production of cubic In_2O_3 nanoparticles (5–50 nm) with indium acetylacetonate as a metallic precursor. In_2O_3 nanoparticles could also be obtained also with another shape, that is, concretely hollow spheres, by the hydrothermal reaction of glucose and $InCl_3$ and further calcination.

Transformation of biomass by nanocatalysts

Biomass derived from different sources includes lignocellulose, oil seed crops, sugar crops, starch crops and aquatic biomass, such as shellfish and micro- and macroalgae. Other biomass materials are biowaste from agricultural, animal fats, urban and domestic wastes and used plant oils (Lam and Luong 2014). The estimated global production of biomass is approximately of 10^{11} tonnes per annum, the 60% being from terrestrial origin and the 40% from an aquatic one (Sheldon 2007). The terrestrial biomass is composed of 75% of carbohydrates (storage carbohydrates and structural polysaccharides), 20% of lignin and 5% of triglycerides (fats and oils), proteins and terpenes (Sheldon 2014) (Fig. 2). Macroalgae contain only 10–15% of dry matter, which is formed mainly (in ca. 60%) by polysaccharides, including cellulose and/or starch. They also contain monosaccharides, as glucose and galactose (Chen et al. 2015).

The components of the biomass can be transformed under the appropriate conditions into added-value chemicals. In this regard, the hydrolysis of cellulose leads to glucose monomers, which can be used for the production of fuel alcohol or other chemicals by fermentation. The hydrolysis of carbohydrates and sugars contained in the hemicellulose produces fructose and C5 monosaccharides, which are the substrates for the production of furfural, 5-hydroxymethylfurfural (HMF) and its derivatives. The lignin is a potential feedstock to obtain higher value fuels and chemicals. The triglycerides (from oil seed feedstocks) can be hydrolytically converted into fatty acids plus glycerol, and finally, the proteins can be transformed into amino acids.

Lignocellulose biomass is much more difficult to process than the first-generation renewable feedstocks (simple sugars, starches and vegetable oils) due to the presence of multiple functional groups and the high oxygen content (Besson et al. 2014); however, it has the advantage that it is not edible and, therefore, does not compete with food. The high oxygen content of carbohydrates of the lignocellulosic materials is a limitation in the processing of the biomass, and therefore it is necessary to lower it. In order to decrease the oxygen content of biomass, three alternative routes can be applied. One way is the fermentative conversion of carbohydrates into ethanol, butanol and CO_2. Hydrogenolysis is another way by which the oxygen is removed in the form of water by combining with one molecule of hydrogen. The third option is the dehydration of carbohydrates leading to the formation of furans and levulinic acid (LA).

The predominant catalysts for many reactions of transformation of biomass are homogeneous Brønsted acids, such as HCl, H_2SO_4 or H_3PO_4 due to their cheap cost and availability. However, they suffer from some disadvantages, such as their corrosiveness and limited reusability and recovery, among others. Additionally,

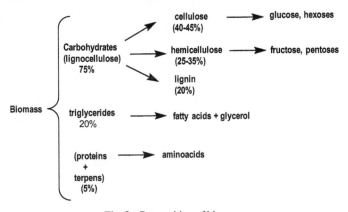

Fig. 2. Composition of biomass.

under the reaction conditions required for the depolymerisation and fragmentation of raw biomass, severe side-reactions often occur, and therefore the use of the mildest conditions possible is necessary. Consequently, the search of alternative highly active heterogeneous solids working under mild conditions is a goal pursued by many scientists. In this context, in the last decade, various modified solid acid catalysts have been developed with tuneable pore architecture for the transformation of raw biomass to chemicals of interest.

Among the solid acids that are intensively studied in the recent times, it is worth mentioning porous metal oxides, sulfonated porous organic polymers (POPs), functionalized zeolites, aluminosilicates (ZSM-5), immobilized ionic-liquids and acid functionalized mesoporous silica materials (Bhanja and Bhaumik 2016). Water is a good medium and a green solvent for the biomass conversion. However, many of the oxide supports used are quite unstable and suffer from deactivation in hot (100–250°C) and pressurized water, a reaction condition commonly employed in this kind of transformation reaction. On the other hand, porous carbon materials are promising supports for heterogeneous catalysis as compared to oxide supports, because of the former's higher structural and chemical stability. In this sense, interesting advancements in the field of catalysts based on carbon as either an organic-inorganic hybrid or carbocatalysts have developed (Matthiesen et al. 2014).

Another group of interest is that of the monometallic and multimetallic catalysts, because many transitional metals are well known catalysts or catalyst components in a wide range of reactions. In particular, they are capable of promoting some important reactions of biomass conversion, such as hydrogenolysis, hydrogenation, oxidation, decarboxylation and decarbonylation (Wang et al. 2016). In this sense, the number of literature references dealing with metal catalysts employed in biomass conversion increased 14-fold in the last decade (Besson et al. 2014). There are many reviews dealing with metal catalysts employed for the conversion of biomass to biofuels. However, less attention has been paid to catalysts applied to the transformation of biomass-to-chemicals.

In this section of the chapter, we will focus on the conversion of biomass into chemicals in the presence of several kinds of catalysts. The different types of reactions involved in the depolymerisation/fragmentation of raw biomass components into platform chemicals, and in the further upgrading of intermediate compounds into value-added end products will be described in different following sub-sections. The production of hydrocarbons, biofuels, and fuel additives will not be considered.

Hydrolysis

The hydrolysis of crystalline cellulose has been carried out using different carbon materials functionalized with acid groups, mainly SO_3H. As an example, an amorphous carbon bearing SO_3H, COOH and OH groups functions as an efficient catalyst for this reaction, due to its ability to adsorb 1,4 glucan. Another carbon material used for the same reaction was an amorphous carbon containing graphene sheets bearing SO_3H, COOH and phenolic OH groups, for which a synergetic effect between the groups in the carbon material, and SO_3H groups bonded to the carbon, was reported (Lam and Luong 2014).

The hydrolysis of cellulose to glucose was achieved in a high yield with a sulfonated activated-carbon catalyst with high hydrothermal stability. A yield of ca. 75% to glucose was also obtained with a sulfonated CMK3 (mesoporous carbon), prepared from sucrose and using SBA-15 as the hard template.

One problem to achieve the hydrolysis of cellulose is its low solubility in water. This can be solved by carrying out the reaction in ionic liquids. In this sense, three different sulfonated catalysts derived consecutively from sucrose-derived carbon, glucose-derived carbon and nut shell activated carbon, were used to perform the cellulose hydrolysis in [BMIM]·Cl (Liu et al. 2013).

The carbohydrates present in the biomass can be processed to furanic aldehydes, such as furfural and 5-hydroxymethylfurfural (HMF). The first is derived from C5 sugars, such as xylose, and the second one is obtained by the dehydration of C6 monomeric and polymeric carbohydrates, mainly glucose, fructose, and cellulose. The furanic aldehydes obtained are key intermediates for the preparation of biofuels and high value-added chemicals (Hu et al. 2012).

Different materials based on functionalized graphene were active for the dehydration of xylose to furfural in water with an average yield of 61%. Better yields to furfural were achieved (67–69%) when the reaction was also catalysed by TiO_2 NPs deposited on graphene oxide or carbon black and activated by microwave irradiation (Russo et al. 2013).

Some amorphous carbon produced by hydrothermal carbonization of cellulose at 200–250°C, were capable of dehydrating the fructose to HMF in ionic liquids with yields comprised between 70 and 83% (Qi et al. 2012). This is a clear example of catalysts obtained from biomass and used for the transformation of the biomass.

CNTs and carbon nanofibres (CNFs) functionalized with poly(*p*-styrenesulfonic acid) and benzenesulfonic acid groups were also successfully tested for the conversion of fructose to 5-HMF at 120°C.

Other different solids from carbon materials, including mesoporous silica, bifunctional catalysts, metal oxides and zeolites, among others, have been also employed as catalysts for the transformation of biomass into HMF (Bhanja and Bhaumik 2016). Thus, the dehydration of fructose to HMF was achieved with a 75% of selectivity and 93% of conversion over a SBA-15-$PrSO_3H$ catalyst in a water-organic biphasic solvent system. Mesoporous TiO_2 prepared using aspartic acid and salicylic acid as soft templates have turned out to be efficient catalysts for the conversion of glucose and fructose into HMF while porous TiO_2 NPs synthesized via biopolymer alginate template transformed some unutilized sugars derivatives to HMF in DMA-LiCl under microwave irradiation. A mesoporous alumina also exhibited a high yield in the dehydration of glucose to HMF. The degradation of cotton cellulose for the synthesis of HMF (81% of conversion) was carried out in the presence of a mixed metal oxide ZnO–ZrO_2 catalyst under mild hydrothermal conditions.

Several zeolites with different Si/Al ratios were tested in the hydrolysis of cellulose, the highest yield to glucose being obtained when the reaction was catalysed by the zeolite with the highest Si/Al ratio (= 75), H-beta, due to the highest hydrophobic character of this zeolite (Onda et al. 2008). Glucose, cellobiose and starch were used as feedstocks for the obtaining of HMF through a one-pot reaction catalysed by a Sn-beta zeolite and hydrochloric acid with a biphasic solvent (Nikolla et al. 2011).

Sulfonated resins have also been employed as catalysts for the hydrolysis of monosaccharides (Bhanja and Bhaumik 2016). Thus, sulfonated Amberlyst-15 transformed the fructose into HMF with a yield of 100% in a water separation reactor. HMF was also obtained from the dehydration of carbohydrates using poly-benzylic ammonium chloride resins as catalysts.

Due to their high porosity and ordered structure, metal-organic frameworks (MOFs) lead to more substrate transfer than conventional carbon materials and inorganic solids. A selective dehydration of fructose to HMF was achieved using MOF and phosphotungstic acid (PTA)-encapsulated MIL-101(Cr), which has a chromium carboxylate cubic structure built up from chromium carboxylate units in a corner-sharing fashion, such as a catalyst. Fructose can be transformed to HMF by using a series of sulfonic acid functionalised MOFs (MOF-SO$_3$H), obtaining full fructose conversion and a yield of 90% to HMF when the reaction was catalysed by MIL-101(Cr)-SO$_3$H.

Metal phosphates are promising catalysts for various dehydration reactions. A phosphate-immobilized anatase TiO$_2$ (phosphate/TiO$_2$) was employed for the selective conversion of glucose to HMF (Nakajima et al. 2014). Other phosphates, such as NbPO, AlPO, TiPO and ZrPO were also successfully applied in glucose dehydration to HMF in the aqueous phase. Fructose, glucose, sucrose, cellobiose and cellulose were converted into HMF with high yield in a water/methyl isobutyl ketone (MIBK) biphasic solvent under microwave activation over a large-pore mesoporous self-aggregated tin phosphate nanoparticles (LPSnP-1) catalyst. The cubic zirconium pyrophosphate showed a very high yield of HMF from carbohydrate biomass.

Hydrogenation

In the heterogeneously catalysed hydrogenation reactions, it is important to consider the phenomenon called "hydrogen spillover", which consists of the dissociative adsorption of the hydrogen on a metal NP and the migration from its surface to the support surface, which can become catalytically active (Filikov and Myasoedov 1986). As a result, the hydrogenation rate is the sum of both reaction rates, on the metal surface and on the support surface. The support not only influences the dispersion of NPs but also on the hydrogen spillover, resulting in a beneficial or detrimental effect to the hydrogenation reaction. An example of the beneficial effect is the conversion of cellulose into sorbitol with a yield of 91.5% over a Pt nanocatalyst loaded on reduced graphene oxide (Pt/rGO) (Wang et al. 2014). The catalyst was prepared using ethylene glycol as a reductant agent under microwave irradiation, and the rGO led to a higher yield than other supports.

Sorbitol is used as an additive in the food, cosmetic and paper industries, as a precursor for the production of fine chemicals. The majority of the sorbitol is produced by the hydrolysis of the starch-containing crops into glucose and its further catalytic hydrogenation.

The industrial reactions for production of sorbitol are carried out in a batch system of stirred tank in the presence of Raney-type nickel catalysts. In the last years, these catalysts have been improved by adding some metal promoters, such as Mo, Cr and Fe. The higher rate in the hydrogenation of glucose is caused by the polarization of the C=O bonds of the aldehyde form of glucose by the electropositive metal promoters. However, the leaching of the metal and the deactivation of the active sites occur, which has provoked several attempts to use supported Ni catalysts as an alternative to the Raney-type nickel (Dechamp et al. 1995; Li et al. 2002). In this regard, the Ni–B/SiO$_2$ amorphous catalyst prepared by reduction with KBH$_4$ aqueous solutions exhibited a higher activity than commercial Raney-type catalysts. When nickel was supported on ZrO$_2$, TiO$_2$, and ZrO$_2$/TiO$_2$, the catalysts obtained were more active and stable than Ni/SiO$_2$, because of the absence of support leaching.

Other catalysts based on supported ruthenium have been developed as an alternative to the nickel catalysts for the production of sorbitol from glucose. Thus, a 5 wt % Ru/MCM-41 catalyst afforded 83% yield to sorbitol. Other Ru supported catalysts were tested in the same reaction, the carbon, ZrO$_2$ and TiO$_2$ being very stable supports in reaction conditions, in contrast to Al$_2$O$_3$ or SiO$_2$. The Ru particles resulted to be very stable to leaching in the reaction media.

Other monosaccharides that can be hydrogenated to alcohols are fructose and the xylose. Specifically, the hydrogenation of the α- and β-furanose forms of D-fructose affords sorbitol and mannitol, respectively. The aim of this process is to maximize the selectivity to the second one, which is a high value-added and low caloric sweetener. The selectivity to mannitol, close to 50%, obtained on platinum-group metal catalysts and Raney-type nickel, was improved to 67% in the case of a 20 wt % Cu/SiO$_2$ catalyst. Other catalysts leading to selectivity values near 65% were Cu/ZnO/Al$_2$O$_3$ and Cu/SiO$_2$.

The second most abundant carbohydrate after D-glucose, xylose, can be hydrogenated to produce xylitol, which is employed in food, cosmetic and pharmaceutical industries. Similar to the case of glucose, Raney-type nickel catalysts have been used industrially for the hydrogenation of xylose; however, they suffer from deactivation, produced by leaching and surface poisoning. As a consequence, other alternative catalysts have been tested in the reaction. Some examples are Ru/SiO$_2$ and Ru/ZrO$_2$, affording a 99.9% yield to xylitol and a 1 wt % Ru/(5 wt % NiO–TiO$_2$) catalyst, with a yield of 99.7%. Pt NPs (1.3 nm) on MWCNTs were also used for the hydrogenation of xylose to xylitol with 100% conversion of xylose and 99.3% selectivity to xylitol, compared to commercial Pt/C, Ru/C and Raney Ni catalysts (Nakagawa and Tomishige 2010).

Xylitol can be converted into ethylene glycol and propylene glycol using Ni/C and Ru/C catalysts in the presence of solid bases such as CeO$_2$ and Ca(OH)$_2$ (Sun and Liu 2014), the first leading to a higher activity than the second one.

The depolymerisation by acid hydrolysis of polysaccharides present in cellulose and hemicellulose leads to hexoses and pentoses, respectively, which can further be hydrogenated to polyols on metal catalysts. The use of bifunctional metal catalysts can combine the two steps in a one-pot process. In this sense, starch was converted to sorbitol by combined hydrolysis–hydrogenation on Ru/HY (Jacobs 1988), where the acid sites promote the hydrolysis of starch to glucose, and the metal sites catalyse the hydrogenation of the latter to sorbitol. In a similar way, the combination of a carbon support with metal nanoparticles of Ru led to an active catalyst that can first promote the hydrolysis of cellulose to glucose and further the hydrogenation of glucose in the presence of H_2 to sugar alcohols, such as sorbitol and mannitol. When the Ru was supported in acid functionalized CNT, the cellulose was converted to C2–C6 polyols (Lam and Luong 2014). Many other examples of this one-pot process can be found in Besson et al. (2014).

Different attempts have been made to replace the combination of diluted acidic solution and supported metal catalysts by acidic solids for the efficient production of polyols (Fig. 3) from cellulose or hemicellulose in one-step. Thus, cellulose was transformed into sorbitol in the presence of Pt/γ-Al_2O_3 and into hexitols (sorbitol, mannitol, sorbitan) with high yield (ca. 70%) over 1 wt % Ru/CNT catalyst. Similarly, the hemicellulose of sugar beet fibres was converted into arabitol at 155°C and 5 MPa of H_2 by using a 2 wt % Ru/C catalyst. Finally, the direct catalytic conversion of raw woody biomass was carried out at 235°C under 6 MPa of H_2 over a carbon supported Ni–W_2C catalyst (Li et al. 2012); the cellulose and hemicellulose fraction was converted to ethylene glycol and other diols whereas the lignin was simultaneously transformed into monophenols.

Many of the platform molecules, obtained from monosaccharides or carbohydrates, can be transformed into a variety of high value chemicals by hydrogenation. One of these platform molecules is the succinic acid, which can be hydrogenated to γ-butyrolactone (GBL), tetrahydrofuran (THF) and 1,4-butanediol (BDO) (Werpy 2006), three solvents commonly used as precursors for downstream chemicals. Thus, NPs of Pd, Pt, Ru and Rh supported on Starbon material were active catalysts for the hydrogenation of succinic acid to THF in aqueous ethanol, with selectivity of 60% in the most favourable case. NPs of Pd located either inside the pores of SBA-15 or outside the pores of MCM-41 also catalysed the hydrogenation of succinic acid, yielding to a higher selectivity to GBL and THF in the first case and to BDO in the second one.

Levulinic acid (LA), obtained by depolymerisation and dehydration of cellulosic biomass, can be catalytically hydrogenated into γ-valerolactone (GVL) and methyltetrahydrofuran (MTHF). GVL was produced mainly in the liquid-phase

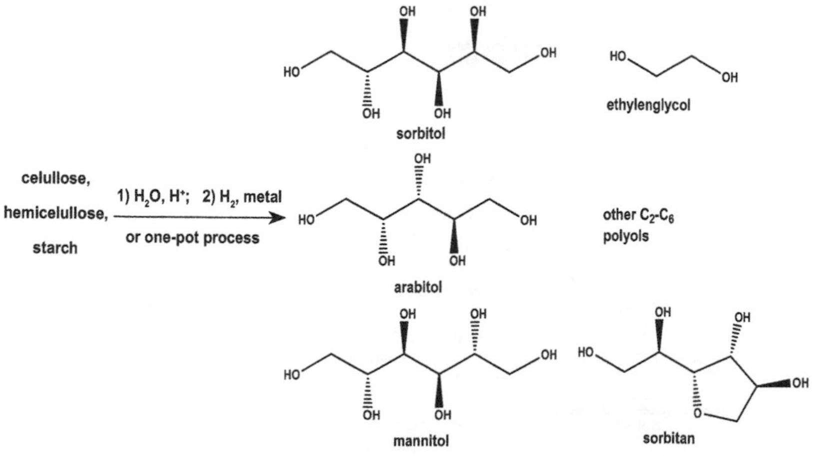

Fig. 3. Some of the polyols produced by hydrolysis-hydrogenation of polysaccharides.

or in supercritical fluids. Thus, GVL was obtained with a yield of 99% when LA hydrogenation was carried out in the presence of a Ru/SiO$_2$ catalyst in supercritical CO$_2$ (Bourne et al. 2007). Other catalysts used for these transformation are based on NPs of Ru or Pt over different supports, such as carbon, titania or silica (Besson et al. 2014).

The lactic acid can be hydrogenated to propylene glycol in the presence of catalysts based on Pd, Ru, Ni and Cu over different supports, Ru supported on activated carbon being the one showing the highest conversion of lactic acid (90%) and selectivity to propylene glycol (95%) (Zhang et al. 2001).

Furfural (FAL) is produced industrially by the hydrolysis and dehydration of agricultural wastes or by the cyclodehydration of xylose on acid catalysts. The hydrogenation of furfural leads to different products. One of these products is the furfuryl alcohol (FOL), which can be obtained by using different catalysts. In this regard, the liquid-phase hydrogenation of FAL dissolved in 2-propanol over a 0.6 wt % Pt/SiO$_2$ catalysts afforded 98.7% selectivity to FOL at 36% conversion (Merlo et al. 2009). A higher conversion (93%), although lower selectivity (89%), was achieved with a CuNi/MgAlO catalyst prepared from a hydrotalcite-like precursor (Xu et al. 2011). The vapour-phase hydrogenation of FAL at atmospheric pressure on a 16 wt % Cu/MgO catalyst yielded FOL with a 98% selectivity and 98% of conversion. Other products resulting from the hydrogenation of furfural are furan, 2 methylfuran and tetrahydrofuran. Several examples of nanocatalysts for these transformations can be found in Besson et al. (2014).

5-hydroxymethylfurfural (HMF) is produced by acid-catalysed dehydration of fructose and glucose or hydrolysis/dehydration of polysaccharides and can be used as platform molecule for organic synthesis (Rosatella et al. 2011; van Putten et al. 2013). Some products of hydrogenation of HMF that are of high interest because of their applications are 2,5-dihydroxymethyltetrahydrofuran (DHMTHF) and 2,5-dimethylfuran (DMF) (Fig. 4). The first one, used as a solvent and a precursor

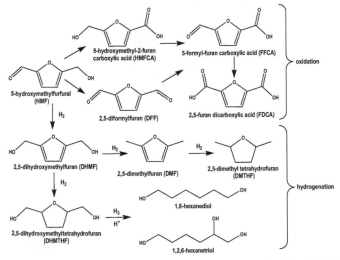

Fig. 4. Hydrogenation and oxidation reactions of 5-hydroxymethylfurfural (HMF).

for polymer synthesis can be obtained in a high yield (around 90%) over Ni-, Pd- and Ru-catalysts, and over a silica-supported Ni-Pd alloy (Ni/Pd = 7) prepared by co-impregnation. DMF was obtained with a 79% yield by vapour-phase hydrogenolysis of HMF with a Cu–Ru/C catalyst (Cu:Ru = 3:2) prepared by incipient wetness impregnation. The catalytic hydrogenation of HMF in supercritical methanol over a Cu-doped porous metal oxide obtained from a hydrotalcite-like material, afforded 48% and 10% yields to DMF and DMTHF, respectively (Hansen et al. 2012).

Hydrodeoxygenation

The hydrodeoxygenation (HDO) reaction plays a key role in the conversion of over-functionalized carbohydrates to unsaturated compounds and chemicals of interest. In order to reduce the oxygen content of bio-based oxygenates, the design of catalysts which effectively favour C-O over C-C bond cleavage and with minimal H_2 consumption is imperative.

One of these catalysts exhibiting high selectivity toward C–O/C=O cleavage with negligible C–C bond is the Mo_2C (Matthiesen et al. 2014), and when it is dispersed over the appropriate support, effective catalysts for HDO reactions can be obtained. In this sense, Mo_2C/activated carbon exhibited a high performance for the conversion of several vegetable oils (olive, soya bean, rapeseed and maize oil) to diesel-range hydrocarbons (Han et al. 2011), and Mo_2C on MWCNTs, produced by the carbothermal hydrogen reduction method, showed HDO capability on vegetable oil (Han et al. 2011). Other material carbons used as supports for Mo_2C are ordered mesoporous carbons (OCM), carbon nanotubes (CNTs) and carbon nanofibres (CNFs) (Matthiesen et al. 2014), this last support being more selective towards the production of un-branched hydrocarbons during the HDO of vegetable bonds, including propanol, 2-propanol and acetone, as compared to the rest.

HDO reactions can be performed over metal sulfides, such as sulfide Co(Ni) Mo/γ-Al_2O_3 supported on alumina, and supported noble metal catalysts, such as Ru, Rh and Pd (Lam and Luong 2014). However, these catalysts are deactivated as a consequence of the coke formation over the acid sites of the alumina support and additionally the products can be contaminated with the sulfiding agent used to regenerate the catalysts. As a consequence, other materials based on Si, Zr and activated carbons have been used as alternative supports (Lam and Luong 2014).

Tungsten carbide (WC) is a highly selective catalyst for the HDO of propanal and propanol to propene (Ren et al. 2014), the selectivity decreasing with the increase of the nanoparticle size. When supported over CNFs with high mesoporosity and surface area, Tungsten carbide (WC) produces the HDO of biomass-derived glyceride (Gosselink et al. 2013).

Lignin has aroused a great interest for the production of biofuels and value-added chemicals due to its abundance and the low oxygen content (Huber et al. 2006; Ma et al. 2014). The selective hydrogenolysis of the aromatic Csp^2–O bonds of the alcohols contained in the lignin is a promising way for lignin valorisation. In order to avoid the undesired hydrogenation of the aromatic rings, some tentative catalysts have been tested. In this regard, the bimetallic Zn/Pd/C system is a good example of synergistic effect for the selective HDO of the β–O–4 linkages of lignin

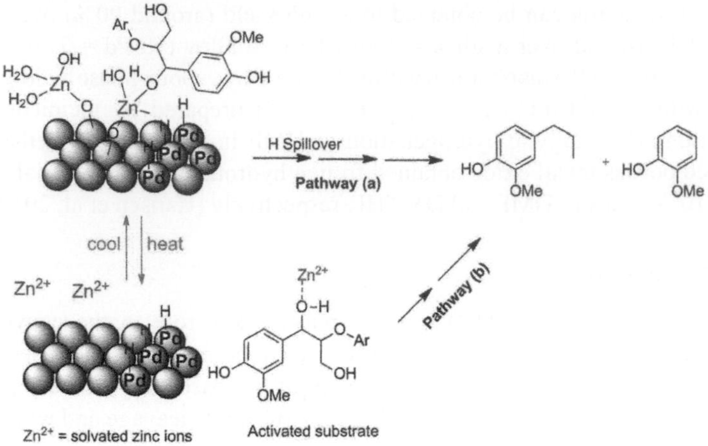

Fig. 5. Proposed pathways for HDO of β-O-4 lignin substrate over Zn/Pd/C catalyst. Pathway (a) involves substrate binding to Zn²⁺ sites and hydrogen spillover from Pd sites. Pathway (b) represents desorption of Zn ions into solution, activation of substrate via binding, and it is the Zn bound substrate that reacts with Pd sites on the catalyst's surface. Reproduced with permission from Parsell et al. 2013. Copyright©2013 The Royal Society of Chemistry.

compounds, which maintained the aromatical functional groups (Parsell et al. 2013) (Fig. 5). Other examples are core-shell bimetallic catalysts, NiM (M = Ru, Rh, Pd, and Au) (Zhang et al. 2014).

The interest of the biorefineries in the conversion of lignin to biofuels has increased in the last years. However, the bio-oils derived from the pyrolysis of woody biomass contain a high amount of oxygen, which confer to them undesired properties, such as high viscosity, low heating value, incomplete volatility and thermal instability. In this sense, the HDO reactions to upgrade bio-oils are imperative. Two major steps are mainly required to achieve this: The first one involves the depolymerisation of lignin into phenolic compounds; the subsequent step transforms the phenols to alkanes (Yan et al. 2008). This step, in turn, consists of four steps, for example, hydrogenation of phenols to ketones and alcohols, further hydrogenation of ketones to alcohols, dehydration of alcohols to alkenes and finally, hydrogenation of these to alkanes.

The conventional multi-step HDO reaction, catalysed by NiMo and CoMo catalysts, requires high temperature. The combination of metal NP_s, capable of hydrogenating aromatic rings, C=O bonds and C=C bonds, with acid functional groups (effective in the dehydration of alcohols to alkenes), leads to the direct conversion of lignin biomass into hydrocarbon fuels. In this sense, some authors have developed catalytic systems of metal NPs (Rh, Ru, Pd and Pt) in ionic liquids functionalized with Brønsted acidic sites, for the one-pot conversion of lignin-derived phenols into alkanes (Yan et al. 2010).

Oxidation

Selective oxidation is a key reaction in introducing desired functionalities in biomass-derived compounds (Davis et al. 2013). The combination of supported metal NPs as catalysts and molecular oxygen appears as a green alternative to classical chemical oxidants for the oxidation of oxygenated compounds, including sugars, glycerol and HMF. The most employed catalysts are noble metals such as Pd, Pt, Au and their alloys (PdAu, PtAu) highly dispersed on appropriate supports (carbon being one of the most used supports), due to its stability in acidic and basic media (Prati et al. 2012; Davis et al. 2013).

Glucose can be oxidised in an aqueous solution in the presence of supported metal catalysts to gluconic, glucaric and 2-keto-gluconic acids and their salts (Singh and Kumar 2007), although it is industrially produced by expensive fermentation processes. The production of gluconic acid and gluconate is of high interest, because they have applications in food and pharmaceutical industries.

The oxidation of glucose solutions under acidic conditions at 70°C has been tested on mono- or bimetallic catalysts (Au, Pt, Pd) as colloidal dispersions and in the form of NPs supported on carbon. Unsupported Au NPs were initially as active as Au/C catalysts at 30°C in a basic solution; however, the Au NPs sintered, causing a significant decrease of the glucose oxidation. The activity of bimetallic catalysts, especially those containing Au, was significantly higher than that of monometallic catalysts (Comotti et al. 2006).

One of the problems in the oxidation of glucose to gluconic acid over monometallic Pd and Pt catalysts is the strong adsorption of oxygen on the metal surface. Thus, gluconic acid was obtained with a selectivity of only 80% over a Pt/C catalyst. This drawback can be solved by carrying out a controlled O_2-diffusion, achieving a selectivity of 95% (Delidovich et al. 2010). A way to increase the selectivity of this type of catalysts is by promoting them with Bi atoms deposited on the surface of Pd particles, preventing the over oxidation of these catalysts (Karski et al. 2003). As a result, a Pd-Bi/C catalyst oxidised the glucose to gluconate (yield of 99%) with a 20-fold higher rate than a Pd/C catalyst. Furthermore, no leaching of Bi was observed and the catalyst could be recycled five times without loss of activity. Other promoters of Pd different from Bi are Tl, Sn and Co; however, they led to worse activity and selectivity.

The use of supported Au catalysts as an alternative to Pd systems in the conversion of glucose to gluconate has aroused a great interest in the last years due to their higher activity and selectivity (Davis et al. 2013). Numerous examples of catalysts of Au over different supports can be found in Besson et al. (2014). Thus, glucose was oxidised to gluconic acid (yield of 99%) over a 0.9 wt % Au/C catalyst, with higher reaction rate than on commercial Pd–Bi/C or Pd–Pt–Bi/C catalysts. However, the catalyst suffered from deactivation, due to the sintering and leaching of Au particles. The confinement of Au nanoparticles into the channels of mesoporous carbons leads to a high turnover frequency. Glucose oxidation with 30% H_2O_2 hydrogen peroxide at 25°C, over a 1 wt % Au/SiO_2 under ultrasonic activation afforded an 85% yield to gluconate.

The selective oxidation of lactose, arabinose and galactose over gold supported catalysts leads to the formation of aldonic acids, which find many applications in the food, cosmetic and pharmaceutical industries (Kusema and Murzin 2013).

Lactose, a disaccharide formed by glucose and galactose units, can be oxidised to lactobionic acid (LBA) and subsequently, to 2-keto-lactobionate. The first one, that is, lactobionic acid, is applied in the food industry and in the synthesis of biodegradable surfactants. Lactose oxidation performed over a 2 wt % Au/Al_2O_3 catalyst at basic pH afforded lactobionate in high yields with less catalyst deactivation than over palladium catalysts (Tokarev et al. 2007). LBA was obtained with a 95% yield on recyclable Pd-Bi/C catalysts (until 15 cycles without significant activity loss) with Bi/Pd ranging between 0.50 and 0.67 by oxidation of lactose at pH = 7–10 (Hendriks et al. 1990). When bimetallic Pd-Bi particles (Bi/Pd = 3) were supported on a mesoporous SBA-15 silica, a yield of LBA of 96% was achieved in the aerobic oxidation of lactose at pH 9 (Belkacemi and Hamoudi 2010). The oxidation of lactose over a Pt–Bi/C catalyst at pH 7 yielded lactobionate, which was subsequently transformed in to 2-keto-lactobionate with a final yield of 80%.

The bimetallic Pd-Au (1 wt %–4 wt %) catalysts supported over Al_2O_3 and CeO_2 were more active and selective for the oxidation of arabinose than the corresponding monometallic catalysts. Pd species were responsible for oxygen activation, whereas Au-species produced the activation of arabinose (Smolentseva et al. 2011).

Au/Al_2O_3 catalysts calcined at different temperatures were highly active for the oxidation of D-galactose to galactonic acid at basic pH and 60°C; however, a decrease of the pH led to a lower selectivity to galactonic acid and to the formation of galactolactone as co-product (Kusema et al. 2011).

The direct oxidation of cellulose has been tested over different Pt or Pd supported on alumina, resulting in malic, acetic and oxalic acids, which are formed in neutral or acidic conditions, in contrast to the acetic acid and glucose obtained as major product under basic media (Besson et al. 2014). A ball-milled crystalline cellulose was oxidised to gluconic acid, with a yield of 60% over an $Au/Cs_{1.2}H_{1.8}PW_{12}O_{40}$ catalyst.

Cellobiose, a dimer of glucose, was oxidised selectively to gluconic acid over Pt-, Pd- and Au-catalysts on various supports (Besson et al. 2014). Some examples of catalysts are Au/TiO_2, Pd/Al_2O_3 and Pt/Al_2O_3 (Mirescu and Pruesse 2007). Au/ CNT has resulted to be a more selective catalyst for the oxidation of cellobiose to gluconic acid (yield of 54%) in an aqueous solution than Au/TiO_2, A/Al_2O_3, Au/ MgO and Au/C, due to the generation of acidic groups on the CNT surface by pre-treatment with HNO_3 (Tan et al. 2009). The presence of acid groups in a bifunctional $Pt/AC-SO_3H$ also produced the oxidation of cellobiose to gluconic acid and glucose, with yields of 46 and 38%, respectively.

The glycerol can be oxidised to glyceric acid or directly to lactic acid in the one-pot reaction by using monometallic Au and Pt and their alloys supported on nanocrystalline CeO_2 in an aqueous solution in the presence of a base and oxygen (Lam and Luong 2014). The bimetallic systems that resulted were highly selective (80%) to lactic acid at 99% of glycerol conversion. The $Au-Pt/CeO_2$ catalyst could be used in five cycles of reaction in a batch system without a significant decrease in the activity and selectivity. No more attention to the oxidation of glycerol is given here because it is revised in another chapter of this book, together with other types of transformations.

5-hydroxymethylfurfural (HMF) can be transformed into high added-value molecules by different reactions (Davis et al. 2013). More concretely, the selective oxidation of the hydroxyl and aldehyde functional groups of HMF produces valuable chemicals, such as 5-hydroxymethyl-2-furancarboxylic acid (HFCA) and 2,5-furandicarboxylic acid (FDCA) (Fig. 4). In particular, FDCA is a potential substitute for terephtalic acid in the production of aromatic polyesters with properties similar to those of polyethyleneterephtalate.

The oxidation of HMF on supported metal NPs has been reviewed (Davis et al. 2013). Experiments of HMF oxidation were usually performed in aqueous solutions with variable amounts of liquid bases to neutralise the acidic functions, the hydroxyl bases being more effective than carbonate bases. Relatively diluted solutions and high catalyst loads were used to achieve a fast oxidation to FDCA. Different metallic salts were added to the reaction medium to act as promoters, but only $PbCl_2$ significantly improved the catalytic activity.

When the oxidation of HMF was performed in 2 M NaOH solutions over a commercial 1 wt % Au/TiO_2 catalyst, a yield of 71% to FDCA was obtained at 25°C (Gorbanev et al. 2009). Different Ru(OH)x catalysts supported on basic solids, as magnesium oxide, magnesium-lanthanum oxide and hydrotalcite were very selective to FDCA without the addition of a liquid base. Another example of catalyst promoting the aerobic oxidation of HMF to FDCA in water without any bases is an Au-Pd alloy catalyst supported on CNTs (Wan et al. 2014). Those CNT containing more carbonyl/quinone and less carboxyl groups enhanced the adsorption of the reactant and reaction intermediates and favoured the FDCA formation.

In order to study the influence of the oxygen nature and content, different carbon materials, such as activated carbon, CNF, CNT, and graphite, were used as supports for Pd, Pt, Au and their alloys to obtain catalysts for the selective oxidation of HMF to HFCA and FDCA (Villa et al. 2013). The least functionalized supports (CNT, graphite) were more selective towards HFCA, while Au/activated carbon was more selective to FDCA.

The mechanism of selective oxidation of HMF to HMFCA and FDCA at high pH has been studied over supported Pt and Au catalysts (Davis et al. 2012). Authors found that the hydroxide anions of water were the source of oxygen atoms. Molecular oxygen played an indirect role during the oxidation process by removing the electrons deposited into the supported metal NPs. An Au/TiO_2 catalyst produced HFMA as a major product, in contrast to the HMF yields of 70–80% obtained over Pd/C and Pt/C catalysts under the same conditions.

One-pot conversion of fructose to FDCA has been tested over different catalysts. In this sense, the combination of a Pt-Bi/C catalyst with a membrane technology afforded a 25% yield to FDCA at 80°C under atmospheric pressure in the fructose transformation, by carrying out the HMF oxidation in methyl-isobutylketone (MIBK). A one-pot conversion of fructose to FDCA was also achieved with a 72% conversion and 99% selectivity under air pressure using cobalt acetylacetonate encapsulated in sol–gel silica as the catalyst (Ribeiro and Schuchardt 2003).

Another product of interest resulting from the oxidation of HMF is 2,5-Diformylfuran (DFF). Different catalysts based on V_2O_5 are active and selective for this transformation (Besson et al. 2014). Thus, a 85% yield to DFF was achieved in the aerobic oxidation of HMF in toluene and methyl isobutyl ketone in the presence of V_2O_5/TiO_2 catalysts and a 84% yield to DFF was reported for the oxidation of HMF in DMSO at 125°C under 1 MPa of O_2 over a 1 wt % V_2O_5/H-beta. The selective oxidation of HMF to DFF in toluene was also performed over Ru/γ-Al_2O_3 catalyst at 130°C, affording 97% selectivity to DFF.

Water solutions of carbohydrates can be oxidised by oxygen or air at atmospheric pressure in the presence of Pd or Pt catalysts at 30–80°C. An oxidative dehydrogenation reaction occurs and primary and secondary alcohol functions are oxidised to carbonyl and carboxylic functions (Besson et al. 2014).

Summary and outlook

In this chapter, we have revised the synthesis methods employed for obtaining nanocatalysts from biomass and the application of the nanocatalysts in the different reactions involved in the transformation of biomass into chemicals.

Among the nanocatalysts obtained from biomass, biomass-derived carbons are very promising, because they can be synthesised with conveniently adjusted structures and surface properties by selecting different synthesis strategies and the appropriate precursor for the sake of specific applications. Apart from the suitable selection of a biomass precursor, the synthesis method is a very important factor to obtain biomass-derived carbons with excellent physicochemical properties. The traditional activation methods can be applied to synthesize microporous carbons, but the necessity of obtaining carbons with well-defined shapes and hierarchical pore structure has driven the development of other synthesis methods. In this sense, the direct hydrothermal carbonization (HTC) of biomass-derived precursors delivers porous structural carbons with tuneable properties and desirable high concentrations of oxygen functionalities, which serves as anchoring places for the active catalyst on their surface. However, the knowledge of the detailed chemical mechanism of low HTC, which could allow a rational design and control of the HTC process for a variety of carbon materials, is still a challenge. Suitable porous carbons with high surface area and hierarchical porosity can be obtained by synthesis assisted by templates, and carbons synthesized from carbohydrates and functionalized with sulphonic groups are appropriate catalysts for a broad range of acid-catalysed reactions. Carbons prepared form biomass can also serve as supports for active metal catalysts for several organic reactions. However, the heterogeneous structure and surface chemistry of these carbons make them unsuitable for complex reactions that require well-defined active sites on the atomic scale, the synthesis of alternative nanocarbons with tailored surface properties, such as graphene, CNFs and CNTs being necessary.

Another type of nanocatalyst obtained from biomass revised here is the group of metal and metal oxide nanoparticles which can be synthesized by green routes involving products derived from plant materials, such as extracts of various plants, tea, coffee, as well as sugars and glucose, which act as reducing and/or capping agents in the formation of nanoparticles. Frequently, the synthesis of these nanoparticles occurs in a single-pot reaction and the stability and longevity of formed nanoparticles is increased, because they are protected from aggregation and sintering reactions.

The environmental benefits of producing catalysts from renewable biomass could be countered if large amounts of solvents or toxic chemicals, high temperature and pressure conditions or expensive commercial additives are necessary for their synthesis. Therefore, significant efforts are still required for the design of catalysts eliminating these conditions and reducing the length, number and cost of processing steps.

In the last two decades, a great effort has been dedicated to the conversion of lignocellulosic biomass into platform molecules, and their further upgrading to lignocellulosic biomass into value added chemicals over a wide range of heterogeneous catalysts. Many studies report the chemocatalytic conversion of cellulose or the component sugars of biomass. However, little effort has been dedicated to the use of multifunctional catalysts for processing the biomass as a whole, not its individual components.

Most of the reactions of biomass processing occur in the condensed phase, generally in water above the boiling point. Therefore, it is necessary to design and synthesize catalysts which are chemically and structurally stable under these conditions. In this sense, carbon-supported and carbon-based catalysts, as well as novel organic-inorganic hybrids are the leading candidates in biomass transformation applications.

Components in biomass feedstock have plenty of functional groups, which reduce their stability at the high temperatures generally used to process petrochemicals. Consequently, the de-functionalization, particularly deoxygenation under mild reaction conditions, is imperative. For this purpose, well-defined nano-scale metal catalysts are very suitable for promoting this kind of reactions, including hydrogenolysis, hydrolysis and decarboxylation. Among the metal nanocatalysts, two metals, ruthenium and gold, are of vital importance in biomass transformation, the first for hydrogenation, hydrolysis and hydrogenolysis/dehydroxylation reactions, and the second one for oxidation reactions. However, the gold catalysts suffer from a dramatic deactivation caused by the sintering of nanoparticles.

In acidic and/or chelating media, necessary to carry out some of the transformations of biomass, carbons or stable oxides, such as CeO_2, TiO_2 or ZrO_2, have replaced alumina and silico-alumina used as traditional supports in hydrocarbon processing and organic synthesis in nonpolar solvents.

Some of the biomass conversion processes based on supported metal catalysts meet many of the green chemistry principles, because metal nanoparticles and supports do not leach in reaction media, and many of the catalysts are recyclable. One example of this is the aerobic oxidation of glucose into gluconic acid in an aqueous solution. However, other biomass transformations involve toxic metals and reaction media, such as the use of chromium catalysts in the production of HMF.

Many of the processes of biomass transformation require too many conversion and separation steps. In addition, the platform molecules, particularly those obtained by depolymerisation and/or fermentation of cellulosic materials, have to be treated with costly purification technologies before being transformed into valuable chemicals. These circumstances exert a negative impact on the overall process economy. Although some of these processes have already been commercialized or are currently being reduced to industrial practice, the production of fuels and chemicals from biomass is still not cost competitive, and further studies on more sustainable technologies based on renewable biomass, and the design of multifunctional catalysts capable of promoting the one-pot biomass transformation in a cascade manner, will be necessary in the future.

Acknowledgements

This work has been supported by the Spanish Ministry of Science and Innovation (CTM2014-56668-R). Dr. Francisco José Plou Gasca (from Institute of Catalysis and Petrochemistry, CSIC) is acknowledged for the revision of this chapter.

Reference cited

Abioye, A.M. and F.N. Ani. 2015. Recent development in the production of activated carbon electrodes from agricultural waste biomass for supercapacitors: A review. Renewable Sustainable Energy Rev. 52: 1282–1293.

Aguayo-Villarreal, I.A., A. Bonilla-Petriciolet and R. Muniz-Valencia. 2017. Preparation of activated carbons from pecan nutshell and their application in the antagonistic adsorption of heavy metal ions. J. Mol. Liq. 230: 686–695.

Balakrishnan, M., V.S. Batra, J.S.J. Hargreaves and I.D. Pulford. 2011. Waste materials—catalytic opportunities: an overview of the application of large scale waste materials as resources for catalytic applications. Green Chem. 13: 16–24.

Belkacemi, K. and S. Hamoudi. 2010. Chemocatalytic oxidation of lactose to lactobionic acid over Pd-Bi/ SBA-15: Reaction kinetics and modeling. Ind. Eng. Chem. Res. 49: 6878–6889.

Bennett, J.A., K. Wilson and A.F. Lee. 2016. Catalytic applications of waste derived materials. J. Mater. Chem. A 4: 3617–3637.

Besson, M., P. Gallezot and C. Pinel. 2014. Conversion of biomass into chemicals over metal catalysts. Chem. Rev. 114: 1827–1870.

Bhanja, P. and A. Bhaumik. 2016. Porous nanomaterials as green catalyst for the conversion of biomass to bioenergy. Fuel 185: 432–441.

Bourne, R.A., J.G. Stevens, J. Ke and M. Poliakoff. 2007. Maximizing opportunities in supercritical chemistry: the continuous conversion of levulinic acid to γ-valerolactone in CO_2. Chem. Commun. 44: 4632–4634.

Budarin, V., J.H. Clark, J.J.E. Hardy, R. Luque, K. Milkowski, S.J. Tavener and A.J. Wilson. 2006. Starbons: new starch-derived mesoporous carbonaceous materials with tunable properties. Angew. Chem., Int. Ed. 45: 3782–3786.

Budarin, V., R. Luque, D.J. Macquarrie and J.H. Clark. 2007. Towards a bio-based industry: benign catalytic esterifications of succinic acid in the presence of water. Chem. Eur. J. 13: 6914–6919.

Budarin, V.L., J.H. Clark, R. Luque and D.J. Macquarrie. 2007. Versatile mesoporous carbonaceous materials for acid catalysis. Chem. Commun. 6: 634–636.

Budarin, V.L., J.H. Clark, R. Luque, D.J. Macquarrie and R.J. White. 2008. Palladium nanoparticles on polysaccharide-derived mesoporous materials and their catalytic performance in C-C coupling reactions. Green Chem. 10: 382–387.

Cagnon, B., X. Py, A. Guillot, F. Stoeckli and G. Chambat. 2009. Contributions of hemicellulose, cellulose and lignin to the mass and the porous properties of chars and steam activated carbons from various lignocellulosic precursors. Bioresour. Technol. 100: 292–298.

Campelo, J.M., D. Luna, R. Luque, J.M. Marinas and A.A. Romero. 2009. Sustainable preparation of supported metal nanoparticles and their applications in catalysis. ChemSusChem. 2: 18–45.

Casas Hidalgo, A.I., M. Roman Aguirre, E. Valenzuela, J.Y. Verde Gomez, A. Camacho Davila, R.S. Varma and V.H. Ramos Sanchez. 2016. Sustainable application of pecan nutshell waste: greener synthesis of Pd-based nanocatalysts for electro-oxidation of methanol. Int. J. Hydrogen Energy 41: 23329–23335.

Chen, H., D. Zhou, G. Luo, S. Zhang and J. Chen. 2015. Macroalgae for biofuels production: Progress and perspectives. Renewable Sustainable Energy Rev. 47: 427–437.

Clark, J.H., V. Budarin, T. Dugmore, R. Luque, D.J. Macquarrie and V. Strelko. 2008. Catalytic performance of carbonaceous materials in the esterification of succinic acid. Catal. Commun. 9: 1709–1714.

Comotti, M., C. Della Pina and M. Rossi. 2006. Mono- and bimetallic catalysts for glucose oxidation. J. Mol. Catal. A: Chem. 251: 89–92.

Cui, X., M. Antonietti and S.-H. Yu. 2006. Structural effects of iron oxide nanoparticles and iron ions on the hydrothermal carbonization of starch and rice carbohydrates. Small 2(6): 756–759.

Davis, S.E., B.N. Zope and R.J. Davis. 2012. On the mechanism of selective oxidation of 5-hydroxymethylfurfural to 2,5-furandicarboxylic acid over supported Pt and Au catalysts. Green Chem. 14: 143–147.

Davis, S.E., M.S. Ide and R.J. Davis. 2013. Selective oxidation of alcohols and aldehydes over supported metal nanoparticles. Green Chem. 15: 17–45.

De, S., A.M. Balu, J.C. van der Waal and R. Luque. 2015. Biomass-derived porous carbon materials: synthesis and catalytic applications. ChemCatChem. 7: 1608–1629.

Dechamp, N., A. Gamez, A. Perrard and P. Gallezot. 1995. Kinetics of glucose hydrogenation in a trickle-bed reactor. Catal. Today 24: 29–34.

Delidovich, I.V., O.P. Taran, L.G. Matvienko, A.N. Simonov, I.L. Simakova, A.N. Bobrovskaya and V.N. Parmon. 2010. Selective oxidation of glucose over carbon-supported pd and pt catalysts. Catal. Lett. 140: 14–21.

Demir-Cakan, R., P. Makowski, M. Antonietti, F. Goettmann and M.-M. Titirici. 2010. Hydrothermal synthesis of imidazole functionalized carbon spheres and their application in catalysis. Catal. Today 150: 115–118.

Demirbas, A. 2009. Agricultural based activated carbons for the removal of dyes from aqueous solutions: a review. J. Hazard Mater. 167: 1–9.

Deng, J., M. Li and Y. Wang. 2016. Biomass-derived carbon: synthesis and applications in energy storage and conversion. Green Chem. 18: 4824–4854.

Deng, J., T. Xiong, H. Wang, A. Zheng and Y. Wang. 2016. Effects of cellulose, hemicellulose, and lignin on the structure and morphology of porous carbons. ACS Sustainable Chem. Eng. 4: 3750–3756.

Dodson, J.R., V.L. Budarin, A.J. Hunt, P.S. Shuttleworth and J.H. Clark. 2013. Shaped mesoporous materials from fresh macroalgae. J. Mater. Chem. A 1: 5203–5207.

Filikov, A.V. and N.F. Myasoedov. 1986. Hydrogen spillover and the rate of heterogeneous catalytic hydrogenation. Quantitative model. J. Phys. Chem. 90: 4915–4916.

Gorbanev, Y.Y., S.K. Klitgaard, J.M. Woodley, C.H. Christensen and A. Riisager. 2009. Gold-catalyzed aerobic oxidation of 5-hydroxymethylfurfural in water at ambient temperature. ChemSusChem. 2: 672–675.

Gosselink, R.W., D.R. Stellwagen and J.H. Bitter. 2013. Tungsten-based catalysts for selective deoxygenation. Angew. Chem., Int. Ed. 52: 5089–5092.

Guiotoku, M., C.R. Rambo, F.A. Hansel, W.L.E. Magalhaes and D. Hotza. 2009. Microwave-assisted hydrothermal carbonization of lignocellulosic materials. Mater. Lett. 63: 2707–2709.

Hadi, P., S.K. Sharma and G. McKay. 2015. Removal of dyes from effluents using biowaste-derived adsorbents. Scrivener Publishing LLC.

Han, J., J. Duan, P. Chen, H. Lou and X. Zheng. 2011. Molybdenum carbide-catalyzed conversion of renewable oils into diesel-like hydrocarbons. Adv. Synth. Catal. 353: 2577–2583.

Han, J., J. Duan, P. Chen, H. Lou, X. Zheng and H. Hong. 2011. Nanostructured molybdenum carbides supported on carbon nanotubes as efficient catalysts for one-step hydrodeoxygenation and isomerization of vegetable oils. Green Chem. 13: 2561–2568.

Hansen, T.S., K. Barta, P.T. Anastas, P.C. Ford and A. Riisager. 2012. One-pot reduction of 5-hydroxymethylfurfural via hydrogen transfer from supercritical methanol. Green Chem. 14: 2457–2461.

Hendriks, H.E.J., B.F.M. Kuster and G.B. Marin. 1990. The effect of bismuth on the selective oxidation of lactose on supported palladium catalysts. Carbohydr. Res. 204: 121–129.

Hill, J.M. 2017. Sustainable and/or waste sources for catalysts: Porous carbon development and gasification. Catal. Today 285: 204–210.

Hirose, T., T. Fujino, T. Fan, H. Endo, T. Okabe and M. Yoshimura. 2002. Effect of carbonization temperature on the structural changes of woodceramics impregnated with liquefied wood. Carbon 40: 761–765.

Hu, B., K. Wang, L. Wu, S.-H. Yu, M. Antonietti and M.-M. Titirici. 2010. Engineering carbon materials from the hydrothermal carbonization process of biomass. Adv. Mater. 22: 813–828.

Hu, L., G. Zhao, W. Hao, X. Tang, Y. Sun, L. Lin and S. Liu. 2012. Catalytic conversion of biomass-derived carbohydrates into fuels and chemicals via furanic aldehydes. RSC Adv. 2: 11184–11206.

Hu, Y. and J. Cao. 2016. Review on adsorbents from plant waste for chromium removal from sewage and wastewater. Anquan Yu Huanjing Gongcheng 23: 51–58.

Huang, C.-H. and R.-A. Doong. 2011. Sugarcane bagasse as the scaffold for mass production of hierarchically porous carbon monoliths by surface self-assembly. Microporous Mesoporous Mater. 147: 47–52.

Huang, J., Q. Li, D. Sun, Y. Lu, Y. Su, X. Yang, H. Wang, Y. Wang, W. Shao, N. He, J. Hong and C. Chen. 2007. Biosynthesis of silver and gold nanoparticles by novel sundried *Cinnamomum camphora* leaf. Nanotechnology 18: 105104/1–105104/11.

Huang, J., C. Liu, D. Sun, Y. Hong, M. Du, T. Odoom-Wubah, W. Fang and Q. Li. 2014. Biosynthesized gold nanoparticles supported over TS-1 toward efficient catalyst for epoxidation of styrene. Chem. Eng. J. 235: 215–223.

Huber, G.W., S. Iborra and A. Corma. 2006. Synthesis of transportation fuels from biomass: chemistry, catalysts, and engineering. Chem. Rev. 106: 4044–4098.

Ioannidou, O. and A. Zabaniotou. 2007. Agricultural residues as precursors for activated carbon production—A review. Renewable Sustainable Energy Rev. 11: 1966–2005.

Jacobs, P.H., H. 1988. Patent EP 329923.

Jain, A., R. Balasubramanian and M.P. Srinivasan. 2016. Hydrothermal conversion of biomass waste to activated carbon with high porosity: A review. Chem. Eng. J. 283: 789–805.

Jia, L., Q. Zhang, Q. Li and H. Song. 2009. The biosynthesis of palladium nanoparticles by antioxidants in Gardenia jasminoides Ellis: long lifetime nanocatalysts for p-nitrotoluene hydrogenation. Nanotechnology 20: 385601/1–385601/10.

Jung, A., S. Han, T. Kim, W.J. Cho and K.-H. Lee. 2013. Synthesis of high carbon content microspheres using 2-step microwave carbonization, and the influence of nitrogen doping on catalytic activity. Carbon 60: 307–316.

Karski, S., T. Paryjczak and I. Witonska. 2003. Selective oxidation of glucose to gluconic acid over bimetallic Pd-Me catalysts (Me = Bi, Tl, Sn, Co). Kinet. Catal. 44: 618–622.

Kharissova, O.V., H.V.R. Dias, B.I. Kharisov, B.O. Pérez and V.M.J. Pérez. 2013. The greener synthesis of nanoparticles. Trends Biotechnol. 31: 240–248.

Knox, J.H., B. Kaur and G.R. Millward. 1986. Structure and performance of porous graphitic carbon in liquid chromatography. J. Chromatogr. 352: 3–25.

Kou, J., C. Bennett-Stamper and R.S. Varma. 2013. Green synthesis of noble nanometals (au, pt, pd) using glycerol under microwave irradiation conditions. ACS Sustainable Chem. Eng. 1: 810–816.

Kubo, S., R.J. White, N. Yoshizawa, M. Antonietti and M.-M. Titirici. 2011. Ordered carbohydrate-derived porous carbons. Chem. Mater. 23: 4882–4885.

Kusema, B.T., B.C. Campo, O.A. Simakova, A.-R. Leino, K. Kordas, P. Maeki-Arvela, T. Salmi and D.Y. Murzin. 2011. Selective oxidation of D-galactose over gold catalysts. ChemCatChem. 3: 1789–1798.

Kusema, B.T. and D.Y. Murzin. 2013. Catalytic oxidation of rare sugars over gold catalysts. Catal. Sci. Technol. 3: 297–307.

Kyzas, G.Z., J. Fu, N.K. Lazaridis, D.N. Bikiaris and K.A. Matis. 2015. New approaches on the removal of pharmaceuticals from wastewaters with adsorbent materials. J. Mol. Liq. 209: 87–93.

Lam, E. and J.H.T. Luong. 2014. Carbon materials as catalyst supports and catalysts in the transformation of biomass to fuels and chemicals. ACS Catal. 4: 3393–3410.

Lee, J.S., X. Wang, H. Luo, G.A. Baker and S. Dai. 2009. Facile ionothermal synthesis of microporous and mesoporous carbons from task specific ionic liquids. J. Am. Chem. Soc. 131: 4596–4597.

Li, C., M. Zheng, A. Wang and T. Zhang. 2012. One-pot catalytic hydrocracking of raw woody biomass into chemicals over supported carbide catalysts: simultaneous conversion of cellulose, hemicellulose and lignin. Energy Environ. Sci. 5: 6383–6390.

Li, H., H. Li and J.-F. Deng. 2002. Glucose hydrogenation over Ni-B/SiO_2 amorphous alloy catalyst and the promoting effect of metal dopants. Catal. Today 74: 53–63.

Lin, X., M. Wu, D. Wu, S. Kuga, T. Endo and Y. Huang. 2011. Platinum nanoparticles using wood nanomaterials: eco-friendly synthesis, shape control and catalytic activity for p-nitrophenol reduction. Green Chem. 13: 283–287.

Liu, M., S. Jia, Y. Gong, C. Song and X. Guo. 2013. Effective hydrolysis of cellulose into glucose over sulfonated sugar-derived carbon in an ionic liquid. Ind. Eng. Chem. Res. 52: 8167–8173.

Liu, R.-L., W.-J. Ji, T. He, Z.-Q. Zhang, J. Zhang and F.-Q. Dang. 2014. Fabrication of nitrogen-doped hierarchically porous carbons through a hybrid dual-template route for CO_2 capture and haemoperfusion. Carbon 76: 84–95.

Lu, A.-H., W. Schmidt, B. Spliethoff and F. Schueth. 2003. Synthesis of ordered mesoporous carbon with bimodal pore system and high pore volume. Adv. Mater. 15: 1602–1606.

Lua, A.C. and T. Yang. 2005. Characteristics of activated carbon prepared from pistachio-nut shell by zinc chloride activation under nitrogen and vacuum conditions. J. Colloid Interface Sci. 290: 505–513.

Luque, R. and J.H. Clark. 2010. Water-tolerant Ru-Starbon materials for the hydrogenation of organic acids in aqueous ethanol. Catal. Commun. 11: 928–931.

Ma, R., M. Guo and X. Zhang. 2014. Selective conversion of biorefinery lignin into dicarboxylic acids. ChemSusChem. 7: 412–415.

Makowski, P., R. Demir Cakan, M. Antonietti, F. Goettmann and M.-M. Titirici. 2008. Selective partial hydrogenation of hydroxy aromatic derivatives with palladium nanoparticles supported on hydrophilic carbon. Chem. Commun. 8: 999–1001.

Matos, I., M. Bernardo and I. Fonseca. 2017. Porous carbon: A versatile material for catalysis. Catal. Today 285: 194–203.

Matthiesen, J., T. Hoff, C. Liu, C. Pueschel, R. Rao and J.-p. Tessonnier. 2014. Functional carbons and carbon nanohybrids for the catalytic conversion of biomass to renewable chemicals in the condensed phase. Cuihua Xuebao 35: 842–855.

Merlo, A.B., V. Vetere, J.F. Ruggera and M.L. Casella. 2009. Bimetallic PtSn catalyst for the selective hydrogenation of furfural to furfuryl alcohol in liquid-phase. Catal. Commun. 10: 1665–1669.

Mirescu, A. and U. Pruesse. 2007. A new environmental friendly method for the preparation of sugar acids via catalytic oxidation on gold catalysts. Appl. Catal., B 70: 644–652.

Mohammad-Khah, A. and R. Ansari. 2009. Activated charcoal: preparation, characterization and applications: a review article. Int. J. ChemTech Res. 1: 859–864.

Nadagouda, M.N., G. Hoag, J. Collins and R.S. Varma. 2009. Green synthesis of Au nanostructures at room temperature using biodegradable plant surfactants. Cryst. Growth Des. 9: 4979–4983.

Nakagawa, Y. and K. Tomishige. 2010. Total hydrogenation of furan derivatives over silica-supported Ni-Pd alloy catalyst. Catal. Commun. 12: 154–156.

Nakajima, K., R. Noma, M. Kitano and M. Hara. 2014. Selective glucose transformation by titania as a heterogeneous Lewis acid catalyst. J. Mol. Catal. A: Chem. 388-389: 100–105.

Nikolla, E., Y. Roman-Leshkov, M. Moliner and M.E. Davis. 2011. One-Pot synthesis of 5-(Hydroxymethyl) furfural from carbohydrates using Tin-Beta Zeolite. ACS Catal. 1: 408–410.

Onda, A., T. Ochi and K. Yanagisawa. 2008. Selective hydrolysis of cellulose into glucose over solid acid catalysts. Green Chem. 10: 1033–1037.

Pan, X., L. Wang, F. Ling, Y. Li, D. Han, Q. Pang and L. Jia. 2015. A novel biomass assisted synthesis of Au-SrTiO$_3$ as a catalyst for direct hydrogen generation from formaldehyde aqueous solution at low temperature. Int. J. Hydrogen Energy 40: 1752–1759.

Paraknowitsch, J.P., J. Zhang, D. Su, A. Thomas and M. Antonietti. 2010. Ionic liquids as precursors for nitrogen-doped graphitic carbon. Adv. Mater. 22: 87–92.

Parsell, T.H., B.C. Owen, I. Klein, T.M. Jarrell, C.L. Marcum, L.J. Haupert, L.M. Amundson, H.I. Kenttaemaa, F. Ribeiro, J.T. Miller and M.M. Abu-Omar. 2013. Cleavage and hydrodeoxygenation (HDO) of C-O bonds relevant to lignin conversion using Pd/Zn synergistic catalysis. Chem. Sci. 4: 806–813.

Prati, L., A. Villa, A.R. Lupini and G.M. Veith. 2012. Gold on carbon: one billion catalysts under a single label. Phys. Chem. Chem. Phys. 14: 2969–2978.

Qi, X., H. Guo, L. Li and R.L. Smith. 2012. Acid-catalyzed dehydration of fructose into 5-hydroxymethylfurfural by cellulose-derived amorphous carbon. ChemSusChem. 5: 2215–2220.

Qian, H.-S., M. Antonietti and S.-H. Yu. 2007. Hybrid "golden fleece": synthesis and catalytic performance of uniform carbon nanofibers and silica nanotubes embedded with a high population of noble-metal nanoparticles. Adv. Funct. Mater. 17: 637–643.

Ren, H., Y. Chen, Y. Huang, W. Deng, D.G. Vlachos and J.G. Chen. 2014. Tungsten carbides as selective deoxygenation catalysts: experimental and computational studies of converting C$_3$ oxygenates to propene. Green Chem. 16: 761–769.

Ribeiro, M.L. and U. Schuchardt. 2003. Cooperative effect of cobalt acetylacetonate and silica in the catalytic cyclization and oxidation of fructose to 2,5-furandicarboxylic acid. Catal. Commun. 4: 83–86.

Richardson, Y., J. Blin, G. Volle, J. Motuzas and A. Julbe. 2010. *In situ* generation of Ni metal nanoparticles as catalyst for H₂-rich syngas production from biomass gasification. Appl. Catal., A 382: 220–230.

Rivera-Utrilla, J., M. Sanchez-Polo, M.A. Ferro-Garcia, G. Prados-Joya and R. Ocampo-Perez. 2013. Pharmaceuticals as emerging contaminants and their removal from water. A review. Chemosphere 93: 1268–1287.

Rosatella, A.A., S.P. Simeonov, R.F.M. Frade and C.A.M. Afonso. 2011. 5-Hydroxymethylfurfural (HMF) as a building block platform: biological properties, synthesis and synthetic applications. Green Chem. 13: 754–793.

Russo, P.A., S. Lima, V. Rebuttini, M. Pillinger, M.-G. Willinger, N. Pinna and A.A. Valente. 2013. Microwave-assisted coating of carbon nanostructures with titanium dioxide for the catalytic dehydration of d-xylose into furfural. RSC Adv. 3: 2595–2603.

Ryoo, R., S.H. Joo and S. Jun. 1999. Synthesis of highly ordered carbon molecular sieves via template-mediated structural transformation. J. Phys. Chem. B 103: 7743–7746.

Ryoo, R., S.H. Joo, M. Kruk and M. Jaroniec. 2001. Ordered mesoporous carbons. Adv. Mater. 13: 677–681.

Saha, D., Y. Li, Z. Bi, J. Chen, J.K. Keum, D.K. Hensley, H.A. Grappe, H.M. Meyer, S. Dai, M.P. Paranthaman and A.K. Naskar. 2014. Studies on supercapacitor electrode material from activated lignin-derived mesoporous carbon. Langmuir 30: 900–910.

Salvador, F., M.J. Sanchez-Montero and C. Izquierdo. 2007. C/H₂O reaction under supercritical conditions and their repercussions in the preparation of activated carbon. J. Phys. Chem. C 111: 14011–14020.

Sen, A., H. Pereira, M.A. Olivella and I. Villaescusa. 2015. Heavy metals removal in aqueous environments using bark as a biosorbent. Int. J. Environ. Sci. Technol. 12: 391–404.

Sevilla, M., G. Lota and A.B. Fuertes. 2007. Saccharide-based graphitic carbon nanocoils as supports for PtRu nanoparticles for methanol electrooxidation. J. Power Sources 171: 546–551.

Sevilla, M., C. Sanchis, T. Valdes-Solis, E. Morallon and A.B. Fuertes. 2007. Synthesis of graphitic carbon nanostructures from sawdust and their application as electrocatalyst supports. J. Phys. Chem. C 111: 9749–9756.

Sheldon, R.A., I. Arends and U. Hanefeld. 2007. Green Chemistry and Catalysis. Weinheim, Wiley-VCH.

Sheldon, R.A. 2014. Green and sustainable manufacture of chemicals from biomass: state of the art. Green Chem. 16: 950–963.

Shin, H.-D. 2012. Utilization of agricultural residues as low cost adsorbents for the removal dyes from aqueous solution. Chawon Rissaikuring 21: 9–16.

Singh, O.V. and R. Kumar. 2007. Biotechnological production of gluconic acid: Future implications. Appl. Microbiol. Biotechnol. 75: 713–722.

Smolentseva, E., B.T. Kusema, S. Beloshapkin, M. Estrada, E. Vargas, D.Y. Murzin, F. Castillon, S. Fuentes and A. Simakov. 2011. Selective oxidation of arabinose to arabinonic acid over Pd-Au catalysts supported on alumina and ceria. Appl. Catal., A 392: 69–79.

Sun, J. and H. Liu. 2014. Selective hydrogenolysis of biomass-derived xylitol to ethylene glycol and propylene glycol on Ni/C and basic oxide-promoted Ni/C catalysts. Catal. Today 234: 75–82.

Tan, X., W. Deng, M. Liu, Q. Zhang and Y. Wang. 2009. Carbon nanotube-supported gold nanoparticles as efficient catalysts for selective oxidation of cellobiose into gluconic acid in aqueous medium. Chem. Commun. 46: 7179–7181.

Tang, S., J. Liu, S. Yao and Y. Chen. 2015. A review of walnut shell based activated carbon for heavy metals removal from aqueous solution. Sichuan Huagong 18: 18–20.

Tian, Z., L. Wang, L. Jia, Q. Li, Q. Song, S. Su and H. Yang. 2013. A novel biomass coated Ag-TiO₂ composite as a photoanode for enhanced photocurrent in dye-sensitized solar cells. RSC Adv. 3: 6369–6376.

Titirici, M.-M., A. Thomas and M. Antonietti. 2007. Replication and coating of silica templates by hydrothermal carbonization. Adv. Funct. Mater. 17: 1010–1018.

Titirici, M.-M., M. Antonietti and N. Baccile. 2008. Hydrothermal carbon from biomass: a comparison of the local structure from poly- to monosaccharides and pentoses/hexoses. Green Chem. 10: 1204–1212.

Tokarev, A.V., E.V. Murzina, J.P. Mikkola, J. Kuusisto, L.M. Kustov and D.Y. Murzin. 2007. Application of *in situ* catalyst potential measurements for estimation of reaction performance: Lactose oxidation over Au and Pd catalysts. Chem. Eng. J. 134: 153–161.

van Putten, R.-J., J.C. van der Waal, E. de Jong, C.B. Rasrendra, H.J. Heeres and J.G. de Vries. 2013. Hydroxymethylfurfural, a versatile platform chemical made from renewable resources. Chem. Rev. 113: 1499–1597.

Varma, R.S. 2012. Greener approach to nanomaterials and their sustainable applications. Curr. Opin. Chem. Eng. 1: 123–128.

Varma, R.S. 2013. Greener routes to organics and nanomaterials: sustainable applications of nanocatalysts. Pure Appl. Chem. 85: 1703–1710.

Varma, R.S. 2016. Greener and sustainable trends in synthesis of organics and nanomaterials. ACS Sustainable Chem. Eng. 4: 5866–5878.

Vilchis-Nestor, A.R., M. Avalos-Borja, S.A. Gomez, J.A. Hernandez, A. Olivas and T.A. Zepeda. 2009. Alternative bio-reduction synthesis method for the preparation of Au(AgAu)/SiO$_2$-Al$_2$O$_3$ catalysts: Oxidation and hydrogenation of CO. Appl. Catal., B 90: 64–73.

Villa, A., M. Schiavoni, S. Campisi, G.M. Veith and L. Prati. 2013. Pd-modified Au on carbon as an effective and durable catalyst for the direct oxidation of HMF to 2,5-Furandicarboxylic acid. ChemSusChem. 6: 609–612.

Virkutyte, J. and R.S. Varma. 2011. Green synthesis of metal nanoparticles: Biodegradable polymers and enzymes in stabilization and surface functionalization. Chem. Sci. 2: 837–846.

Wan, X., C. Zhou, J. Chen, W. Deng, Q. Zhang, Y. Yang and Y. Wang. 2014. Base-free aerobic oxidation of 5-hydroxymethylfurfural to 2,5-furandicarboxylic acid in water catalyzed by functionalized carbon nanotube-supported Au-Pd alloy nanoparticles. ACS Catal. 4: 2175–2185.

Wang, D., W. Niu, M. Tan, M. Wu, X. Zheng, Y. Li and N. Tsubaki. 2014. Pt nanocatalysts supported on reduced graphene oxide for selective conversion of cellulose or cellobiose to sorbitol. ChemSusChem. 7: 1398–1406.

Wang, Y., S. De and N. Yan. 2016. Rational control of nano-scale metal-catalysts for biomass conversion. Chem. Commun. 52: 6210–6224.

Werpy, T., J. Frye and J. Holladay. 2006. *In:* Kamm, B., P.R. Gruber and M. Kamm (eds.). Biorefineries-Industrial Processes and Products. Wiley-VCH: Weinheim.

White, R.J., V. Budarin, R. Luque, J.H. Clark and D.J. Macquarrie. 2009. Tuneable porous carbonaceous materials from renewable resources. Chem. Soc. Rev. 38: 3401–3418.

Wohlgemuth, S.-A., F. Vilela, M.-M. Titirici and M. Antonietti. 2012. A one-pot hydrothermal synthesis of tunable dual heteroatom-doped carbon microspheres. Green Chem. 14: 741–749.

Xu, C., L. Cheng, P. Shen and Y. Liu. 2007. Methanol and ethanol electrooxidation on Pt and Pd supported on carbon microspheres in alkaline media. Electrochem. Commun. 9: 997–1001.

Xu, C., L. Zheng, J. Liu and Z. Huang. 2011. Furfural hydrogenation on nickel-promoted Cu-containing catalysts prepared from hydrotalcite-like precursors. Chin. J. Chem. 29: 691–697.

Xue, Z., X. Sun, Z. Li and T. Mu. 2015. CO$_2$ as a regulator for the controllable preparation of highly dispersed chitosan-supported Pd catalysts in ionic liquids. Chem. Commun. 51: 10811–10814.

Yahya, M.A., Z. Al-Qodah and C.W.Z. Ngah. 2015. Agricultural bio-waste materials as potential sustainable precursors used for activated carbon production: A review. Renewable Sustainable Energy Rev. 46: 218–235.

Yan, N., C. Zhao, P.J. Dyson, C. Wang, L.-t. Liu and Y. Kou. 2008. Selective degradation of wood lignin over noble-metal catalysts in a two-step process. ChemSusChem. 1: 626–629.

Yan, N., Y. Yuan, R. Dykeman, Y. Kou and P.J. Dyson. 2010. Hydrodeoxygenation of lignin-derived phenols into alkanes by using nanoparticle catalysts combined with bronsted acidic ionic liquids. Angew. Chem., Int. Ed. 49: 5549–5553, S5549/1–S5549/13.

Yilmaz, M., H. Turkdemir, M.A. Kilic, E. Bayram, A. Cicek, A. Mete and B. Ulug. 2011. Biosynthesis of silver nanoparticles using leaves of Stevia rebaudiana. Mater. Chem. Phys. 130: 1195–1202.

Yu, J.-S., S. Kang, S.B. Yoon and G. Chai. 2002. Fabrication of ordered uniform porous carbon networks and their application to a catalyst supporter. J. Am. Chem. Soc. 124: 9382–9383.

Yuan, J., C. Giordano and M. Antonietti. 2010. Ionic liquid monomers and polymers as precursors of highly conductive, mesoporous, graphitic carbon nanostructures. Chem. Mater. 22: 5003–5012.

Zhang, J., J. Teo, X. Chen, H. Asakura, T. Tanaka, K. Teramura and N. Yan. 2014. A series of nim (M = Ru, Rh, and Pd) bimetallic catalysts for effective lignin hydrogenolysis in water. ACS Catal. 4: 1574–1583.

Zhang, S., K. Dokko and M. Watanabe. 2015. Carbon materialization of ionic liquids: from solvents to materials. Mater. Horiz. 2: 168–197.

Zhang, Z., J.E. Jackson and D.J. Miller. 2001. Aqueous-phase hydrogenation of lactic acid to propylene glycol. Appl. Catal., A 219: 89–98.

Zhao, L., Z. Bacsik, N. Hedin, W. Wei, Y. Sun, M. Antonietti and M.-M. Titirici. 2010. Carbon dioxide capture on amine-rich carbonaceous materials derived from glucose. ChemSusChem. 3: 840–845.

Part IV
Environmental Applications

Chapter 7

Nanocatalysis and their Application in Water and Wastewater Treatment

Hanna S. Abbo,[1,2,*] *Nader Ghaffari Khaligh*[3] *and Salam J.J. Titinchi*[2]

INTRODUCTION

The global water crisis has become an emerging dilemma for humans in the recent years. Water is becoming scarcer in many countries and various strategies are being formulated to try and preserve water. Various factors, such as population growth, urbanization and industrialization (associated with an increase in production and consumption), have continually stressed hydrological resources due to the release of vast volumes of wastewater (United Nations Educational, Scientific and Cultural Organization 2015). The world demand for water has increased several fold due to the growing global population and in turn, agriculture, industry, and energy demands on water resources has increased and will continue to grow. This will result in billions of people living in regions described as "Water scarce". The world's growing population, expansion of urban areas and pollutants, all adversely impact water resources, especially in areas where natural resources are limited. The world's growing demand for clean water requires the protection of water resources more than ever. Therefore, better water treatment technology is vital. Several traditional treatment methods and techniques have been applied for remediation of waste water. These methods and techniques are not efficient enough, are generally expensive and

---

[1] Department of Chemistry, University of Basrah, Basrah, Iraq.
[2] Department of Chemistry, University of the Western Cape, Bellville, Cape Town, South Africa.
[3] Nanotechnology and Catalysis Research Center, 3rd Floor, Block A, Institute of Postgraduate Studies, University of Malaya, 0603 Kuala Lumpur, Malaysia.
* Corresponding author: hsabbo@gmail.com

produce by-products causing further substantial pollution and hence, are not viable. Several new concepts and technologies for water purification are fast replacing the traditional methods in order to address these challenges. These methods are more cost-effective, less time and energy consuming and result in less waste generation than conventional methods. Nanotechnology holds great potential in handling water treatment problems and clearly makes further improvements in treatment efficiency all of which have advantages over traditional treatment processes. These can include lower capital outlay, operations and maintenance costs, higher efficiency, easier operation, better effluent water quality and lower waste production.

Advanced green nanotechnology proved to be superior to the conventional methods in terms of enhancing the environmental sustainability of water treatment. Nowadays, there is a continuously increasing worldwide concern for the development of cost-effective, environmental friendly and sustainable wastewater treatment technologies on a large-scale.

Removal of organic compounds from wastewater is a very important subject of research in the field of environmental chemistry. In particular, heterogeneous photocatalysis has been demonstrated to have tremendous promise in water purification and the treatment of several pollutant materials that include naturally occurring toxins, pesticides and other deleterious contaminants. In this sense, photocatalysis is a handy promising technology which is very attractive for wastewater and potable water treatment. The photocatalytic activity is dependent on the surface and structural properties of the semiconductor such as crystal composition, surface area, particle size distribution, porosity, band gap and surface hydroxyl density.

A variety of semiconductor powders acting as photocatalysts have been used (oxides, sulfides, etc.). TiO_2 is preferred due to its chemical and biological inertness, photodurability, high photocatalytic activity having maximum quantum yields, its resistance to photo-corrosion, its biological immunity and low cost.

The advanced oxidation processes based on the photo-activity of semiconductor-type materials can be successfully used in wastewater treatment for destroying the persistent organic pollutants, resistant to biological degradation processes. TiO_2 is an attractive semiconductor because of its higher photocatalytic activity and can be used suspended into the reaction medium (slurry reactors) or immobilized as a film on solid material. A very promising method for solving problems concerning the photocatalyst separation from the reaction medium is to use the photocatalytic reactors in which TiO_2 is immobilized on a support. The immobilization of TiO_2 onto various supporting materials has largely been carried out via physical or chemical processes. Environmental applications of photocatalysis using TiO_2 have attracted an enormous amount of interest over the last three decades (Park 2011).

It is well established that slurries of TiO_2 illuminated with UV light can degrade to the point of mineralization of almost any dissolved organic pollutant. Nevertheless, photocatalysis, particularly in aqueous media, has still not found widespread commercial implementation for environmental remediation. The main hurdle appears to be the cost, which is high enough to prevent the replacement of existing and competing technologies by photocatalysis.

TiO_2 is the most used semiconductor because of its higher photocatalytic activity, resistance to the photocorrosion processes, absence of toxicity, biological immunity and relatively low cost. TiO_2 crystalline powder can be used as slurry (slurry reactors) or immobilized on a film on a carrier material. The anatase form of TiO_2 is reported to give the best combination of photoactivity and photostability. An ideal photocatalyst for oxidation is characterized by the following attributes: photostability; a chemically and biologically inert nature; availability and low cost and capability to adsorb reactants under efficient photonic activation. The support should have the immobilized TiO_2 on different supporting materials and should have a relatively high surface area, small particle size and a porous structure. In particular, the magnetic properties of the powders make them easily recoverable by magnetic separation technology after adsorption or regeneration, which overcomes the disadvantage of separation difficulty of common powdered adsorbents.

The impact of organic pollution is severely harmful due to it resisting biodegradation and it is thus persistent in the environment. Many methods, including chemical precipitation, ion exchange, membrane technologies, coagulation, reduction, biosorption, filtration, adsorption, reverse osmosis, granular ferric hydroxide, electrolysis and surface adsorption have been attempted to address the issue of organic pollution. Most of these methods have economic and technical limitations to achieve the discharge standards.

Advanced oxidation processes (AOPs) are good alternatives for the removal of toxic compounds from wastewater. The AOPs can be successfully applied in wastewater treatment to degrade the persistent organic pollutants, the oxidation process being determined by the very high oxidative potential of the HO· radicals generated in the reaction medium by different mechanisms. AOPs can be applied to fully or partially oxidized pollutants, usually using a combination of oxidants. Photo-chemical and photo-catalytic advanced oxidation processes including UV/H_2O_2, UV/O_3, $UV/H_2O_2/O_3$, $UV/H_2O_2/Fe^{2+}(Fe^{3+})$, UV/TiO_2 and $UV/H_2O_2/TiO_2$ can be used for oxidative degradation of organic contaminants. A complete mineralization of the organic pollutants is not necessary, as it is more worthwhile to transform them into biodegradable aliphatic carboxylic acids followed by a biological process. The efficiency of the various AOPs depend both on the rate of generation of the free radicals and the extent of contact between the radicals and the organic compound.

Photocatalytic oxidation in water treatment has proved its efficiency at many pilot-scale applications. However, wide marketing of commercially available solar detoxification systems is curtailed by the general market situation: a new water treatment procedure has an opportunity to be implemented only when its cost is at least two-fold lower than the cost of a procedure currently in use. Photocatalysis, also called the "green" technology, represents one of the main challenges in the field of treatment and decontamination systems, especially for water and air. Its operating principle is based on the simultaneous action of light and a catalyst (semi-conductor), which causes pollutant molecules to be destroyed without damaging the surrounding environment.

In recent years, applications to environmental remediation has been one of the most active subjects in photocatalysis. However, application of these methods is still limited and in the early stages and more research in the field is certainly necessary.

On the other hand, water pollutants such as heavy metals, detergents, organic pollutants, viz. pesticides, fertilizers and fuel combustion products are posing a threat to nature, humans, animals and plants. During the dyeing process for instance, about 10–15% of the synthetic dyes and other additives do not bind to the product and are released into the waste stream to pollute the environment. Research has shown that over 15% of the total production of dyes is lost as untreated in effluent, thus posing a threat to the health and general well-being of man and animals. These pollutants are difficult to remove and may undergo degradation to form products that are highly toxic and carcinogenic. Thus, contaminants are potential hazards to living organisms and severely affect the quality of water (The United Nations world water development report 2016).

Due to the low biodegradability of dyes, conventional biological treatment processes are inefficient to effectively treat dye wastewater. The same dyes, pharmaceuticals and hormones designed to stimulate a physiological response in humans, plants, and animals pose a significant threat to the aquatic environment due to their intrinsic lipophilic nature, tendency to interact with living tissues and continued comprehensive use.

Nanomaterials in water treatment

Nanocatalysts for organic pollutants degradation

Nanotechnology

Nanotechnology was introduced by the Nobel laureate, Richard P. Feynman, during his famous 1959 lecture *"There's Plenty of Room at the Bottom"* (Feynman 1960; Drexler 2000). Since then, there have been many revolutionary developments in physics, chemistry and biology that have demonstrated Feynman's ideas of manipulating matter at the level of molecules and atoms. Nanotechnology refers to matter observed on a nanometer scale which offers novelty to materials compared to those observed on a bulk scale. It holds the promise of immense improvements in manufacturing technologies, electronics, telecommunications, health and even environmental remediation. It involves the production and utilization of a diverse array of nanomaterials, which include structures and devices ranging in size from 1 to 100 nm and displays unique properties not found in bulk-sized materials. Nanotechnology offers materials with great potential in tunable properties varying from electrical, magnetic, conductive catalytic, mechanical and chemical properties, thereby, allowing materials to be applied in a variety of fields.

Nanotechnology is considered an effective technology in solving water problems. Nano-catalysts are receiving extensive attention in wastewater treatment due to their exceptional water treatment capabilities for the degradation of organic pollutants. Nanotechnology has the potential to not only overcome major water treatment challenges faced by the current technologies, but also to develop and provide new solutions for water treatment.

Advanced oxidation processes

Advanced oxidation processes and their reliable application in degradation of many contaminants have been reported as a potential method to reduce and/or alleviate this problem.

Photocatalysis has shown great potential as a low-cost, environmental friendly and sustainable water treatment technology.

The high surface areas of nanomaterials, upgraded membrane technologies and the catalytic properties of some nanomaterials have potential for removing toxic metal ions, disease causing microbes and inorganic and organic solutes from waste water. It covers details about various nanotechniques like Nanosensors, Nanosorbents, Nanofilteration and Reverse Osmosis. The positive and negative effects of nanotechnology have also been discussed. Nanotechnology has led to various efficient ways for the treatment of waste water in a more precise and accurate way on both the small and large scale. Nanotechnology is able to treat water for daily use and industrial purposes which is a high-priority of an eco-friendly system.

In this respect, nanotechnologies can play an important role. Due to unique properties vis-à-vis extremely high surface area, high absorbsion, interacting and reacting capabilities due to their extremely small size, being lighter, stronger and more efficient nanomaterials are considered as new classes of materials. Nanomaterials can offer a wide range of applications such as catalytic membranes, nanosorbents, bioactive nanoparticles and metal nanoparticles such as iron, silver, titanium oxides and many others. In water research, nanotechnology is applied to develop more cost-effective and high performance waste water treatment systems, as well as to provide rapid and continuous ways to monitor water quality.

Advanced Oxidation Processes (AOP) has been reported to have been effectively applied to purify fractory pollutants found in industrial wastewater. According to Ollis (1993), AOPs aqueous phase oxidation processes which are based primarily on the intermediary of the hydroxyl radical in the mechanism(s) results in the destruction of the target pollutant, that is, the xenobiotic or contaminant compounds include drugs, industrial chemicals, etc. AOPs have been reported to degrade the high loading of organic compounds in high saline conditions over the years (Neyens and Baeyens 2003).

Photocatalysis

Metal oxides

Photocatalytic activities of fabricated nanostructured flowerlike p-BiOI/p-NiO heterostructures for the degradation of acid orange 7 at room temperature under visible light illumination were measured. The study showed that the 10% of NiO in the BiOI/NiO composites exhibited an excellent performance. The morphological and optical studies revealed that the activity enhancement of BiOI/NiO enhanced photocatalytic activity of the composites was attributed to the efficient separation of photoinduced electrons and holes caused by heterostructure due to suitable heterojunction formation. Furthermore, the results showed that the composite is highly active in an acidic environment. The reusability of the nanophotocatalyst and the effects of various parameters such as pH, dye concentration and the catalyst's dosage were also investigated (Yosefi and Haghighi 2018).

The photocatalytic degradation of methyl violet, phenol and rhodamine B (RhB) was investigated using an Ag-doped ZnO catalyst under UV light irradiation. Kumaresan et al. (2017) synthesized various concentrations of silver (Ag) doped zinc oxide (ZnO) nanoparticles by a one step microwave irradiation method. The Ag/ZnO possessed a hexagonal wurtzite type structure with average crystal size of 21–16 nm by powder X-ray diffraction studies of these catalysts. They also possessed a spherical shaped morphology with an average diameter of 32–13 nm as determined by Transmission Electron Microscope analysis. Subsequently, the Ag-ZnO catalyst was investigated for its photocatalytic degradation of methyl violet (MV), phenol and rhodamine B (RhB) under UV light irradiation. The result showed that the photocatalytic property was significantly improved by Ag doping (Kumaresan et al. 2017).

The highly-dispersed carambola-like SnO_2/C composite microparticles have been synthesized through a novel electrospun approach using functional additives. The carbon matrix was used to reduce agglomeration, decrease migration and control the sizes of SnO_2 nanocrystals. The researchers also investigated the influence of the calcination temperature on the microstructures and photocatalytic properties of SnO_2/C products and the relevant photocatalytic mechanism. SnO_2/C photocatalysts showed a superior photocatalytic performance for the degradation of rhodamine B with a high efficiency of 94.6% as the irradiation time was increased to 210 min. Also, it showed good photocatalytic cycle stability (Chen et al. 2017).

A new hierarchical photocatalyst material was thus synthesized with high efficiency. ZnO seeds were uniformly deposited on carbon fibres (CFs) by atmospheric layer deposition (ALD) followed by hydrothermal growth of ZnO nanorods (NRs). Then, Pt particles were dispersed on ZnO NRs using the magnetron sputtering method. The grain size of Pt particles was in the range of 2–5 nm. Significantly, the resultant Pt@ZnO NRs/CFs composites exhibited better photocatalytic performance than the parent material. This photocatalyst was studied for the degradation of methyl orange. The degradation rates of methyl orange dye solution were 72% for ZnO NRs/CFs and 99.8% for Pt@ZnO NRs/CFs after a period of light irradiation. In addition, this novel photocatalyst can be easily recycled with good performance and stability (Gu et al. 2017).

Nanocomposite oxides (NCOs) ZrO_2/CuO was synthesized by utilizing oxalyl dihydrazide as fuel. The XRD and TEM analysis reveal the presence of cubic ZrO_2 and CuO in the nanocomposite oxides. The study showed that the ZrO_2/CuO (2:1) composite exhibited excellent photocatalytic activity towards the degradation of Indigo carmine dye under sunlight with enhanced photocatalytic activity of 97%. This represented improvements of 3.3, 2.4 and 1.5 times higher than that of pure ZrO_2, CuO and commercial P25. This improvement was attributed to the balance between the parameters, band gap, nature of morphology, crystallite size, defects and surface area all which cause a slow electron-hole pair recombination rate with fast electron transfer ability (Renuka 2017).

Typical spinel structured nickel ferrites ($NiFe_2O_4$) were prepared via thermal decomposition of metal oxalates. The products exhibited small nanoparticle size (ca. 12 nm), high BET surface (53.5 m^2 g^{-1}) and a good magnetic response (19.3 emu g^{-1}). $NiFe_2O_4$ was applied in heterogeneous catalysis to generate powerful radicals from peroxymonosulfate (PMS) for the removal of recalcitrant pollutants such as benzoic acid (BA). $NiFe_2O_4$/PMS were found to degrade the pollutants by 82.5% in 60 min and maintain the catalytic efficiency for four recycling experiments. This study confirmed that sulfate and hydroxyl radicals were the main reactive species in the $NiFe_2O_4$/PMS system. On the other hand, XPS spectra revealed that Ni sites on the surface of $NiFe_2O_4$ were the primary active sites while, Raman spectra suggested that inner-sphere complexation between PMS and Ni sites produced peroxo species on the surface, which were further responsible for the efficient generation of radicals (Wang et al. 2016).

Another group of researchers developed a direct preparation method to synthesize a core-shell structure with a similar spherical morphology of Fe_3O_4 nanoparticles as the core and WO_{3-x} as the shell. The XRD patterns showed that a cubic spinel structure of the Fe_3O_4 core and the WO_{3-x} shell were obtained. The nanoparticles showed strong magnetic and microwave properties for enhancing the rate of heating, which lead to development of nanoparticles with substantial potential for various applications such as drug targeting, water treatment, etc. This finding may lead to the development of nanoparticles with great potential for applications in water treatment as well as drug targeting delivery, controlled drug release and photo- and microwave-thermal combination therapy (Peng 2017).

Although Cu_2O is a low-cost semiconductor with a narrow band gap and high absorption coefficient, it suffers from low charge mobility and poor quantum yield and therefore exhibits a low efficiency in catalytic performance under visible light irradiation. A group of researchers worked on Cu_2O to overcome the technical challenges of photochemical water treatment which resulted in drastically improving the catalytic capacity by a simple and effective strategy due to the advantage of the synergistic effects between photocatalysis and Fenton. Thus, an efficient Cu_2O/Nano-C hybrid photocatalyst was synthesized. This catalyst exhibits a significant superiority for the removal of two typical fractory pollutants in wastewater such as Rhodamine B and p-nitrophenol. The superiority of the photocatalysis-driven Fenton system, PFC, is mainly attributed to: the rapid photo-electron transfer driven by Schottky-like junction; the superior *in situ* generation activities for both H_2O_2 and •OH; and the accelerated Fe^{2+}/Fe^{3+} cycling and robust Fe^{2+} regeneration via two additional pathways. This finding opens a new direction to overcome the intrinsic challenges of both photocatalysis and the Fenton reaction, as well as develop novel technologies for advanced water treatment. For an insight into the synergistic PFC on the Cu_2O/N-rGO hybrid under visible light irradiation for efficient fractory pollutant degradation, a mechanism was proposed (Zhang et al. 2017).

On the other hand, another group of researchers synthesized different CuO based immobilized nanoparticles. Collagen-CuO nanoparticles composite beads were synthesised as adsorbent for water treatment applications. Adsorption studies were conducted under optimum conditions (such as initial concentration, contact time, adsorbent dose and pH) for the removal/adsorption of a pesticide, that is, cypermethrin. The results illustrated that the bionanocomposite material removed 94% of cypermethrin from the solution (25 mL, 0.125 mgL^{-1}) at room temperature and pH 7 (Nejaei et al. 2017).

A series of Fe_2O_3 modified $BiOIO_3$ photocatalysts were prepared using a simple hydrothermal method. By introducing Fe, the absorption range of $BiOIO_3$ extended to the visible light region and the apparent band gap is dramatically decreased from 3.00 eV for pure $BiOIO_3$ to 1.81 eV for 30% $Fe_2O_3/BiOIO_3$ nanosheets. The photocatalytic oxidative degradation of *p*-nitrophenol under visible light was improved and has been attributed to the synergistic effect of heterojunction semiconductor photocatalysis and heterogeneous Fenton catalysis caused by $BiOIO_3$ and Fe_2O_3. The intrinsic relationship between structure and catalytic activity was fully studied and a possible catalytic mechanism was proposed. The researchers concluded that this highly active heterogeneous catalyst has the potential for efficient utilization of solar energy in the visible range and can be applied in environmental remediation and waste water treatment. On recycling the catalyst, it was found that it was still stable without any obvious deactivation (Zeng et al. 2017).

Nanoparticle-based immobilized biocatalysts are gathering interest in bioremediation efficiency. In this study, laccase (lac) was immobilized on both ZnO and MnO_2 nanomaterials. The catalytic potential of lac-ZnO and lac-MnO_2 was examined for *in vitro* degradation of alizarin red S dye in simulated-water. The degradation study of the dye at pH (~ 7.0) indicated that lac-ZnO is a better catalyst than ZnO being free from lac. It was assumed that MnO_2 enhanced the stability and activity of lac-MnO_2 over free lac. Moreover, a combination of nanomaterial and enzyme is proposed for achieving biocompatibility in inert conditions without denaturing the enzyme. Hence, lac was immobilized on ZnO followed by the addition of MnO_2. It was found that the degradation activity in this situation was twice that of free lac. Thus, the results clearly showed that ZnO and MnO_2 nanomaterials are important adsorbents in waste water treatment and show the potential for large-scale lac immobilization with improved properties and reuse (Rani et al. 2017).

Recently, photocatalytic membranes have been developed by coupling membrane filtration and photocatalysis. A novel visible-light-driven photocatalytic membrane based on graphitic carbon nitride (g-C_3N_4) for water treatment was developed in which the g-C_3N_4, with a band gap of about 2.70 eV, was found to be an attractive metal-free photocatalyst for organic pollutant degradation. In this study, fabricated quantum dots (g-C_3N_4) were assembled into a TiO_2 nanotube array (TNA) membrane for the first time. A facile approach was used to achieve a vertically oriented and free-standing g-C_3N_4/TNA membrane by immobilizing graphitic carbon nitride Quantum dots g-C_3N_4 QDs in a highly ordered TNA membrane via potentiostatic anodization in a two-electrode electrochemical cell. The electrodeposition of g-C_3N_4 QDs was performed using an electrophoresis device where the pre-anodized Ti foil was connected to the negative pole of the Pt plate while the positive pole was connected to the electrophoresis apparatus. The electrolyte solution contained isopropyl alcohol, magnesium nitrate and g-C_3N_4 QDs. Electrodeposition was maintained at 20 V for 1–3 h duration. Subsequently, the g-C_3N_4/TNA substrate was anodized for a second time at 60 V and at room temperature for 4 h. Finally, a larger anodic voltage (150 V) was applied for 5 min to get a through-hole g-C_3N_4/TNA membrane.

A major benefit from the synergistic effect of membrane filtration and photocatalysis led to a unique photogenerated charge separation. The photo degradation study showed that more than 60% of rhodamine B was removed from water under visible light irradiation. Meanwhile, the g-C_3N_4/TNA membrane presented an enhanced anti-fouling ability during water filtration containing *Escherichia coli* under visible light irradiation, and a permeate flux of twice that of filtration alone was obtained by this integrated process. This study is one of the promising approaches for the potential application of the visible-light-driven membranes in water treatment (Zhang et al. 2017).

Other workers synthesized different types of graphitic carbon nitride based metal oxide nanohybrids, which comprised of strontium oxide/graphitic carbon nitride (SrO_2/g-C_3N_4) nanohybrids, by a facile simple dry protocol. These nanohybrids were tested for photocatalytic degradation against a rhodamine B (RhB) dye solution under visible light irradiation. The SrO_2/g-C_3N_4 nanohybrids displayed superior crystallinity with a crystallite size of ~ 7.1 nm. Moreover, the synthesized nanohybrids displayed excellent activity and good stability towards the photo-degradation of the RhB dye compared to the pure parent vis-á-vis SrO_2 or g-C_3N_4. The excellent photocatalytic activity of the SrO_2/g-C_3N_4 nanohybrid was attributed to the improved charge separation and the complete degradation of the dye solution was achieved within 60 min. A possible mechanism for the charge separation in the photo degradation process has been proposed by the authors. Finally, the synthesized nanohybrids exhibited excellent stability during a recycling photocatalytic experiment. Therefore, the nanohybrids demonstrate potential interest in waste water treatment under visible light irradiation (Prakash et al. 2017).

On the other hand a different type of graphitic carbon nitride, that is, porous CeO_2/sulfur-doped g-C_3N_4 (CeO_2/CNS) composites, with different CeO_2 contents have also been synthesized. The CeO_2(x)/CNS composites were synthesized by a one-pot thermal condensation of thiourea and cerium nitrate as starting materials. CeO_2(x)/CNS composites displayed a nanoporous structure with uniform pore size of ~ 40 nm, while CNS exhibited platelet-like morphology. The CeO_2(9.5)/CNS nanocomposite, exhibited the highest visible light photocatalytic activity (91.4% in 150 min) toward methylene blue (MB) degradation which was greater compared to either CNS or CeO_2 photocatalysts.

This enhanced photocatalytic performance originated from heterojunctions formed between CeO_2 and CNS, which improved the effective charge transfer through the interfacial interactions between both components.

Isopropanol, benzoquinone and ethylenediaminetetraacetate were used in the trapping tests for the ˙OH, ˙O_2 and H^+ scavengers, respectively. The results verified that the ˙OH and ˙O_2 are the major species that directly attacked the MB molecules while H^+ showed a negligible role. The authors concluded that the simultaneous doping of sulphur and CeO_2 within the g-C_3N_4 structure using a simple one-pot synthetic process produced very active photocatalysts and demonstrated their potential for practical applications in industrial water treatment purposes (Jourshabani et al. 2017). Another approach for improving the stability and photocatalytic performance of Ag_3PO_4 nanoparticles in photocatalysis is decorating carbon nitride material ($g-C_3N_4$) with Ag_3PO_4 nanoparticles. The results demonstrated that photocatalytic performance for water purification under visible light irradiation was largely enhanced (~ 7 folds) in comparison to the performance of pure Ag_3PO_4. These nanoparticles show exceptional stability due to the protection effect of the decorated the $g-C_3N_4$ sheet. In addition, the introduction of $g-C_3N_4$ altered the photocatalytic mechanism slightly which resulted in the $\cdot O^{-2}$ radical playing a more important role. These findings were obtained from radical trapping experiments. While the catalytic performance enhancement was attributed to the larger surface area, controllable particle size and the synergistic effect between Ag_3PO_4 and $g-C_3N_4$, are also contributing factors due to their promoting the separation efficiency of the photogenerated electron-hole pairs (Ren et al. 2017).

Graphene/NiO nanocomposite (GNP) was synthesized by the solvothermal method and subsequently tested for the photodegradation of methyl orange and inactivation of bacteria. NiO-nanoflakes grafted graphene shows outstanding photocatalytic activity with 99% degradation compared to pure NiO (34%). This makes it an innovative material to achieve the depollution of dye contaminated water. In addition, the graphene/NiO nanocomposite displays exceptional antibacterial activity with 100% growth inhibition of the pathogenic microorganisms of both Gram-(+) and Gram-(–) bacteria. Therefore, graphene/NiO nanocomposite is an innovative promising bactericidal material that can achieve complete pathogen control and thus, have potential as an economic solution for water treatment. The interplay between solar excitations, charge trapping by graphene and carbon purity enables the photocatalyst to perform in an excellent manner. The NiO and graphene nanocomposite also meets the economic requirements by being solar-light active and recyclable (Arshad et al. 2017a).

A further nanocomposite synthesized by the combination of cupric oxide (CuO) with graphene nanoplatelets (GNPs) has the formula CuO/GNPs and this nanocomposite enhanced the photocatalytic properties without tweaking the inherent properties of the GNPs, that is, the conductive nature of the GNPs. This nanocomposite photodegraded methylene blue by 99.44% within 80 min of exposure to solar light compared to 75% achieved by pure CuO. The improved photocatalytic performance of the nanocomposite is a result of the low bandgap energy of CuO/GNPs and the conducting nature of GNPs that causes fast photo excitation of electrons and lengthens the recombination lifetime of charge carriers (Arshad et al. 2017b).

Cu/Cellulose (C) and Cu_2O/C composites have been synthesized by calcining the precursors (Cellulose/CuO) at different temperatures in a N_2 atm. These precursors have been fabricated by the microwave-assisted ionic liquid method. The photocatalytic degradation results of methylene blue (MB) in aqueous solution under visible-light irradiation demonstrated that Cu_2O/C composites showed the highest degradation rate ~ 99%, and thus provide promising applications for dye removal and wastewater treatment fields (Liu et al. 2016).

Mixed metal oxides supported on reduced graphene oxide (rGO-ZnTi-MMO-x, x presents wt. percentage of GO) were studied for water treatment, that is, the degradation of bisphenol A (BPA). These catalysts were obtained by the thermal treatment of Zn-Ti layered double hydroxides-graphene oxide (GO-ZnTi-LDHs) composites. The photocatalytic activity of the obtained photocatalysts showed significantly enhanced activities compared with the pristine ZnTi-MMOs catalyst for 3 h under solar light irradiation.

Photo-generated holes, OH and singlet oxygen radicals were demonstrated to be the predominant active species responsible for the photodegradation of BPA. The spectro-photo spectroscopic techniques used confirmed that the enhanced photocatalytic activity of rGO-ZnTi-MMOs composites was attributed to the extended visible light absorption region and efficient transportation and separation of photo-induced electron-hole pairs of rGO-ZnTi-MMOs with unique hetero-nanostructure (Yang et al. 2016).

Series palladium-doped-zirconium oxide-multiwalled carbon nanotubes (Pd-ZrO_2-MWCNTs) representing nanocomposites with various percentage of Pd were synthesized and tested for their photocatalytic degradation of organic pollutants, vis-à-vis acid blue 40 dye in water under simulated solar light. These Pd-ZrO_2-MWCNTs nanocomposites were prepared by the homogenous co-precipitation method. The nanocomposites showed enhanced photocatalytic activity toward the degradation compared with ZrO_2 and ZrO_2-MWCNTs. The nanocomposite with 0.5% Pd was the most efficient photocatalyst with 98% degradation after 3 h with corresponding Ka and band gap values of 16.8×10^{-3} m^{-1} and 2.79 eV, respectively (Anku et al. 2016).

At the same time, a similar approach of synthesizing a nanocomposite consisting of silver, silver oxide (Ag_2O), zinc oxide (ZnO) and graphene oxide (GO) was published. The Ag-Ag_2O-ZnO nanostructure was synthesized by a co-precipitation method and calcined at 400°C. Then, after it was functionalized using a silane linker, that is, 3-aminopropyl triethoxysilane, it was further anchored to the carboxylated graphene oxide via the formation of an amide bond to give the Ag-Ag_2O-ZnO/GO nanocomposite. These nanocomposites were screened for their photocatalytic degradation of organic pollutants such as acid blue 74 dye, in water under visible light irradiation. The results indicated that the Ag-Ag_2O-ZnO/GO nanocomposite has a higher photocatalytic activity (90% removal) compared to Ag-Ag_2O-ZnO (85% removal) and ZnO (75% removal) which could thus be applied in water treatment and the removal of organics pollutants (Umukoro et al. 2016).

The photocatalytic activity of activated charcoal (AC) supported bismuth (Bi)-doped zinc oxide (ZnO) nanocomposite material, synthesized by a precipitation method, was studied. Production of hydroxyl radicals (•OH) on the surface of the UV-irradiated photocatalysts was detected by a photoluminescence technique using coumarin as a probe molecule. The photocatalytic activity of the AC-Bi/ZnO material is demonstrated through the photodegradation of methyl violet (MV) under UV-light irradation. AC-Bi/ZnO has an increased absorption in the UV region; moreover, it shows excellent UV-light driven photocatalytic performance. The photocatalyst of AC-Bi/ZnO reveals enhanced photocatalytic activities as compared to ZnO and Bi/ZnO for the degradation of the dye. The enhanced photocatalytic activity of AC-Bi/ZnO is attributed to the low recombination rates of photoinduced electron hole pairs, caused by the transfer of electrons and holes between ZnO and AC supported Bi^{3+} ions (Chandraboss et al. 2015).

Titanium dioxide (TiO$_2$)

Titanium dioxide (TiO$_2$) is a semiconductor material extensively used due to its superior photoreactivity, long-term stability, non-toxicity, low operating cost and radically low level of energy consumption. The photocatalytic activity of TiO$_2$ depends on various parameters, including surface area, crystallinity, impurities and density of the surface hydroxyl groups. Generally, TiO$_2$ could be used as a photocatalyst in anatase and rutile crystal structures (the anatase phase displaying much higher activity than the rutile structure). Titanium shows a novel photoinduced superhydrophilic phenomenon and hence, the interest in its application in water treatment and purification systems (Fujishima et al. 2000). Since the pioneering work of Honda and Fujishima (1972) on considering TiO$_2$ as a possible alternative for sustainable water treatment, numerous TiO$_2$ based materials have been extensively developed and TiO$_2$ has been the subject of intense investigations. This is due to titanium dioxide being a semiconductor material with superior photoreactivity, long-term stability, non-toxicity, high inertness even in a corrosive environment, abundance and a radically low level of energy consumption. However, its wide band gap ($\approx$ 3.2 eV) in the UV region of the spectrum makes it ineffectual under solar illumination. The photocatalytic activity of TiO$_2$ depends on various parameters, that is, surface area, crystallinity, impurities and surface hydroxyl group density. TiO$_2$ exists in different polymorphs; the most common forms are the anatase- and rutile-crystal structures. These two forms could be used as a photocatalyst; however, the anatase phase displaying much higher activity than the rutile structure. Titanium shows a novel photoinduced superhydrophilic phenomenon and hence, the interest in its application in water treatment and purification systems.

Self-organized TiO$_2$ nanotube arrays (TNAs) as immobilized catalysts were evaluated for photocatalytic degradation of the β-blocker metoprolol (MTP) in an aqueous solution under an UV-LED light source. High and rapid MTP degradation were achieved over a wide range of pH (3–11). The total MTP degradation was reduced from ~ 87% to 62% on using tap water samples, demonstrating reasonable efficacy for practical applications. The degradation mechanism suggests that formic acid and *tert*-butanol need to be added as scavengers for photo-generated holes (h+) and hydroxyl radicals (•OH), respectively. Results demonstrate that primary degradation occurred in the liquid phase with participation of hydroxyl radicals (•OH liquid), while a smaller portion of MTP was degraded on the catalyst surface via reaction with h+ and hydroxyl radicals adsorbed on the catalyst surface (•OH surface). A minor role in the degradation process was suggested via other reactive species, for example, photo generated electrons and superoxide radical anions. The mechanistic aspect was further confirmed by the identification of degradation products by LC-MS/MS. The TNAs exhibited good stability after repeated use under varied operation conditions (Ye et al. 2017).

Core-shell magnetic nanoparticles Fe$_3$O$_4$-SiO$_2$-TiO$_2$ were reported as highly active and recyclable efficient photocatalysts for water treatment. The magnetic core of Fe$_3$O$_4$ was synthesized via carbon reduction. A SiO$_2$ layer was coated on the magnetic core, while the final layer was TiO$_2$ following the sol-gel method. It was reported that the methyl orange solution (10 mg/L) was decomposed completely within 240 min by the core-shell nanoparticles (Wang et al. 2016).

A TiO_2/agarose hybrid gel photocatalyst was synthesized by uniformly dispersing TiO_2 nanoparticles (NPs) in an agarose hydrogel matrix. The report indicated that the size, uniformity and concentration of the hybrid gel and the contents of the constituent ingredients exhibit significant effects on the photocatalytic activity.

The smaller and uniform size of the hybrid gel exhibited superior photocatalytic performance in both photodegradation of methylene blue (MB) under UV light and TiO_2 leakage. The hybrid gel showed excellent recycling performance over several reuses (Mai et al. 2017).

Another type of TiO_2 based photo-catalyst using alginate as a non-toxic, biocompatible pore-directing template and binder was investigated in the adsorption and removal of different organic dyes from water. Alginate-TiO_2 porous beads were prepared from commercial type II anatase powders through ionotropic gelation and alginate as pore-directing template and binder, respectively. Both anionic (methyl orange, MO) and cationic (methylene blue, MB) dyes were utilized as a model for organic pollutants. The photocatalytic activity of titania in solution is slightly reduced compared with that of the nanopowders before assembly. However, the formation of the beads brings about many direct advantages. Thanks to their large size (0.5–2 mm) and good mechanical stability, the beads can be easily dispersed in polluted solutions and promptly recovered when stirring is stopped. The beads exhibited an enhanced adsorption of MB in comparison to nanopowder samples (55% vs. 6.5%). Upon recovery, the adsorbed dye can be completely extracted and removed from the beads either by UV or ozone-UV cleaner treatment. The latter procedure resulted in a further increment of adsorption efficiency (up to 64%) as a function of different adsorption-removal cycles (Gjipalaj and Alessandri 2017).

The main challenge and limitation with titanium dioxide (TiO_2) is that it absorbs only ultraviolet light. Thus, only 5% of the solar radiation can be utilized for water decontamination. To overcome this issue, several methodologies have been developed in the last years in order to narrow the band gap of TiO_2 to enhance its photocatalytic performance under visible (solar) irradiation. Most of these methodologies are achieved by varying the chemical composition of TiO_2 by the use of dopants such as N, C and transition metals (Asahi 2001; Litter 1999; Khan 2002) grafting of metals (Yu et al. 2010; Liu et al. 2014; Dette et al. 2014) and coupling with semiconductor quantum dots (Robel et al. 2006) or plasmonic metal nanostructures (Linic et al. 2011; Hou and Cronin 2012; Scuderi et al. 2014).

Chen and his team (Chen et al. 2011) demonstrated an alternative conceptual approach to improve visible and infrared optical absorption. Their findings were published in Science magazine in 2011. In this approach, they developed a disorder-engineered nanophase TiO_2. The disorder-nanophase TiO_2 consists of two phases: a crystalline TiO_2 quantum dot or nanocrystal as a core, and a highly disordered surface layer. The disorder of nanophase TiO_2 surface layers was introduced through a hydrogenation process. On introducing lattice disorders in hydrogenated anatase TiO_2 nanocrystals, mid-gap electronic states were generated, accompanied by band gap reduction. The disordered TiO_2 nanocrystal model, in which one H atom is bonded to an O atom while another H atom is bonded to a Ti atom, yields electronic band structures consistent with the valence band XPS measurements. However, this synthetic route requires high hydrogen pressure and long annealing treatment. Later, an Italian team presented an industrially scalable synthesis of black TiO_x-based material using laser irradiation. The photocatalyst is composed of a nanostructured titanate film (TiO_x) on a titanium foil with a layer of Pt nanoparticles (PtNps) deposited on the rear side of the foil. The result is a monolithic photochemical diode with a stacked, layered structure (TiO_x/Ti/PtNps). The resulting high photo-efficiency is ascribed to both the scavenging of electrons by Pt nanoparticles and the presence of traped surface states for holes in an amorphous hydrogenated TiO_x layer. The resulting black TiO_x shows high activity and adsorbs visible radiation, overcoming the main concerns related to the use of TiO_x under solar irradiation (Zimbone et al. 2017).

Randomly oriented arrays of SiO_2 nanowires (NWs) coated by a thin film of TiO_2 was synthesised and tested as photocatalytic material for water treatment. The highly disordered TiO_2/SiO_2 NWs were obtained via thermal oxidation of Si NWs, grown by plasma enhanced chemical vapour deposition (PECVD) and covered by a TiO_2 film deposited by atmospheric layer deposition (ALD). The materials showed a significant photocatalytic performance for the degradation of methylene blue and phenols in water. The superior photocatalytic activity is attributed to the synergetic effect of the morphology and optical properties of the SiO_2 NWs (Convertino et al. 2016).

In order to improve the visible-light photoactivity of TiO_2 for degradation of pollutants, Zhang and his group developed a synergistic hybrid photocatalyst vis-a-vis $ZnO/CdS/TiO_2$ which exhibited excellent potential for the degradation of refractory pollutants and could act as a robust and self-protected photocatalyst for water purification without additional sacrificial reagents. The superiority of the hybrid photocatalyst $ZnO/CdS/TiO_2$ over the traditional CdS/TiO_2 hybrid in both photocatalytic activity and anti-photocorrosion capacity was demonstrated in the degradation of Atrazine and Rhodamine B, two typical refractory organic pollutants, and the treatment of real textile wastewater under solar light irradiation (Zhang et al. 2016).

A highly efficient visible light photolysis catalyst was synthesized through a sol-gel method using C/C composites (expanded graphite dipping in the phenol formaldehyde resin, EGC) coated with Ag, N and co-doped TiO_2.

The analysis revealed that the photocatalysts were composed of mesoporous structures mainly with anatase TiO_2. The study considered the effects of the Ag content and calcination temperature on the photocatalytic activity of Ag-N-TiO_2/EGC photocatalysts. The results showed that under visible light, the photocatalytic efficiencies of $Ag_{1.0}$-N-TiO_2/EGC (550°C) obtained for the degradation of Rhodamine B (RB) and diesel are 7.4 and 3.9 times higher, respectively, than those of N-TiO_2/EGC with high mineralization. The enhanced photocatalytic activity of the Ag-based composites was attributed to the porosity and flexible electron transport paths for formation of the superoxide radicals (Zhang et al. 2016).

An environmentally friendly one-port *in situ* microwave method was used to synthesize Graphene-titanium oxide (G-TiO_2) nanocomposites by a novel surfactant free protocol. The photocatalytic activity of pure TiO_2 and TiO_2 nanoparticles distributed on the graphene sheets were studied under UV and visible light irradiation sources with methylene blue dye. The rate constant and half life time were calculated from the kinetic studies of the degradation. The highest degradation efficiency achieved using G-TiO$_2$ catalyst was 97% and 96% in UV light and under visible light irradiation, respectively (Shanmugam et al. 2016).

Polymeric nanocomposites have attracted extensive attention in recent years in several fields of science and technology. This class of material joins structural flexibility and relatively simple processing of the polymers with the properties of the nanomaterials, such as the photocatalytic activity of TiO$_2$. A new type of polymeric nanocomposites embedded PMMA titanium dioxide nanotubes can be prepared by a sonication and solution casting method. The evaluation of the photocatalytic activity was studied by bleaching Methylene Blue (MB) dye in an aqueous medium. These materials were also tested for their antibacterial activity using *Escherichia coli* as a model organism (Cantarella et al. 2016).

Fe-doped TiO$_2$ materials were prepared by the solgel auto-combustion technique. Then gold nano-particles (AuNPs) were deposited on the support (Fe/TiO$_2$) using the conventional deposition-precipitation method. The resulting catalysts were subsequently modified with thermal (450°C under vacuum) and plasma (ambient temperature under Argon atm. for 20 min) treatments. The photocatalytic activity was evaluated by assessing the degradation of the MB in water under UV irradiation while the parameters affecting the photocatalytic process such as the catalyst crystallinity, light absorption efficiency, the dosage of catalyst, dopant and MB concentrations were well controlled. Thermal/plasma-treated samples showed significant enhancement in the photocatalytic activity compared to untreated samples by changing the morphology, increasing the number of AuNPs, improving the Au-doped Fe/TiO$_2$ interface and decreasing the band-gap energies thus tuned the Au-doped Fe/TiO$_2$ catalyst to higher efficiency. The AuNPs deposited on the Fe/TiO$_2$ showed good thermal stability as well (Mahmood et al. 2015).

Zinc Oxide Nano Wires (ZNWs) have been considered as a promising material for purification and disinfection of water and remediation of hazardous waste owing to its high activity and relatively low cost. Three-dimensional (3D) structured palladium (Pd)/ZNWs were synthesized on the fabricated electrospun nanofibers and explored for enhancement of organic matter removal efficiency in water by suppressing electron-hole recombination during photocatalytic activity and increased surface area. The densely populated ZNWs were fabricated on the electrospun nanofiber by electroless plating and hydrothermal synthesis. In order to improve photocatalytic efficiency, a thin layer of Pd was coated prior to the ZNWs growth to induce suppression of electron hole recombination produced during the catalyst activity. The creation of a highly porous network of nanofibers decorated with ZNWs resulted in an increase of the specific removal rate of organic matter from 0.0249 to 0.0377 mg CODCr removed/mg ZNWs-hr when ZNWs were grown on a Pd layer. It is believed that the demonstration of OM removal in the water through Pd/ZNW membranes and enhanced photocatalytic activity under UV irradiation from the layered structure can broaden the potential applicability of Pd/ZNWs membranes for various photocatalytic water treatments (Choi et al. 2016).

Photo-catalytic noble metallic and bimetallic nano-composites (Ag or Pd/-TiO$_2$/CNT) were synthesized using multi-walled carbon nanotubes via a modified dry-mix metal-organic chemical vapour deposition method (MO-CVD). The titania loading was varied from 10–40 wt %, and the optimum TiO$_2$/CNT photo-catalyst was determined using methylene blue degradation as a model probe reaction. Furthermore, acid-treated nanotubes and non-acid treated nanotubes were compared as a substrate for the synthesis of various titania nano-composites, and it was found that the acid treatment decreased the photo-catalytic activity of the titania CNT nano-composites. 20 wt % Titania on CNT samples were then further modified with silver, palladium and a combination of both metals using the MOCVD technique. It was found that the silver-titania CNT nano-composites were the most effective photo-catalyst for the degradation of methylene blue. The deposition of 2% Ag on 20% TiO$_2$/MWCNT resulted in 92% degradation of 50 mg/L MB in 4 h with 1 g/L of photo-catalyst. The addition of plasmonic Ag enhanced the activity of 20 wt % TiO$_2$/CNT by about 10%. This was due to the surface plasmon resonance effect observed with silver, either via an electric field enhancement or a charge transfer mechanism. Palladium had little effect in altering the photo-catalytic activity of the titania CNT nano-composites, and the combination of both metals suppressed the photo-catalytic activity of the titania CNT nano-composites (Hintsho et al. 2014).

Other nanomaterials

A nanocomposite chitosan-copper (CS-Cu) was synthesized using biocompatible chitosan without using reducing agents. These nanocomposite photocatalysts were employed for the removal of Rhodamine B (RhB) and Conge red (CR) dyes under visible light irradiation. The nanocomposite demonstrated their efficiency as photocatalysts for removal and decolourisation of both cationic and anionic dyes by varying the catalyst concentration and the pH of the dye solution (Arjunan et al. 2017).

Ag/AgBr nano-particles were fabricated by a solution reduction strategy. Results illustrated that Ag/AgBr nano-particles exhibited intense light absorbance in the whole visible region, high photoinduced charge separation efficiency and visible light driven PC performance. Five degradation intermediates were detected; therefore, two pathways for PC degradation of acetaminophen were proposed viz. hydroxylation addition to parent compound and CN cleavage of side chain from the acetaminophen attacked by photoinduced holes (Ma et al. 2017).

Taguchi's robust design followed by solid state thermal decomposition was used to synthesize holmium carbonate and holmium oxide nanoparticles (Rahimi-Nasrabadi et al. 2017). The photocatalytic behaviour of the prepared nanoparticles in water treatment was evaluated. The photochemical behaviour study revealed that holmium carbonate and oxide nanoparticles can be effectively utilized for the degradation of methyl orange with conversion values of 99.6 and 99.3%, respectively. This confirms the potential of the fabricated products for the efficient elimination of organic pollutants.

A novel MoS$_2$/Ag$_2$WO$_4$ nanohybrid was prepared via a one-step hydrothermal approach. The photocatalytic properties of the nanohybrid material were studied for the degradation of methyl-orange under stimulated irradiation. The nanohybrid shows enhanced efficiency in dye degradation compared to the bare Ag_2WO_4 nanorods. The use of MoS_2 is to prevent the photocorrosion of Ag_2WO_4 and also to diminish the number of photogenerated electron-hole recombinations. The results suggest that the MoS_2/Ag_2WO_4 nanohybrid could be an efficient and suitable photocatalyst for wastewater treatment and remedial applications (Thangavel et al. 2017).

In order to increase the catalytic systems' efficiency while ensuring their optimized reactivities and large-scale development and implementation, Dong et al. 2016 developed graphene-based hybrid composites to increase the catalysts' ability to absorb visible light, in order to retain high corrosion-resistance properties and offer energy levels that match their reduction and oxidation half-reactions (Dong et al. 2016). They firstly constructed hybrids of platinum/tungsten trioxide conjugates and followed this by decorating such conjugates onto graphene surfaces. The results demonstrated that the synthesized hybrids can be degraded by a model azo dye experiment and further, that graphene plays an important role in delaying electron transfer at its interface.

Gold-palladium (Au-Pd) bimetallic nanoparticles immobilized MgZnAl-Cl layered double hydroxides (LDH) were prepared by the co-precipitation and colloidal sol immobilization methods. The LDH and Au-Pd@LDH were investigated for their photocatalytic degradation of orange II dye. The influence of parameters such as catalyst loading and Orange II concentration were also evaluated. It was found that after 50 min irradiation time under the tested conditions, conversions were 58% and 65% for MgZnAl-LDH and Au-Pd@MgZnAl-LDH, respectively. Despite the fact that LDH is photocatalytically active in degrading orange II dye, adsorption of orange II on the LDH surface (36%) predominates compared to its degradation (22%) (Sobhana et al. 2016).

A heterogeneous Fenton-like catalyst, Fe_3O_4-graphite composite, was prepared via a one-step solvothermal method and studied for its degradation of levofloxacin (LEV) in an aqueous solution. The study revealed that the Fe_3O_4-graphite composite exhibited excellent properties for the degradation of LEV with nearly complete degradation in 15 min and 48% of total organic carbon removal in 60 min under optimal conditions. This is due to the fast production of •OH radical species by the easy reduction of Fe(III) to Fe(II). In the work mentioned above, they observed that the graphite can degrade LEV in the presence of H_2O_2. Due to the synergistic results of the large electronic conjugation in graphite structures and Fe_3O_4 magnetic nanoparticles (MNPs), have led to contribute to high catalytic activity. On the other hand, less iron leaching of the Fe_3O_4-graphite composite was observed during the degradation compared with pure MNPs. The degradation efficiency of LEV remained ~ 80% at the fifth recycling run (Wang et al. 2015).

Nanomaterials and nanocatalysts for heavy metals removal

Introduction

Water pollution and the supply of clean water have become major worldwide problems and challenges (The Millennium Development Goals Report 2008). Currently, the quality of water resources is deteriorating owing to population growth, industrial and agricultural activities as well as other ecological and environmental phenomena (Chong et al. 2010). The continuous release of different types of contaminants such as heavy metals and organic dyes into clean water systems is causing a growing concern to many countries (Zeng et al. 2013). Heavy metals have a density greater than 5 g/mL and atomic weights of between 63.5–200.6 g/mol. Heavy metals are particularly problematic because, unlike most organic contaminants, they are non-biodegradable and can accumulate in living tissue, thus, posing a serious threat to both human health and the ecological environment.

The most common heavy metals include mercury, cadmium, lead, chromium, arsenic, zinc, copper, nickel and cobalt. These metal ions are toxic and cause serious side effects toward human health if ingested. For example, copper has universally been considered to be very toxic at high concentrations causes copper poisoning in humans such as gastrointestinal problems, kidney damage, hair loss, nausea, anemia, hypoglycemia, severe headaches and even death (Rahman and Islam 2009). Cadmium is a toxic heavy metal of significant environmental and occupational concern. It has been identified as a human carcinogen and teratogen substance severely impacting lungs, kidneys, liver and reproductive organs (Waalkes 2000).

Chromium is widely used in mining and pigment industries. Both Cr(III) and Cr(VI) exist in the environment. Cr(III) is highly toxic to humans even at trace levels and the Environmental Protection Agency (EPA) set the permissible limits for Cr(III) as 0.1 mg/L in drinking water (National Primary Drinking Water Regulation, US). However, Cr(VI) is five hundred times more toxic than Cr(III) (Selvi 2001) as well as being extremely mobile in the environment and is a proven carcinogen to living organisms (Depault et al. 2006).

Pb(II) is mostly associated with aqueous effluents and the EPA has set the permissible limits for Pb(II) at 0.015 mg/L in drinking water (Lingamdinne et al. 2015). The consumption of heavy metal ions leads to various diseases, such as skin irritation, lung cancer, renal abnormalities, DNA damage, kidney and liver failure, allergic reactions, gastric damage and reduced fertility. Consequently, the need for removal of heavy metals has become a matter of extreme importance for a healthy livable environment.

Copper is an important strategic raw material for domestic and international economic development and together with cobalt is widely applied in alloy manufacturing and in aerospace materials. Co(II) and Cu(II) cause acute and chronic poisoning, with adverse effects organs such as the kidney, liver and heart. These metals also cause asthma and degradation in the functioning of the thyroid gland and damages both the vascular and immune systems (Boyd 2010).

To date, various methods have been proposed for efficient heavy metal removal from waters, including but not limited to coagulation, chemical precipitation, membrane filtration, reverse osmosis, solvent extraction, flotation, ion exchange, electrochemical treatments, evaporation, oxidation, adsorption and biosorption (Hua et al. 2012; Fu and Wang 2011; Wang et al. 2003; Ali 2012). In spite of the successes of these applications certain drawbacks have been encountered such as precipitation or coagulation demands a variety of chemicals and includes high sludge volumes and filtration or electrochemical technologies need relatively large capital investments and electricity supplies (Kurniawan et al. 2006). Among the aforementioned techniques, adsorption/ion exchange has been demonstrated to be a most important method for the removal of heavy metal ions due to great removal performance, simple and easy process, cost effectiveness and the considerable choice of adsorbent materials.

In the last two decades, nanotechnology has developed with its applications in almost all branches of science and technology (Kaur and Gupta 2009; Savage and Diallo 2005). With the rapid development of nanotechnology, there has been a great deal of interest in environmental applications of nano materials. Nano materials are excellent adsorbents and catalysts (Khin 2012). Since nanomaterials offer significant improvement with extremely high specific surface area, numerous associated sorption sites, low temperature modification, short intraparticle diffusion distance, tunable pore size and surface chemistry compared to other materials. Extensive research has been carried out to remove heavy metals from wastewater by developing and using various nanomaterials. The adsorption capacity and the desorption property are two key parameters to evaluate an adsorbent. The adsorption parameters, such as the amount of absorbent used, temperature, pH, ionic strength, metal ion concentration and competition among metal ions, are often studied and optimized.

This chapter highlights recent developments for the removal of heavy metals by various nanomaterials, mainly including carbon-based nanomaterials, iron-based nanomaterials and photocatalytic nanomaterials in batch and flow systems. Finally, future perspectives are offered to inspire more exciting developments in this promising field.

Removal of heavy metals using various nanomaterials

Carbon based nano-adsorbents

Carbon nanotubes

Carbon nanotubes (CNTs) are known as one of the allotropes of carbon and are fundamentally constituted of cylindrical shape rolled up tube-like structures. The single walled carbon nanotubes (SWCNTs) and multi walled carbon nanotubes (MWCNTs) are two types of CNTs where single walled carbon nanotubes are composed of single graphene sheets rolled up tubes and multi walled carbon nanotubes are comprised of multiple graphene rolled up sheets (Zhao and Stoddart 2009).

CNTs have been widely applied in the removal of heavy metal ions from aqueous solutions (Mubarak et al. 2013; Upadhyayula et al. 2009). Due to their high surface active site to volume ratio, light mass density, high porous and hallow structure, controlled pore size distribution, CNTs have an exceptional sorption capability and high sorption efficiency compared to conventional granular and powder activated carbon (Santhosh 2016).

In recent years, CNTs have been used to remove cadmium (Ihsanullah et al. 2015a; Vuković et al. 2010), chromium (Ihsanullah et al. 2015b; Gu et al. 2013), lead (Shao et al. 2010), nickel (Yu et al. 2013), copper (Ge et al. 2014; Ren et al. 2013; Li et al. 2010), mercury (Shadbad et al. 2011; Tawabini et al. 2010), arsenic (Tawabini

Table 1. Application of carbon nanotubes (CNTs), and modified carbon nanotubes as adsorbents.

Adsorbent	Target contaminant	Adsorption capacity (mg/g)	References
Oxidized MWCNT/SDBS	Pb(II)	66.95	Li et al. 2011
Diethylenetriamine MWCNT	Pb(II), Cd(II)	58.26 for Pb(II) and 31.45 for Cd(II)	Vuković et al. 2011
Functionalized MWCNTs	As(III)	109.5	Mishra and Ramaprabhu 2010
MWCNT/Fe$_3$O$_4$ based electrodes	As(III), As(V)	39 for As(III), 53 for As(V)	Mishra and Ramaprabhu 2012
Purified CNTs	Zn(II)	43.66	Lu and Chiu 2006a
Modified MWCNTs	Zn(II)	32.68	Lu and Chiu 2008a
CNT sheets	Zn(II)	74.63	Tofighy and Mohammadi 2011
MWCNTs/polyacrylamide composites	Pb(II)	37.44	Yang et al. 2011
Functionalized MWCNTs based electrodes	As(III)	109.457	Mishra and Ramaprabhu 2012
Oxidized CNT sheets	Pb(II), Cd(II), Co(II), Zn(II), Cu(II)	117.65 for Pb(II), 92.59 for Cd(II), 85.74 for Co(II), 74.63 for Zn(II), 64.93 for Cu(II)	Tofighy and Mohammadi 2011
CNTs	Pb(II), Cu(II), Cd(II), Hg(II)	1.406 for Pb(II), 1.219 for Cu(II), 1.291 for Cd(II), 1.068 for Hg(II)	Anitha et al. 2015
CNT-OH	Pb(II), Cu(II), Cd(II), Hg(II)	2.07 for Pb(II), 1.342 for Cu(II), 1.513 for Cd(II), 1.284 for Hg(II)	Anitha et al. 2015
CNT-CONH$_2$	Pb(II), Cu(II), Cd(II), Hg(II)	1.907 for Pb(II), 1.755 for Cu(II), 1.563 for Cd(II), 1.658 for Hg(II)	Anitha et al. 2015
CNT-COO$^-$	Pb(II), Cu(II), Cd(II), Hg(II)	4.672 for Pb(II), 3.565 for Cu(II), 3.325 for Cd(II), 3.300 for Hg(II)	Anitha et al. 2015
CNTs	Pb(II)	17.44	Stafiej and Pyrzynska 2008
CNTs (HNO$_3$)	Pb(II)	49.95	Li et al. 2002
MWCNTs	Ni(II)	7.53	Lu and Liu 2006b
SWCNTs	Ni(II)	9.22	Lu and Liu 2006b
MWCNTs (HNO$_3$)	Pb(II)	97.08	Li et al. 2003
NaOCl-MWCNTs	Ni(II)	38.46	Lu et al. 2008b
NaOCl-SWCNTs	Ni(II)	47.85	Lu et al. 2008b
CS/PVA/MWCNTNH$_2$	Cu(II)	20.1	Salehi et al. 2012
CS–MWNT–PAA–PADPA/FG	Cr(VI)	2000.0	Kim et al. 2015
CPMP	Cu(II)	35.1	Salehi et al. 2013

CS–MWNT–PAA–PADPA/FG: Chitosan based functional gel, comprising multiwall carbon nanotube-poly(acrylicacid)-poly(4-aminodiphenylamine); CPMP: chitosan/polyvinyl alcohol thin adsorptive membranes containing combined MWCNT-NH$_2$ and PEG.

et al. 2011; Addo Ntim et al. 2011), zinc (Shin et al. 2011; Vellaichamy and Palanivelu 2011) and cobalt (Wang et al. 2011; Chen et al. 2012; Gupta et al. 2015).

The studies showed that the adsorption capacity of CNTs depends on the surface functional groups, their purity, porosity, surface area, site density, type of CNTs and the nature of the sorbate. The adsorption behaviors of CNTs mainly involve chemical and physical interaction between adsorbent and heavy metal ions on possible surface sites including internal sites, interstitial channels (ICs), grooves and outside surface. The maximum removal efficiency of CNTs was reported at pH 7–10 and the ionization and competition between ions could occur in this pH range (Table 1). CNTs are expensive but their adsorption and desorption cycles are more efficient compared to conventional activated carbon. Various carbon nanotubes (CNTs), and modified carbon nanotubes as adsorbents for the removal of different heavy metals and their adsorption capacity is summarized in Table 1.

The adsorption of Zn(II) was investigated using purified SWCNTs and MWCNTs (Lu and Chiu 2006a). The maximum adsorption capacity of Zn(II) onto CNTs was obtained in the pH range of 8–11. A comparative study on the adsorption of Zn(II) showed that the maximum adsorption capacities calculated by the Langmuir model were 43.66, 32.68 and 13.04 mg/g with SWCNTs, MWCNTs and commercial powdered activated carbon (PAC), respectively.

Removal of Cr(III) was studied on an absorbent by combining the magnetic properties of iron oxide with the adsorption properties of MWCNTs. Acid treatment (HNO_3) of MWCNTs provided the additional adsorbing sites by the oxygen atoms of iron oxide nanoparticles on the surface of MWCNTs which enhanced the removal capacity for Cr(III).

The composite was effective for Cr(III) removal in the pH range of 3.0–7.0 due to the presence of $Cr(OH)^{2+}$ species of chromium at pH 4.0–7.0. Also, at pH values of 4.0–7.0, the net negative surface charge on the former allowed increased adsorption of chromium species on MWCNTs. The results of the fixed bed experiments revealed that lower flow rates favoured Cr(III) removal due to the increased contact time between Cr(III) and adsorbent. Increase in the number of fixed bed layers revealed an increase in Cr(III) uptake which was attributed to the availability of more adsorption sites. Several researchers have also modified the carbon nanotubes to evaluate the efficiency of the former with the unmodified CNTs for the removal of various contaminants.

Cyclodextrin/MWCNT/iron oxide was synthesized by a novel low-temperature plasma technique to graft β-cyclodextrin (β-CD) on the surfaces of magnetic MWCNT/iron oxide particles that exhibited high saturation magnetization and good physicochemical stability in solution (Dong et al. 2014). The surface-coated β-CD improved the dispersion property of CD/MWCNT/iron oxides and therefore, increased their removal capacity for Ni(II). X-ray absorption fine structure (XAFS) analysis suggests that Ni(II) can bind on the hydroxyl sites on the surface-coated β-CD and also the FeO_6 octahedra of iron oxides in an edge-shared mode, forming strong inner-sphere complexes with high thermodynamic stability. The sorption kinetics of Ni(II) on CD/MWCNT/iron oxides achieved equilibrium within 4 h and maximum adsorption capacity 38.24 mg/g for Ni(II) was calculated by the Langmuir model at pH 6.5 and 25°C.

Multi-walled carbon nanotubes (MWCNTs) modified with 8-hydroxyquinoline were used for the removal of heavy metal ions such as Cu(II), Pb(II), Cd(II) and Zn(II) from aqueous solutions (Kosaa et al. 2012). Results showed that most of the metals were removed from aqueous solution using 250 mg of MWCNTs at pH 7.0 and 25°C in 0.01 M KNO_3 after 10 min of adsorption. The maximum adsorption capacities were calculated at 0.080 mg/g for Cu(II), 0.064 mg/g for Pb(II), 0.011 mg/g for Cd(II) and 0.063 mg/g for Zn(II) when pristine MWCNTs (10–20) were used, and 0.080 mg/g for Cu(II), 0.076 mg/g for Pb(II), 0.032 mg/g for Cd(II) and 0.075 mg/g for Zn(II), when 8-HQ-MWCNTs were applied. These values were lower than many others reported for carbon nanotubes (Rao et al. 2007) which was attributed to the lower specific surface area of the pristine MWCNTs (10–20) (69.1 m^2/g) and 8-HQ-MWCNTs (76.2 m^2/g) compared with carbon nanotubes described in previous reports. The results showed that the competition between the target heavy metals was in the order of Cu(II) > Pb(II) $\approx$ Zn(II) > Cd(II) for % adsorption. The MWCNTs could be reused for up to three cycles of adsorption/desorption without losing efficiency. Both pristine MWCNTs (10–20) and 8-HQ-MWCNTs were capable of removing heavy metals from two real samples collected from the Red Sea and a wastewater treatment plant.

Carbon NanoTubes (CNTs) have shown higher efficiency than activated carbon on adsorption of various organic chemicals (Pan and Xing 2008a). CNTs have shown high adsorption capacity for removal of both organic and inorganic pollutants from the liquid phase because of the large specific surface area and the diverse contaminant-CNT interactions via electrostatic bonding and van der Waals forces (Tofighy and Mohammadi 2011). The effective surface area of CNTs reduced in the aqueous medium due to the hydrophobicity of the graphitic surface and formation of loose bundles/aggregates in the aqueous phase. These bundles/aggregates contain significant interstitial spaces, pores and grooves which are high potential adsorption sites for bulky organic molecules such as sulfamethoxazole, tetracycline, and tylosin (Pan et al. 2008b; Ji et al. 2009).

Oxidized CNTs have shown a high capacity for adsorption of various metal cations with fast kinetics. Its high adsorption capacity mainly stems from the surface functional groups (e.g., carboxyl, hydroxyl and phenol) of CNTs which are the major adsorption sites for metal cations, mainly through electrostatic attraction and chemical bonding (Rao et al. 2007). Therefore, surface oxidation can significantly increase the adsorption capacity of CNTs. Based on the results of several studies, CNTs are better adsorbents than activated carbon for heavy divalent metals such as Cu^{2+}, Pb^{2+}, Cd^{2+} and Zn^{2+} (Li et al. 2003; Lu et al. 2006c). The adsorption kinetics are fast on CNTs owing to the highly accessible adsorption sites and the short intraparticle diffusion distances.

Carbon nanotube (CNT) sheets were prepared by chemical vapour deposition of cyclohexanol and ferrocenein nitrogen atmosphere at 750°C and oxidized with concentrated nitric acid at room temperature and then employed as adsorbent for removal of heavy metal ions such as Cu(II), Zn(II), Pb(II), Cd(II) and Co(II) from aqueous solutions. The results showed that kinetics of adsorption varies with initial concentration of heavy metal ions and the preference of adsorption onto the oxidized CNT sheets was ordered as Pb(II) > Cd(II) > Co(II) > Zn(II) > Cu(II) (Tofighy and Mohammadi 2011). The maximum adsorption capacities calculated using the Langmuir isotherm were 117.65, 92.59, 85.74, 74.63 and 64.93 mg/g for Pb(II), Cd(II), Co(II), Zn(II) and Cu(II), respectively.

Amino functionalized multi-walled carbon nanotubes (MWCNT-NH$_2$) were synthesized from raw MWCNTs and utilized to prepare novel chitosan/PVA thin adsorptive membranes for copper ion removal from water (Salehi et al. 2012). Finger-like nanochannels generated in the compact structure of the CS/PVA membrane by the addition of 0.5 wt. % of the MWCNTs. However, fibril-shaped and dense structures were observed at 1 and 2 wt. %, respectively. Adsorption capacity of the membranes containing 2 wt. % MWCNTs was 20.1 mg/g and for the plain membrane 11.1 mg/g at 40°C. However, the adsorption capacity showed no significant increase when the MWCNTs' content was increased from 1 to 2 wt. % (< 3 mg/g). However, kinetic studies showed that membranes with higher MWCNTs' content could support faster adsorption rates. Thermodynamic studies revealed that the adsorption process is spontaneous and endothermic in nature with an entropy generation at the solid-liquid interface. A slight adsorption capacity loss (~ 3%) was observed for the membrane containing MWCNT-NH$_2$ in comparison to the capacity loss of the plain membrane (~ 10%) after four successive adsorption/regeneration cycles.

Overall, CNTs may not be a good alternative for activated carbon as wide-spectrum adsorbents. Rather, as their surface chemistry can be tuned to target specific contaminants, they may have unique applications in polishing steps to remove recalcitrant compounds or in pre-concentration of trace organic contaminants for analytical purposes. These applications require small quantities of materials and hence, are less sensitive to the material cost.

Of the various nanomaterials based adsorbents, carbon based materials have emerged as superior adsorbents for removal of inorganic and organic pollutants. Since the discovery of carbon nanotubes (CNTs) and fullerene, these materials have been extensively used as effective adsorbents but their large scale application is limited on economic grounds and hence, designing the adsorbents at a lower cost remains a great challenge. Multiwalled carbon nanotubes show considerable removal efficiency of inorganic metal ions with the help of magnetic nanomaterials (Yang et al. 2006).

Graphene, graphene oxide and reduced graphene oxide

Graphene (G) is a single carbon layer with a graphite structure ordered in a honeycomb network structure and often shows excellent thermal and electrical conductivity (Zhang et al. 2016). Graphene oxide (GO) is an oxidative form of graphene which comprises a variety of oxygen-containing functional groups including hydroxyl, carboxyl, carbonyl and epoxy groups. Reduced graphene oxide (rGO) is more defective and less conductive than graphene but is relatively easy to be modified by other functional groups (Avouris and Dimitrakopoulos 2012). rGO is poorly water-soluble with good conductivity while GO has low conductivity with high solubility in water (Park et al. 2009).

Graphene based materials have been widely applied in environmental remediation because graphene (G) and graphene oxide (GO) have large surface areas, rich π-bonds and oxygen-containing functional groups. In general, graphene based materials can be used as adsorbents for decontamination of heavy metals from water via five possible interactions including hydrophobic effect, π-π bonds, hydrogen bonds, covalent and electrostatic interactions (Zhu et al. 2010).

Graphene based-nanomaterials were continuously reported as adsorbents for their capability to remove heavy metals such as Pb(II), Zn(II), Cu(II), Cd(II), Hg(II) and As(III/V). The effect of pH, adsorbent dosage, foreign ions, temperature, contact time and adsorption isotherm on the adsorption of Zn(II) on GO was studied and the results showed that the maximum adsorption capacity for Zn(II) was up to 246 mg/g based on the Langmuir model at pH around 7.0 within a short equilibrium time (Wang et al. 2013).

The adsorption capacity for Pb(II) could be increased to 479 ± 46 mg/g at pH 6.8 when GO was linked with EDTA due to the added chelation ability of EDTA (Madadrang et al. 2012). An ideal adsorbent should possess two properties such as a higher adsorption capacity along with a better desorption property, which will significantly reduce the overall cost for the adsorbents. The results showed that Pb(II) desorption for EDTA-GO absorbent increased with decreasing pH values and reached about 90% at pH < 2.0.

The reduced graphene oxide/poly(acrylamide) (rGO/PAM) composite was prepared and its adsorption was investigated for Pb(II) by Yang and coworkers (Yang et al. 2013). The rGO/PAM showed a very high adsorption capacity of up to 1000 mg/g for Pb(II).

The hybrids of monolayer GO with manganese ferrite magnetic nanoparticles were applied for efficient decontamination of Pb(II), As(III) and As(V) from water. The maximum capacity of 673 mg/g for Pb(II), 146 mg/g for As(III) and 207 mg/g for As(V) were obtained (Kumar et al. 2014). This absorbent was promising for co-removal of multiple heavy metals or metalloids from water due to its high capacity and ease of magnetic separation.

The aggregation problem of rGO could be minimized or prevented by the incorporation of nanoparticles into rGO (Wei et al. 2012) and these nanocomposites show superior properties in comparison with bare nanoparticles (Song et al. 2011). A simple solvothermal strategy was reported to prepare $rGO-Fe_3O_4$ nanocomposite for removal of Cr(VI) (Zhou et al. 2013). The size of $rGO-Fe_3O_4$ hybrid particles was larger than 100 nm which can reduce cell toxicity in water treatment processes. The results revealed that the $rGO-Fe_3O_4$ composite with the highest loading of Fe_3O_4 showed the fastest removal of 500 g/L Cr(VI) which could reach 85% within 5 min at neutral pH. Although, the large saturation magnetization of $rGO-Fe_3O_4$ non-nanoparticles allowed fast separation of the adsorbent from water, the removal efficiency of Cr(VI) strongly decreased when common hazardous ions such as Cu(II), Zn(II), Pb(II) and As(III) were added in water.

Magnetite-rGO (M-rGO) composites were applied for the removal of As(III) and As(V) from water (Chandra et al. 2010). The composites showed > 99.9% arsenic removal within 1 ppb.

Fan et al. 2013 prepared magnetic chitosan/graphene oxide (MCGO) materials through a facile and fast process and demonstrated their ability as excellent adsorbents for metal ions (Fan et al. 2013). The SEM and TEM showed that magnetic chitosan had been assembled on the surface of graphene oxide layers with a high density. The XRD and VSM revealed the MCGO had enough magnetic response to meet the need of magnetic separation. The results indicated that adsorption of Pb(II) onto MCGO was mainly dependent on pH and the abundant functional groups on the surfaces of MCGO played an important role on Pb(II) removal. The maximum adsorption capacity for Pb(II) was estimated to be 76.94 mg/g based on the Langmuir model. The MCGO was stable and easily recovered.

A simple chemical bonding method to synthesize magnetic cyclodextrin-chitosan/graphene oxide (CCGO) was reported by Luo and coworkers (Li et al. 2013). The adsorption behaviors of Cr(VI) in an aqueous solution on CCGO were systematically investigated. Due to the high surface area, abundant hydroxyl and amino groups of CCGO, and the magnetic property of Fe_3O_4, the Cr(VI) was easily and rapidly adsorbed from the water. The adsorbent exhibited better Cr(VI) removal efficiency in solutions at low pH and it was demonstarted that the Cr(VI) adsorption performance of CCGO strongly depends on their surface charge concentration and specific surface area. Chitosan contains two types of reactive functional groups, that is, amino groups and hydroxyl groups, which act as chelation sites. Cyclodextrin can enable them to bind metal ions into their cavities to form stable host-guest inclusion complexes. Strong surface complexation between the graphene oxide and metal ions occurs through the Lewis acid base interaction, which also contributes to metal ion sorption on graphene oxide nanosheets. The maximum adsorption capacity of Cr(VI) on CCGO was 67.66 mg/g.

A one-pot simple method was developed to produce the rGO-Fe_3O_4 hybrid nanocomposite for removal of Pb(II) and the maximum adsorption capacities for Pb(II) were 30.68 mg/g (Cao ct al. 2015).

Sitko et al. 2016 prepared two types of pressed and non-pressed graphene oxide/cellulose membranes in order to perform effective adsorption of heavy metal ions such as Co(II), Ni(II), Cu(II), Zn(II), Cd(II) and Pb(II). The pressed membranes were highly durable at different pH values and were applied in the separation/removal of heavy metal ions during vigorous shaking in an aqueous solution. The non-pressed membranes were less stable however but were successfully applied in the filtration process at high flow-rates. Results revealed that the maximum adsorption was achieved at pH 4–8. Adsorption isotherms and kinetic studies indicated that the sorption of the metal ions on the membranes is a chemical adsorption. The maximum adsorption capacity values of Co(II), Ni(II), Cu(II), Zn(II), Cd(II) and Pb(II) on the graphene oxide/cellulose membranes at pH 4.5 were 15.5, 14.3, 26.6, 16.7, 26.8, 107.9 mg/g, respectively and the affinities of prepared membranes for the metal ions were in the order of Pb > Cu > Cd > Zn ≥ Ni ≥ Co.

Al Nafiey et al. 2017 reported a one-step synthesis of reduced graphene oxide-cobalt oxide nanoparticles (rGO-Co_3O_4) nanocomposite under mild conditions (Al Nafiey et al. 2017). A maximum adsorption capacity of 208.8 mg/g was obtained for the removal of Cr(VI). An effective separation and recyclability of the material was achieved by the simple application of an external magnet due to the ferromagnetic properties of the nanocomposite.

Zhang and coworkers reported a room-temperature approach to synthesizing reduced graphene oxide/NiO (rGO/NiO) nanocomposites (Zhang et al. 2018). The

Table 2. Application of graphene (G), graphene nanosheets (GNs), graphene oxide (GO), reduced graphene oxide (rGO) and modified graphene as adsorbents.

Adsorbent	Target contaminant	Adsorption capacity (mg/g)	References
GO	Pb(II)	328	Madadrang et al. 2012
EDTA-rGO		204	
EDTA-GO		479 ± 46	
GO	Au(III), Pd(II), Pt(IV)	108.342 for Au(III), 80.775 for Pd(II), 71.378 Pt(IV)	Liu et al. 2013
GO	Zn(II)	246	Wang et al. 2013
rGO/PAM	Pb(II)	1000	Yang et al. 2013
PVP-rGO	Cu(II)	1689	Zhang et al. 2014a
GO-MnFe$_2$O$_4$	Pb(II), As(III), As(V)	673 for Pb(II), 146 for As(III) and 207 for As(V)	Kumar et al. 2014
Few layered GO	Pb(II)	824 (20°C), 1150 (40°C), 1850 (60°C)	Zhao et al. 2011
MCGO	Pb(II)	76.94	Fan et al. 2013
GNs	Pb(II)	35.5	Huang et al. 2011
GO/chitosan	Pb(II)	99	He et al. 2011
GO−gelatin/chitosan	Pb(II)	100	Zhang et al. 2011
Functionalized GNs based electrodes	As(III), As(V)	138.8 for As(III), 142 for As(V)	Mishra et al. 2011
M-rGO	As(III), As(V)	13.10 for As(III), 5.83 for As(V)	Chandra et al. 2010
GO/Fe$_3$O$_4$	Cu(II)	18.3	Li et al. 2012
GO/silica/Fe$_3$O$_4$	Pb(II), Cd(II)	333.3 for Pb(II), 166.7 for Cd(II)	Wang et al. 2013
GO/Fe$_3$O$_4$/sulfanilic acid	Cd(II)	55.4	Hu et al. 2014
GO/Fe$_3$O$_4$/sulfanilic acid	Cu(II)	50.7 and 56.8	Hu et al. 2013
rGO/CoFe$_2$O$_4$	Pb(II)	299.4	Zhang et al. 2014b
Graphene/Mn-doped Fe(III) oxide	Cd(II), Cu(II)	87.2 for Cd(II), 129.7 for Cu(II)	Nandi et al. 2013
GO/Fe$_3$O$_4$	Cu(II), Pb(II), Cd(II)	23.1 for Cu(II), 38.5 for Pb(II), 4.4 for Cd(II)	Hur et al. 2015
MnO$_2$/GNs	Hg(II)	10.8	Sreeprasad et al. 2011
SiO$_2$−GNs	Pb(II)	113.6	Hao et al. 2012
GNs	Pb(II)	22.4	Huang et al. 2011
GNs-500	Pb(II)	35.2	Huang et al. 2011
TiO$_2$/GO	Zn(II), Cd(II), Pb(II)	88.9 ± 3.3 for Zn(II), 72.8 ± 1.6 for Cd(II), 65.6 ± 2.7 for Pb(II)	Lee and Yang 2012

Table 2 contd. ...

...Table 2 contd.

Adsorbent	Target contaminant	Adsorption capacity (mg/g)	Ref.
Colloidal GO	Zn(II), Cd(II), Pb(II)	30.1 ± 2.5 for Zn(II), 14.9 ± 1.5 for Cd(II), and 35.6 ± 1.3 for Pb(II)	Lee and Yang 2012
GO/cellulose	Pb(II), Cd(II)	107.9 for Pb(II), 26.8 for Cd(II)	Sitko et al. 2016
CCGO	Cr(VI)	67.66	Li et al. 2013
MCGO-IL	Cr(VI)	145.35	Li et al. 2014a
CS/GO-SH	Cu(II), Pb(II), CD(II)	425.0 for Cu(II), 447.0 for Pb(II), 177.0 for Cd(II)	Li et al. 2015
MCts-DETA	U(VI)	177.93	Mahfouz et al. 2015
MCGS	Hg(II)	361.0	Zhang et al. 2014c
Magnetic EDTA-modified chitosan/SiO$_2$/Fe$_3$O$_4$			
β-cyclodextrin/GO	Cr(VI)	103.4	Fan et al. 2012
Fe$_3$O$_4$@TiO$_2$@GO	Cr(VI)	117.94	Li et al. 2014b
ED-rGO	Cr(VI)	80	Ma et al. 2012
ED-DMF-rGO	Cr(VI)	92.15	Zhang et al. 2013
Fe$_3$O$_4$/GO	Cr(VI)	32.33	Liu et al. 2013
Fe$_3$O$_4$/GS	Cr(VI)	17.29	Guo et al. 2014
rGO/NiO	Cr(VI)	198	Zhang et al. 2018

MCGO-IL: magnetic ionic liquid/chitosan/GO composite; CS/GO-SH: chitosan/sulfydryl-functionalized graphene oxide composite; MCts-DETA: magnetic nano-based particles of diethylenetriamine-functionalized chitosan; MCGS: amino-functionalized magnetic composite of CoFe$_2$O$_4$–chitosan–graphene composite; MCGO: magnetic chitosan/graphene oxide composites.

maximum adsorption capacity of Cr(VI) over rGO/NiO nanocomposites at pH 4.0 and 25°C was reported 198 mg/g based on the Langmuir isotherm model. Table 2 summarizes the maximum adsorption capacity of different heavy metals using various graphene based materials.

Based on the principle of "using waste to treat waste" and "waste resource recovery", Al$_2$O$_3$/C composites were prepared from two wastewater facilities such as the alkaline wastewater of an oil refinery and the acid wastewater from an aluminum anodizing factory (Chen et al. 2018). The maximum adsorption capacity of S-600 (Al$_2$O$_3$/C composites with calcinating temperature 600°C) was calculated to be 709.2 mg/g for Pb(II) by the Langmuir model and 1299.4 mg/g for Cd(II) by the Freundlich model.

Synthesis of a hybrid graphene oxide based inverse spinel nickel ferrite (GONF) nano-composite material and its applications in heavy metal removal from an aqueous solution was developed (Lingamdinne et al. 2016a). GONF was successfully used for the removal of Pb(II) and Cr(III) by batch adsorption techniques. The maximum sorption capacity of GONF were calculated to be 25.0 mg/g for Pb(II) at pH 5.5

and 45.5 mg/g for Cr(III) at pH 4.0 and 25 ± 2°C by the Langmuir isotherm model which revealed that the sorption of Pb(II) and Cr(III) onto GONF occurred through monolayer chemisorptions on the homogeneous surface of GONF. The maximum adsorption capacity of metal ions was enhanced with increasing temperature, which was evident for endothermic chemisorption by inner-sphere surface complexation. The GONF was regenerated and reused for up to three cycles.

A flower-like TiO_2-graphene oxide (GO-TiO_2) hybrid was synthesized through the hydrothermal method by stirring a titanium oxide precursor in isopropyl alcohol with a graphene oxide colloidal solution (Lee and Yang 2012). The maximum removal capacities for the GO-TiO_2 hybrid, after 12 h of hydrothermal treatment at 100°C and pH 5.6, were 88.9 ± 3.3 mg/g for Zn(II), 72.8 ± 1.6 mg/g for Cd(II) and 65.6 ± 2.7 mg/g for Pb(II). Under identical conditions, the colloidal GO showed removal capacities of 30.1 ± 2.5 mg/g for Zn(II), 14.9 ± 1.5 for mg/g Cd(II) and 35.6 ± 1.3 mg/g for Pb(II).

The graphene-cobalt oxide nanoparticles (G-Co_3O_4) nanocomposite was successfully utilized for the retention/separation of Pb(II), Cu(II) and Fe(III) ions from environmental water and food samples prior to flame atomic absorption detection (Yavuz et al. 2013). The maximum adsorption capacity of the G/Co_3O_4 composite was reported to be 58, 77 and 78 mg/g for Pb(II), Cu(II) and Fe(III), respectively. This method was used for the separation/preconcentration of trace heavy metal ions in a variety of water samples including tap water, wastewater, dam water, well water, kiwi and wheat samples.

Amino-functionalized magnetic composites of $CoFe_2O_4$-chitosan-graphene (MCGS) have been successfully synthesized by Zhang and coworkers (Zhang et al. 2014). Results showed that optimal adsorption efficiency reached at pH 7.0 for the Hg(II) adsorption kinetics on MCGS was consistent with the pseudo second-order process as well as that the Langmuir model fitted better with a maximum adsorption capacity of 361.0 mg/g at pH 7.0 and 50°C. Thermodynamic studies illustrated that the adsorption process was endothermic and spontaneous.

Recently, reduced graphene oxide-cobalt oxide nanoparticles (rGO-Co_3O_4) nanocomposite was synthesized in a one-step protocol under mild conditions (Al Nafiey et al. 2017). A maximum adsorption capacity of 208.8 mg/g for Cr(VI) removal was reported for this nanocomposite which was higher than that obtained using many other magnetic adsorbents. Owing to its ferromagnetic properties, this nanocomposite was simply separated and recycled by the application of an external magnet.

Carbon nanotubes (CNTs) modified with a fourth generation poly-amidoamine dendrimer (PAMAM, G4) was used to remove Cu(II) and Pb(II) heavy metals from an aqueous solution in single and binary component systems (Hayati et al. 2017). The high adsorption capacities such as 3333 for Cu(II) and 4870 mg/g for Pb(II) were obtained at pH 7.0. Results showed that the adsorption process for Cu(II) and Pb(II) in single and binary component systems follow the Langmuir and extended Langmuir models, respectively, and PAMAM/CNTs act as super-adsorbents. PAMAM/CNTs with numerous specific surface and amino terminal groups can adsorb heavy metal ions by chelating, encapsulating, hydrogen and van der Waals bonding (Zhang et al. 2014). By increasing the nanocomposite loading in industrial wastewater, the adsorption capacity for Cu(II) and Pb(II) is constantly being improved due to the increased nanocomposite surface area and accessibility of many adsorption sites and maximum removal was achived at a dose of 0.03 g/L and above. Desorption results showed that a maximum metal ion release of > 80% for both Cu(II) and Pb(II) was obtained in an aqueous solution at pH 2.0. By decreasing the pH of the system, the number of positively charged sites increased, which favoured the desorption of Cu(II) and Pb(II) as previously reported (Han et al. 2016). PAMAM/CNTs nanocomposites have advantages compared to other adsorbents such as high adsorption capacity, good dispersion, excellent porosity, high adsorption rate and recyclability.

Vilela and coworker reported graphene oxide-based microbots (GOx-microbots) as active self-propelled systems for the capture, transfer and removal of toxic heavy metals such as Pb(II) from contaminated water through an adsorption process (Vilela et al. 2016). The structure of microbots consists of nanolayers of graphene oxide (GOx), Pt/Ni layers, Ni magnetic layers and Pt catalytic inner layers, all of which provide different functionalities. Hydrogen peroxide decomposed by a platinum layer into water and oxygen microbubbles which provide enough force for the self-propulsion of GOx-microbots. After Pb(II) decontamination, GOx-microbots were easily removed from the water using a magnet due to the magnetic properties of the layers of Pt/Ni and Ni. Recovery of the adsorbed Pb(II) ions on GOx-microbots was performed via acid pH adjustment thus allowing them to be recycled and reused for further decontamination processes. Results showed that mobile GOx-microbots remove Pb(II) 10 times more efficiently than non-mobile GOx-microbots, decontaminating Pb(II) ions in water from 1000 ppb down to below 50 ppb during a period of 60 min.

Iron-based nanomaterials

Zero-valent iron nanomaterials

Nano zero valent iron (nZVI) was used as an effective tool for water remediation due to their large specific areas (Tosco et al. 2014). The advantages of nZVI are its high reactivity towards a broad range of contaminants and the possibility of it being injected in aqueous slurries for a targeted remediation of contaminated areas. However, the stability against aggregation, mobility in subsurface environments and longevity are important problems which should be addressed. Modification of the surface of the particles and viscosity of the dispersant fluid were found to be effective approaches in improving both colloidal stability and the mobility of nZVI.

nZVI particles can be synthesized by bottom-up processes such as the generation of iron nanoparticles from ions or smaller particles via nucleation, deposition, precipitation and agglomeration. Other factors such as size reduction of larger particles by milling and ablation also play a role (Crane and Scott 2012). Furthermore, it was demonstrated that ultrasound in the presence of sodium borohydride established positive results in reducing the nZVI particles size and modifying their shape (Jamei et al. 2014). Previously, studies showed strong particle-particle attractive interactions and that the dendritic aggregates observed in aqueous dispersions of nZVI are 10–100 times larger than primary particles (10–100 nm) (Tiraferri et al. 2008).

Oxidation is extremely fast in the first few days after synthesis; thus, fresh nZVI particles are mainly composed of Fe(0) with a thin layer of oxides which increases with time. After the formation of the oxide shell, the Fe(0) core is partly protected from undesired reactions and corrosion is slower (Yan et al. 2013). A variety of surface analysis techniques, including X-ray photoelectron spectroscopy (XPS), extended X-ray absorption spectroscopy (EXAFS), X-ray absorption near edge structure spectroscopy (XANES), X-ray diffraction (XRD) and scanning transmission electron microscopy (STEM) with energy dispersive X-ray (EDX), can be used to investigate the metal removal mechanisms associated with nZVI (O'Carroll et al. 2013).

Kanel et al. 2006 reported that by manipulating the size of ZVI from micron to nanoscale, the rate constant for As(V) removal was increased by 1–3 orders of magnitude owing to its drastically enhanced surface area and active sites in comparison with bulky ZVI. nZVI exhibited a high surface area and excellent magnetic and biocompatible properties (Xu et al. 2012).

EDX analysis demonstrated the presence of cadmium ions on the nZVI surface (Boparai et al. 2011). Results revealed that the adsorption process was chemisorption, endothermic and spontaneous in nature as well as the maximum adsorption capacity of nZVI for Cd(II) was found to be 769.2 mg/g at 24°C.

Yu et al. 2014 demonstrated that reduction of Cr(VI) onto nZVI particles depends mainly not only on the nZVI dosages but also the surface characteristics of nZVI particles caused by different synthetic methods. It was found that monitoring of the oxidation–reduction potential (ORP), dissolved oxygen (DO) and pH in the reactor and presented good correlations with the Cr(VI) removal efficiencies. The experimental results revealed that the Cr(VI) removal efficiencies could reach 99% when the proper nZVI dosages and contact times were used.

Both of the aforementioned results showed that heavy metal ions, such as Cd(II) and Cr(VI), formed inner-sphere complexes with the iron oxide layer surrounding the nZVI core.

Crane et al. 2011 reported that U(VI) was removed by nZVI to < 10 μg/L (> 98% removal) within 2 h. The partial chemical reduction of U(VI) to U(IV) concurrent with Fe oxidation was confirmed by X-ray photoelectron spectroscopy analysis. In contrast, nano-Fe_3O_4 failed to achieve > 20% U removal from the water sample. While the outer surface of both nano-Fe(0) and nano Fe_3O_4 was initially near-stoichiometric magnetite, the greater performance displayed by nano-Fe(0) was attributed to the presence of a Fe(0) core for enhanced aqueous reactivity, sufficient to achieve near-total removal of aqueous U despite any competing reactions within the carbonate-rich water sample.

A team of researchers evaluated the Chitosan zerovalent Iron Nanoparticle (CIN) for As(III/V) removal (Gupta et al. 2012). The stability of Fe(0) nanoparticles was increased in the presence of chitosan. The maximum adsorption capacity was found to be 94 ± 1.5 mg/g for As(III) and 119 ± 2.6 mg/g for As(V) at pH 7.0 based on the Langmuir monolayer model. Anions such as sulfate, phosphate and silicate cause no significant interference in the adsorption behavior of As(III) and As(V). The adsorbent was recycled five times and applied to the removal of total inorganic arsenic from real life groundwater samples.

Li et al. 2014 demonstrated that the average removal capacity of Cu(II) by nZVI was 343 mg/g and Cu(II) was reduced to metallic copper or cuprite (Cu_2O) after adsorption. Pilot experiments were performed at a printed-circuit board manufacturing plant, treating 250 000 L of wastewater containing 70 mg/L Cu(II) with a total of 55 kg of nZVI. The average Cu(II) removal efficiency was greater than 96% with 0.20 g/L nZVI and a hydraulic retention time of 100 min. The average removal capacity of nZVI for Cu(II) was 343 mg/g. The end product was a valuable composite of iron and copper (20–25%), which could partially offset the treatment costs.

Ramos et al. 2009 demonstrated that reduction to elemental arsenic by nZVI was an important process for arsenic immobilization. They presented clear evidence of As(0) species on nZVI surfaces after reactions with As(III) or As(V) species in solution using high-resolution X-ray photoelectron spectroscopy (HR-XPS) and it was proved that reduction to elemental arsenic by nZVI was an important mechanism for arsenic immobilization. Additionally, the formation of As(0), As(III) and As(V) on the nZVI surface after reaction indicated that both the reduction and oxidation of As(III) had occurred when reacting with nZVI. Therefore, nZVI could have performed the dual redox functions by its core-shell structure containing a metallic core with a highly reducing characteristic and a thin amorphous iron (oxy)hydroxide layer promoting As(III) coordination and oxidation.

Removal of heavy metal ions could be performed by reduction and subsequent precipitation or co-precipitation if the redox potential of heavy metal ions is significantly higher than that of Fe(0) and if it is significantly lower, the heavy metal ions can be eliminated through adsorption on the iron particles (O'Carroll et al. 2013).

The nZVI interactions with various heavy metal ions were categorized by O'Carroll et al. as:

1. Reduction—Cr, As, Cu, U, Pb, Ni, Se, Co, Pd, Pt, Hg, Ag.
2. Adsorption—Cr, As, U, Pb, Ni, Se, Co, Cd, Zn, Ba.
3. Oxidation/reoxidation—As, U, Se, Pb.
4. Co-precipitation—Cr, As, Ni, Se.
5. Precipitation—Cu, Pb, Cd, Co, Zn.

Furthermore, they reviewed some of the heavy metal ions that interacted with nZVI by more than one mechanism in detail (O'Carroll et al. 2013).

The factors such as nZVI chemical properties and structure, the presence of more than one contaminant species, pH, redox potential and natural dissolved species can be effective on the reaction processes and pathways occurring at the iron surface (Yan et al. 2013).

To sum up, nZVI is a highly reactive, cost-effective, environment-friendly material and exhibits multifaceted removal pathways for wastewater treatment.

Iron oxide nanomaterials

The synthesis of magnetite nanoparticles and their applications as nano-sorbents for water treatment has attracted considerable attention. Bare magnetite nanoparticles are easily aggregated in aqueous systems due to air oxidation (Maity and Agrawal 2007). Thus, surface modification needs to be performed for the stabilization of magnetite nanoparticles. The magnetic properties of nanoparticles and nanocrystals strongly depend upon the structure of the surface layer and the dimension of the nanoparticles (Zhang et al. 2006). Thus, only magnetite particles with a size of less than 30 nm have a large enough surface area allowing them to exhibit super paramagnetic properties making them prone to magnetic fields (Zhou et al. 2009).

Magnetic nanomaterials demonstrate their advantage as nano-sorbents over other nanocomposites for water treatment applications due to their large surface area, chemical stability, high adsorption capacity as well as being easily recovered and separated using an external magnetic force and thus, need neither filtration nor centrifugation (Li et al. 2013; Lingamdinne et al. 2016a; Xu et al. 2012; Lingamdinne et al. 2016b; Lv et al. 2009; Mahmoud et al. 2013; Mallakpour and Khadem 2016; Tu et al. 2015; Zhao et al. 2016). However, synthesis of these magnetic nanoparticles requires the use of chemicals which cause secondary pollution to the environment (Chen et al. 2011). In addition, the synthesis of these superparamagnetic nanoparticles is difficult owing to their colloidal nature (Laurent et al. 2008).

A variety of chemical methods have been developed to synthesize magnetic nanoparticles including co-precipitation, reverse micelles and micro-emulsion technology, sol-gel syntheses, sonochemical reactions, hydrothermal reactions, hydrolysis and thermolysis of precursors, flow injection syntheses and electrospray syntheses (Laurent et al. 2008; Teja and Koh 2009; Wu 2016; Ali et al. 2016).

For metal removal applications, surface modification of the magnetic nanoparticles plays an important role regarding the stability and selectivity of these materials in an aqueous medium. A variety of organic and inorganic functionalized magnetic nanoparticles have been reported in literature. Herein, we present a brief description of some methods more widely applied for preparing magnetic nanoparticles.

Co-precipitation

The low-cost, reasonably quick and simple chemical method to synthesize magnetic nanoparticles involves co-precipitation. A stoichiometric mixture of (Fe^{2+}/Fe^{3+}) and a base as precipitant are stirred in an aqueous medium at 70–90°C in an inert atmosphere and Fe_3O_4 is precipitated at a pH 8–14. Temperature, pH, Fe^{2+} and Fe^{3+} ratio, ratio of hydroxide ions to iron ions, the base addition rate, ionic strength and nature of salts are factors that control the size, shape and particle size distribution of the nanoparticles (Ali et al. 2017; Schwarzer et al. 2004). Particles with sizes ranging from 5–100 nm can be obtained using this method. The addition of chelating organic anions or polymer surface complexing agents during the formation of the magnetite can help to control the size of the nanoparticles (Wu 2016; Ali et al. 2016).

Reverse micelles and micro-emulsion technology

In this method, the nanometer sized aqueous cores of the reverse micelles establish an appropriate stabilized medium for the formation of nanoparticles of fairly uniform size, through chemical reactions occurring in the core as well as inhibit the aggregation of nanoparticles formed. The main advantage of the reverse micelle or emulsion technology is the diversity of nanoparticles that can be formed by varying the nature and amount of surfactant and co-surfactant, the oil phase or the reacting conditions. The size of the magnetite particle can be controlled by the temperature and the surfactant concentration (Deng et al. 2003).

Thermolysis of precursors

Organic solution-phase thermodecomposition of the iron precursor at temperatures > 200°C was used in the preparation of new types of iron oxide nanoparticles (Laurent et al. 2008). The control of the mean size, size distribution and crystallinity of the magnetic iron nanoparticles was improved by adjusting a variety of parameters such as solvent, reaction time and temperature. The monodispersed magnetite nanoparticles with size of 3 to 20 nm were synthesized by reaction of iron acetylacetonate, $Fe(acac)_3$, in phenyl ether in the presence of alcohol, oleic acid and oleylamine at high-temperature (Sun and Zeng 2002).

Hydrothermal reactions

The hydrothermal synthesis of Fe_3O_4 nanoparticles are performed in aqueous media in reactors or autoclaves at high pressures and temperatures (Daou et al. 2006; Mizutani et al. 2008). Formation of magnetite can be performed by hydrolysis and oxidation or neutralization of mixed metal hydroxides by the hydrothermal method in which the particle size can be controlled mainly by the nucleation and grain growth rates.

Magnetic nanocomposites

A novel sensitive and recyclable surface-enhanced Raman scattering (SERS) reagent was synthesized as uniform Fe_3O_4@Ag nanoparticles which adsorbed Cr(VI) in water and substantially enhanced the Raman signal (Du and Jing 2011). The closely spaced Fe_3O_4@Ag substrate with a core–shell structure exhibited a 25 nm surface roughness. The sensitivity and reproducibility of Fe_3O_4@Ag NPs was demonstrated by using rhodamine 6G. SERS spectra of Cr(VI) in simulated and real contaminated water showed that the symmetric stretching vibrations of Cr–O occurred at 796 cm^{-1} which showed a linear dependence ($R^2 = 0.9992$) on the Cr(VI) concentration between 5 and 100 μg/L. Coexisting anions such as sulfate, nitrate, chloride, carbonate and humic acid decreased the sensitivity of the SERS analysis which could be attributed to the competitive occupancy for the available surface sites of the SERS substrate. However, the adverse effect of the competing ions was eliminated by controlled dilution of the raw sample.

Ou et al. 2012 reported the synthesis of bifunctional Fe_3O_4@MgSiO$_3$ via a solvothermal method using Fe_3O_4@SiO$_2$ as a self-template with mean diameter of 220 nm which displayed a superparamagnetic property and high removal capacity for Pb(II). The maximum adsorption capacity of 242.1 mg/g for Pb(II) was calculated based on the Langmuir model. The Fe_3O_4@MgSiO$_3$ was regenerated effectively by a NaOH or Mg^{2+} solution.

Ren et al. developed a magnetic EDTA-modified chitosan/SiO_2/Fe_3O_4 (EDCMS) adsorbent for removal of heavy metal ions from aqueous solution (Ren et al. 2013). This absorbent was synthesized by surface modification of chitosan/SiO_2/ Fe_3O_4 (CMS) with EDTA using water-soluble carbodiimide as the cross-linker in a buffer solution. Results illustrated that the adsorption kinetics for EDCMS and CMS followed the mechanism of a pseudo-second order kinetic model, and the equilibrium data were mainly fitted with the Langmuir isothermal model. EDCMS had better tolerance at low pH and exhibited maximum adsorption capacities of 44.41 mg/g (0.699 mmol/g) for Cu(II), 123.5 mg/g (0.596 mmol/g) for Pb(II) and 63.28 mg/g (0.563 mmol/g) for Cd(II) at pH 5.0 and 25°C in comparison with CMS which showed maximum adsorption capacities of 31.45 mg/g (0.495 mmol/g) for Cu(II), 9.32 mg/g (0.045 mmol/g) for Pb(II), 4.5 mg/g (0.040 mmol/g) for Cd(II) ions under the same conditions. EDCMS showed about 25% loss in the adsorption capacity for heavy metal ions after being used 12 times.

A nanosorbent was synthesized using the Fe_3O_4 magnetic core-shelled by mesoporous silica and cetyltrimethylammonium bromide (CTAB) as surfactant template through a sol-gel process. The then obtained nanomagnetic material was further modified with bis(3-triethoxysilylpropyl)tetrasulfide (MSC MNPs-S_4) (Vojoudi et al. 2017). The potential of the resultant mesoporous magnetite nanomaterials was investigated as a convenient and effective adsorbent for the removal of toxic heavy metal ions from aqueous solutions in a batch system. The effect of essential parameters on the removal efficiency including initial pH of the solution, adsorbent amount, metal ion concentration, contact time and type and quantity of the eluent on the adsorption characteristics of the MSCMNPs-S4 were studied. Under optimized conditions, the proposed nanosorbent exhibited a high adsorption capacity of 303.03, 256.41 and 270.27 mg/g and maximum removal percentages of 98.8%, 96.4% and 95.7% for Hg(II), Pd(II) and Pb(II) ions, respectively. The mesoporous MSC MNPs-S4 magnetic adsorbent was then applied for the removal of metal ions in three real water samples such as tap water, well water and lake water. Results showed that heavy metal ions were removed with > 90% removal efficiency from the environmental water samples using this adsorbent.

A continuous method for the preparation of magnetic Fe_3O_4/Chitosan nanoparticles (Fe_3O_4/CS NPs) was developed and applied for the efficient removal of heavy metal ions from aqueous solution (Fana et al. 2017). Using a novel impinging stream-rotating packed bed, the continuous preparation of Fe_3O_4/CS NPs reached a theoretical production rate of 3.43 kg/h. The Fe_3O_4/CS NPs revealed better adsorption capacity and faster adsorption rates for Pb(II) and Cd(II) than those of pure Fe_3O_4 due to the strong metal chelating ability of chitosan. The maximum adsorption capacities of Fe_3O_4/CS NPs were calculated 79.24 mg/g for Pb(II) and 36.42 mg/g for Cd(II) and additionally, the Fe_3O_4/CS NPs showed excellent reusability after five adsorption-desorption cycles.

Lee et al. synthesized iron oxide@carbonate (IO@$CaCO_3$) magnetic adsorbent via a hydrothermal method which could remove both anionic [As(V) and Cr(VI)] and cationic [Pb(II)] heavy metal ions in a very short time, that is, the toxic heavy metal ions were completely removed from waste water in only 9 min (Islama et al. 2017). Removal capacities of 184.1, 251.6 and 1041.9 mg/g were reported for As(V),

Cr(VI) and Pb(II) over a wide pH range, respectively. The ion-exchange reaction between the cationic/anionic heavy metal ions and the positively/negatively charged groups on the adsorbent was proposed as the mechanism for heavy metal removal. This absorbent was also used by column filtration which exhibits the potential of IO@CaCO$_3$ for industrial applications. Moreover, this adsorbent was separated by an external magnetic field and was effectively reused. However, the removal efficiency decreased by 4.6% over the five runs.

Magnetized iron-oxide-impregnated Lonicera japonica flower biomass (IO-LJFP) was synthesized and utilized to remove Pb(II), Co(II) and Cu(II) from aqueous solutions (Lingamdinne et al. 2016b). The structure of IO-LJFP was characterized using Fourier transform infrared (FT–IR) spectroscopy, scanning electron microscopy (SEM) and X-ray diffractometer (XRD). The kinetics, isotherms and thermodynamic studies revealed that sorption of heavy metal ions on IO-LJFP was a rate-limiting step, with pseudo-second-order kinetics involved monolayer chemi-sorption on a homogeneous surface and this sorption was spontaneous and endothermic in nature. In addition, IO-LJFP could be reused with no significant reduction in its initial sorption capacity.

It was recently reported that magnetic inverse spinel iron oxide nanoparticles (MISFNPs) were efficient in removing Pb(II) and Cr(III) in batch adsorption and that Pb(II) and Cr(III) follow pseudo-second-order kinetics as well as an endothermic process during adsorption onto the homogeneous surface of MISFNPs (Lingamdinne et al. 2017). MISFNPs can be recycled after removal of heavy metals without loss of its stability.

The synthetic protocol of chitosan-coated magnetic nanoparticles (CCMNPs) was modified with α-ketoglutaric acid (α-KA) and applied as a magnetic nanoadsorbent to remove toxic Cu(II) from the aqueous solution (Zhou et al. 2009). The dimension of multidispersed circular particles was about 30 nm and no marked aggregation occurs. Maximum adsorption capacity for Cu(II) was estimated to be 96.15 mg/g and 60.606 mg/g for α-KA-CCMNPs and CCMNPs by the Langmuir model, respectively.

Jiang et al. 2015 prepared magnetic attapulgite/fly ash/poly(acrylic acid-co-acrylamide) (ATP/FA/Poly(AA-co-AM)) ternary nanocomposite microgels via an inverse suspension and copolymerization of acrylic acid (AA) and acrylamide (AM) in liquid paraffin with span-80 as stabilizer, in which attapulgite (ATP) and fly ash (FA) were used to develop a rigid inorganic skeleton, while the soft and elastic poly(acrylic acid-co-acrylamide) (Poly(AA-co-AM)) blocks were grafted onto the rigid inorganic skeleton to form the beadlike ATP/FA/Poly(AA-co-AM) microgels with a 3-D network, validating the expected performance. These microgels showed a certain magnetic characteristic and good adsorption capacity vis-à-vis heavy metal ions, especially Pb(II). The ATP/FA/Poly(AA-co-AM) microgel showed an adsorption capacity of about 40 mg/g for Pb(II) at pH 6 in 24 h and the adsorbed Pb(II) was completely eluted with a 0.2 mol/L aqueous HCl in 2.0 h.

Adsorption and desorption of Mo(VI) from water and wastewater were investigated using maghemite (γ-Fe$_2$O$_3$) which was prepared via a reductive coprecipitation method followed by aeration oxidation (Afkhami and Norooz-Asl 2009). The results illustrated that adsorption was independent of initial concentration of Mo(VI) and the maximum Langmuir adsorption was 33.4 mg/g at pHs between 4.0 and 6.0.

Superparamagnetic iron oxide (Fe_3O_4) nanoparticles modified with dimercaptosuccinic acid (DMSA) were prepared and applied as sorbent material for the removal of Hg(II), Ag(I), Pb(II), Cd(II) and Tl(I) from an aqueous solution (Yantasee et al. 2007). The chemical affinity, capacity, kinetics and stability of the magnetic nanoparticles were compared to those of conventional resin based sorbents (GT-73), activated carbon, and nanoporous silica (SAMMS) of similar surface chemistries in river water, groundwater, seawater and human blood plasma. Fe_3O_4/ DMSA showed a capacity of 227 mg/g for Hg(II) which is 30-fold greater than that of GT-73.

Thinh et al. 2013 described a simple method to prepare magnetic chitosan nanoparticles by co-precipitation via an epichlorohydrin cross-linking reaction. The average size of magnetic chitosan nanoparticles was ~ 30 nm. The maximum adsorption capacity was calculated to be 55.80 mg/g at pH 3.0 and room temperature based on the Langmuir isotherm model.

Meng and coworkers prepared chitosan-modified Mn ferrite nanoparticles by a one-step microwave-assisted hydrothermal method and employed these to absorb Cu(II) from water (Meng et al. 2015). Chitosan modification caused no phase change of $MnFe_2O_4$. FTIR and zeta potential curves revealed that chitosan was successfully coated on the Mn ferrites and TEM characterization showed that the chitosan-modified $MnFe_2O_4$ nanoparticles have a cubic shape with a mean diameter of ~ 100 nm. The results showed that increasing pH and extending contact time improved adsorption efficiency. Optimal adsorption was achieved after 500 min adsorption at pH 6.5 for the solutions with initial Cu(II) concentration of 50 mg/L. According to the Langmuir isotherm model, the maximum adsorption capacity was calculated to be 65.1 mg/g.

Zhu et al. 2017 reported on *in situ* oxidized Fe_3O_4 membranes using a 316 L porous stainless steel filter tube for adsorbing Cd(II) and Pb(II). The membranes showed maximum adsorption capabilities of 0.800 mg/g for Cd(II) and 2.251 mg/g for Pb(II) at 45°C by the Langmuir model. The results displayed the existence of electrostatic attraction and chemisorption. Competitive adsorption of Cd(II) and Pb(II) revealed a preferential adsorption for Pb(II) in binary solutions. Table 3 summarizes the maximum adsorption capacity of different heavy metals using magnetite nanoparticles (MNTs), and various modified magnetite nanoparticles.

Titania nanomaterials and photocatalytic removal

TiO_2 exhibits photocatalytic activity due to its excellent chemical properties, high quantum size effect and strong adsorption in the ultraviolet or visible region which has lead to it being widely utilized in the photocatalytic reduction of Cr(VI) (Qiu et al. 2012). Titanate nanotubes (TNTs) were used as adsorbents for the removal of heavy metals due to their high specific area, irreversible and selective ion exchange properties (Li et al. 2012; Li et al. 2011; Xiong et al. 2011). The photocatalytic activity of TNTs is rather weak due to fast recombination of the created electron-hole pairs after excitation (Kim et al. 2012). In the conventional methods, reduction of Cr(VI)

Table 3. Application of magnetite nanoparticles (MNTs) and modified magnetite nanoparticles as adsorbents.

Adsorbent	Target contaminant	Adsorption capacity (mg/g)	References
$MnFe_2O_4$ nanoparticle	Pb(II), As(III), As(V)	488 for Pb(II), 97 for As(III), 136 for As(V)	Kumar et al. 2014
Hematite nanoparticles	Zn(II)	8.56	Grover et al. 2012
Thiol@magnetic mesoporous silica	Pb(II), Hg(II)	91.5 for Pb(II), 260 for Hg(II)	Li et al. 2011
Fe_3O_4@APS@AA-co-CAMNP	Pb(II), Cd(II), Cu(II), Zn(II)	166.1 for Pb(II), 29.6 for Cd(II), 126.9 for Cu(II), 43.4 for Zn(II)	Ge et al. 2012
Cyclodextrinpolymer@MNP	Pb(II), Cd(II), Ni(II)	64.5 for Pb(II), 27.7 for Cd(II), 13.2 for Ni(II)	Badruddoza et al. 2013
Fe_3O_4@graphenes composite	Cr(VI), 17.29, Pb(II), Hg(II), Cd(II), Ni(II)	17.29 for Cr(VI), 27.95 for Pb(II), 23.03 for Hg(II), 27.83 for Cd(II), 22.07 for Ni(II)	Guo et al. 2014
EDTA@magnetic GO	Pb(II), Hg(II), Cu(II)	508.4 for Pb(II), 268.4 for Hg(II), 301.2 for Cu(II)	Cui et al. 2015
Fe_3O_4@SiO_2-SH	Hg(II)	132	Wang et al. 2016b
MSC MNPs-S_4	Pb(II), Hg(II), Pd(II)	270.3 for Pb(II), 303 for Hg(II), 256.4 for Pd(II)	Vojoudi et al. 2017
IO@$CaCO_3$	As(V), Cr(VI), Pb(II)	184.1 for As(V), 251.6 for As(III), 1041.9 for Pb(II)	Islama et al. 2017
Fe_3O_4@ DAPF CSFMNRs	Pb(II)	83.3	Venkateswarlu and Yoon 2015
Fe_3O_4@$MgSiO_3$ sub-microsphere	Pb(II)	242.1	Ou et al. 2012
γ-Fe_2O_3@MgSNTs	Pb(II), Cd(II)	430 for Pb(II), 200 for Cd(II)	Cao et al. 2013
Fe_3O_4 nanoparticles	As(V)	16.6	Feng et al. 2012
Fe_2O_3/activated carbon fiber (Fe_2O_3/ACF)	As(V)	20.3	Chen et al. 2016
CCMNPs	Cu(II)	60.606	Zhou et al. 2009
α-KA-CCMNPs	Cu(II)	96.15	Zhou et al. 2009
Fe_3O_4/DMSA	Hg(II)	227	Yantasee et al. 2007
Chitosan zerovalent iron nanoparticle CIN	As(III), As(V)	94.00 ± 1.50 for As(III), 119.00 ± 2.6 for As(V)	Gupta et al. 2012
Chitosan-modified $MnFe_2O_4$ nanoparticles	Cu(II)	65.1	Meng et al. 2015
Magnetic chitosan nanoparticles	Cr(VI)	55.80	Thinh et al. 2013
CTS/MMT–Fe_3O_4	Cr(VI)	74.2	Chen et al. 2013
γ-Fe_2O_3-chitosan bead	Cr(VI)	106.5	Jiang et al. 2013

to Cr(III) is the first step of remediation for Cr pollution and in the second step, Cr(III) is removed by precipitation by adding lime or NaOH. Reduction Cr(VI) to Cr(III) over TiO_2 photocatalyst is more effective in the presence of electron donation organic compounds (Ku and Jung 2001).

Reduction of Cr(VI) to Cr(III) is an effective remediation method of wastewater since Cr(VI) exhibites a higher toxicity and mobility than Cr(III) for environment and human beings. Among reduction technologies, that is, electroreduction (Tian et al. 2012) and chemical reduction (Lai and Lo 2008; Lo et al. 2006; Lu et al. 2006), reduction of Cr(VI) to Cr(III) by a photocatalytic reduction process is an efficient and cost-effective method (Mu et al. 2010).

The photocatalytic reduction of Cr(VI) over TiO_2 as a semiconductor photocatalyst was shown to occur as a sequence of one electron transfer steps according to the experimental evidence in literature (Testa et al. 2001; Testa et al. 2004). On the other hand, the influence of TiO_2 phase composition and its surface area on the photocatalytic reduction of Cr(VI) in an aqueous suspension at pH 3.7 under UV–visible light irradiation was investigated (Dozzi et al. 2012). The effect of deposition of gold nanoparticles on the TiO_2 surface on Cr(VI) photocatalytic reduction was also studied since it was expected to enhance its photocatalytic activity by increasing the separation of photogenerated electron–hole pairs (Kamat 2002). Their results illustrated that the rate of Cr(VI) reduction was totally independent of the photocatalyst surface area, and was mainly dependant on the intrinsic efficiency in photoproduced charge separation.

Reduction of Cr(VI) over TiO_2 was studied in an aqueous solution by the UV photocatalysis process by Ku and Jung (2001). Their results revealed that reduction rates of Cr(VI) were higher in acidic medium than in alkaline solutions and the optimal pH range was reported to be 3–6 for the adsorption of Cr(VI) by TiO_2 particles. Increasing the light intensity promoted the reduction rate of Cr(VI) which implied that the extent of the excitation of the particular wavelengths of light plays an important role (Chen et al. 2010). In general, there are two methods to achieve visible light-induced photocatalytic activity of TiO_2 for Cr(VI) reduction, that is, dye sensitization of the ligand-to-metal charge transfer (LMCT) mechanism (Wang et al. 2010; Kim and Choi 2010; Zhang et al. 2014), and doping TiO_2 with metal/nonmetal elements or modifying it with another sensitizer which can be directly excited by visible light (Qiu et al. 2012; Sajjad et al. 2010).

The synthesis of low-cost adsorbents of titanate nanofibers with the formula Na_xH_2–$xTi_3O_7 \cdot nH_2O$ for Cu(II) removal from aqueous solutions was described by Li et al. 2011. The results suggested that the nanofibres with a high sodium content were effective adsorbents for Cu(II) removal. Results showed that the maximum adsorption capacity for Cu(II) calculated to be 167.224 mg/g based on the Langmuir isotherm model and the time needed for equilibrium was 180 min. Thermodynamic studies illustrated that adsorption was spontaneous and endothermic. EDTA-2Na solutions exhibited a high efficiency for Cu(II) desorption from adsorbents and the adsorbents can be used repeatedly.

Xiong and coworkers synthesized titanate nanotubes (TNTs) with specific surface areas of 272.31 m^2/g and pore volumes of 1.264 cm^3/g by the alkaline hydrothermal method for removal of Pb(II) and Cd(II) from aqueous solutions (Xiong et al. 2011).

FT-IR analysis indicated that Pb(II) and Cd(II) adsorption were mainly assigned to the hydroxyl groups in the TNTs. It was shown that adsorption equilibrium was reached after 180 min and the maximum adsorption of Pb(II) and Cd(II) were found to be 520.83 and 238.61 mg/g at pH 5.0–6.0, respectively. Moreover, more than 80% of Pb(II) and 85% of Cd(II) adsorbed onto TNTs were desorbed with 0.1 M HCl within 3 h.

Ni and coworkers studied the regeneration of desorbed titanate nanotubes (TNTs) with cycled Cd(II) adsorption and desorption processes (Wang et al. 2013). They found that desorption of Cd(II) from TNTs was possible using 0.1 M HNO_3, and that TNTs could be regenerated with 0.2 M NaOH at ambient temperature. The regenerated TNTs displayed similar adsorption capacities of Cd(II) even after six recycles. The ion-exchange mechanism with Na^+ in TNTs was confirmed by the change of $-TiO(ONa)_2$ by FTIR spectroscopy. TEM and XRD demonstrated that the damaged tubular structures of TNT were recovered. Furthermore, the cost-effective regeneration of the tubular structures was ascribed to be possibly related to the complex formed by TNTs-OCd⁺OH⁻ onto the adsorbed TNTs, which was identified using X-ray photoelectron spectroscopy.

Liu et al. studied the adsorption of Pb(II), Cd(II), Cu(II) and Cr(III) onto titanate nanotubes (TNTs) in multiple systems from aqueous solutions (Liu et al. 2013). The adsorption mechanism was an ion-exchange between metal ions and H^+/Na^+ located in the interlayers of TNTs. The results of binary or quaternary competitive adsorption indicated that the adsorption capacity of the four heavy metal ions onto TNTs followed the sequence of Pb(II) 547.0 mg/g (2.64 mmol/g) $\gg$ Cd(II) 239.41 mg/g (2.13 mmol/g) > Cu(II) 122.0 mg/g (1.92 mmol/g) $\gg$ Cr(II) 71.22 mg/g (1.37 mmol/g), which followed the reverse order of their hydration energies. In addition, the cations Na^+, K^+, Mg^{2+} and Ca^{2+} inhibited the adsorption of heavy metal ions on TNTs since they competed for adsorption sites and promoted the aggregation of TNTs. However, Al^{3+} and Fe^{3+} generally enhanced adsorption because the resulting hydroxyl-Al/Fe intercalated or coated TNTs and thus capture metal ions. Due to formation of the stable complex of HOCr(OTi)2` with TNTs, desorption of Cr(III) proved to be challenging.

Ni et al. reported a one-step simultaneous adsorption of Cr(VI) and Cr(III) by a mixture of TiO_2 and titanate nanotubes (TNTs) (Liu et al. 2014a) which significantly reduced reaction time. The instant transfer of the reduced Cr from TiO_2 surface to the TNTs interlayer greatly promoted the release of photocatalytic sites of TiO_2, which in turn increased the photocatalytic activity of TNTs by inhibiting electron-hole pair recombination. The optimal pH was 5.0 for the whole process and higher pH lead to precipitation of $Cr(OH)_3$ onto TNTs.

Remove Cu(II) and Cd(II) from aqueous solution was investigated by a series of titanate nanomaterials (TNMs) which were synthesized under different NaOH concentrations (4–15 mol/L) via a hydrothermal reaction (Liu et al. 2014b). TEM and XRD analysis showed that the morphology of the TNMs presented as nanogranules, nanoplates, nanotubes, nanosheets and nanoblocks with increasing NaOH concentration. Furthermore, the crystal phase of the TNMs changed as mixtures of TiO_2 and crystalline titanate, pure titanate and amorphous phase with increasing NaOH concentration. The maximum adsorption capacity of ca. 120 mg/g and 210 mg/g for Cu(II) and Cd(II), respectively, were found for the TNMs which had the titanate phase. A good correlation of the adsorption capacity with the sodium content of TNMs was observed for Cu(II) and Cd(II). This lead to the suggestion that the primary adsorption mechanism was an ion-exchange between Na^+ and metal ions and that the ONa groups were the main adsorption sites. Other mechanisms, that is, ion-exchange with H^+ and complexation by OH were also considered. No definite relationship between the specific surface area and average pore diameter with adsorption capacity for the TNMs with the titanate phase was found.

Thiourea-modified magnetic ion-imprinted chitosan/TiO_2 (MICT) was synthesized and used as a composite adsorbent for Cd(II) removal (Chen et al. 2012). The maximum adsorption capacity for cadmium was 256.41 mg/g at pH 6.0–7.0 according to the Langmuir model. The used sorbent was reusable and no significant decrease of adsorption capacities was observed after five cycles.

Cong and coworkers prepared AgI/TiO_2 using a dissolution-precipitation method, followed by calcination at a temperature range 100–700°C (Wang et al. 2016). It was found that calcination temperature significantly affected the visible light absorption of AgI/TiO_2 along with a shift from metastable γ-AgI to relatively stable β-AgI. Results revealed that the highest photocatalytic reduction rate for Cr(VI) and β-AgI are produced at a calcination temperature of 350°C. The pseudo first order rate constant of the Cr(VI) reduction reaction for a photocatalyst calcined at 350°C was five times that for a photocatalyst calcined at 100°C. This enhancement was attributed to enhanced visible light absorption and greatly reduced charge transfer resistance, which facilitates a more efficient separation and easier transfer of photogenerated electron–hole pairs to the catalyst surface. Little change was observed in the activity of AgI/TiO_2 after five cycles.

Na et al. synthesized Mn-doped TiO_2 grown on reduced graphene oxide (rGO) by a one-pot hydrothermal method and investigated the photocatalytic removal of Cr by the material under sunlight (Chen et al. 2016). As the initial concentration of Cr(VI) was 20 mg/L, Cr (total) removal efficiency of the material proved to be 97.32% in 30 min and 99.02% in 60 min under sunlight irradiation. rGO plays an important role in the synergetic effect of adsorption and photocatalysis to sustain the high and efficient removal of Cr(VI) and Cr(III). The high photocatalytic activity under visible light was considered to be mainly due to the Mn-doping. Cr(VI) adsorbed on the surface of rGO was reduced to Cr(III) by photo electrons which are transported through rGO and the reaction product Cr(III) continues to be adsorbed. The process contributes to the release of abundant photocatalytic sites of Mn-TiO_2 and thereby improves photocatalytic efficiency.

The chitosan/TiO_2 composite nanofibrous adsorbents were prepared by two techniques including TiO_2 nanoparticles coated chitosan nanofibers (coating method) and electrospinning of chitosan/TiO_2 solutions (entrapped method) (Razzaz et al. 2016). Investigation of these chitosan/TiO_2 composite nanofibers for removal of Pb(II) and Cu(II) ions in a batch system indicated that pseudo-first-order and Redlich–Peterson isotherm models for the chitosan/TiO_2 nanofibers were more suited. The maximum adsorption capacities of Cu(II) and Pb(II) ions using entrapped and coating methods were found to be 710.3, 579.1 and 526.5, 475.5 mg/g at an equilibrium time of 30 min and 45°C, respectively. Reusability studies indicated that the chitosan/TiO$_2$ nanofibers prepared by the entrapped method could be reused frequently with no significant loss in adsorption performance after five adsorption/ desorption cycles, whereas nanofibres prepared by the coating method showed lower than 60% of total adsorption for metal ions sorption after the first cycle. In a binary sorption system, the selectivity of metal sorption using chitosan/TiO$_2$ nanofibres was in the order of Cu(II) > Pb(II).

Photocatalytic removal of Cr(VI) from wastewater was investigated using TiO$_2$, ZnO and CdS (Joshi and Shrivastava 2011). Removal of chromium was a maximum at pH 2.0 for different amounts of photocatalysts which may be attributed to the large number of H$^+$ ions being neutralized. The chromium removal percentage and mass removal rate increased with an increase in photocatalyst loading. These results illustrated that the nanocrystalline powder form of TiO$_2$ was the most effective for degradation of Cr(VI) compared to the TiO$_2$ thin film prepared by the sol-gel method, and that ZnO was a less effective photocatalyst than TiO$_2$. It also illustrated that CdS is a very poor photocatalyst for removing Cr(VI) from aqueous solutions.

Other metal oxide nanomaterials

Oxide based nanomaterials are inorganic nanoparticles which are extensively applied for the removal of heavy metal ions from wastewater and include composites such as titaniumoxide/dendrimers (Barakat 2013).

Other nano-sized metal oxides (NMOs), including manganese oxides, zinc oxides, aluminum oxides, magnesium oxides and cerium oxides, provide high removal capacities of heavy metals due to their high surface areas, minimum adverse environmental impact, low solubility, no secondary pollutants and specific affinities for the removal of heavy metal ions (Gupta et al. 2015; Ma et al. 2010; Su et al. 2010).

Nagarajah and coworkers prepared magnesium oxide cores with silica coated nano-magnetite (MTM) by a simple method for the removal of heavy metal ions such as Pb(II), Cd(II) and Cu(II) (Nagarajah et al. 2017). The designed morphology of the magnetite was octahedral magnetite, which enhanced the magnetic separation strength due to its high magnetic saturation and coercivity force. Isothermal, kinetic and intraparticle diffusion studies showed that the mechanism of heavy metal removal was ion exchange, followed by the metal precipitation for Pb(II), while the mechanism for Cd(II) and Cu(II) removal was mainly ion exchange and partial precipitation. The MTM had maximum adsorption capacities of 238 mg/g (1.15 mmol/g) for Pb(II), 85.1 mg/g (0.75 mmol/g) for Cd(II) and 33.5 mg/g (0.52 mmol/g) for Cu(II). Although, MTM has a low reusability, the adsorbed metals could be recovered and no leaching of Fe ions was observed at pH > 2.0.

Other miscellaneous nanomaterials

Dual-layer nanofiltration (NF) hollow fibre membranes were prepared by the simultaneous co-extrusion of polybenzimidazole (PBI) and polyethersulfone (PES)/polyvinylpyrrolidone (PVP) dopes through a triple-orifice spinneret using a dry-jetwet phase inversion process. Removal of heavy metal ions, that is, Cd(II), Cr(VI) and Pb(II) was studied on this absorbent from model wastewater (Zhu et al.

2014). PBI was chosen as the outer selective layer because of its superior chemical resistance and unique charge characteristics, while a PES/PVP blend was employed as the support layer due to its reasonable cost, superior spinnerability, hydrophilic nature, good mechanical properties and easy formation of porous membranes. The newly developed dual-layer NF membrane showed superior rejection to various salts. The adsorption amounts of the membrane for Cd(II), Cr(VI) and Pb(II) were only 0.014 mg/g, 0.051 mg/g and 0.031 mg/g, respectively. While the membrane rejection rate to Mg(II) and Cd(II) were 98% and 95%, respectively. By changing the pH of the solution, the rejections to Cr(VI) and Pb(II) can reach more than 98% and 93%, respectively. Experimental results illustrated that the high rejections are due to the following factors: (a) a narrow pore size distribution membrane with a mean effective pore radius of 0.32 nm and a molecular weight cut off (MWCO) of 249 Da; (b) an enhanced Donnan exclusion effect due to the amphoteric PBI charge property and (c) low adsorption of heavy metals on the PBI surface due to its hydrophilic nature.

2,4-Dinitrophenylhydrazine (DNPH) immobilized on sodium dodecyl sulfate coated nano-alumina was developed for the removal of the heavy metal ions: Pb(II), Cd(II), Cr(III), Co(II), Ni(II) and Mn(II) from water samples (Afkhami et al. 2010). The results illustrated that the adsorbent has the highest capacity for Pb(II), Cr(III) and Cd(II) in an ionics mixture of the three. The experimental data were interpreted by the Freundlich adsorption isotherm equation for Mn(II), Pb(II), Cr(III) and Cd(II) ions and by the Langmuir isotherm equation for Ni(II) and Co(II) ions. The maximum adsorption capacities calculated by the Langmuir equation were 100.0 mg/g for Cr(III), 83.33 for Cd(II) mg/g and 100.0 mg/g for Pb(II) with modified alumina nanoparticles. Desorption experiments showed that the modified alumina nanoparticles could be reused without significant loss of its initial properties even after three adsorption-desorption cycles.

An amino-modified MCM-41/poly(vinylalcohol) nanocomposite (M-MCM-41/ PVOHNC) was synthesized by an ultrasonic-assisted procedure for the adsorption of Cd(II) from aqueous media (Soltani et al. 2018). The kinetic results revealed that the adsorption of Cd(II) onto M-MCM-41/PVOHNC was a chemical adsorption and followed the PSO kinetic model. The maximum adsorption capacity of M-MCM-41/ PVOHNC at 25°C was calculated to be 46.73 mg/g for Cd(II) by the Langmuir isotherm at 25°C.

Table 4 summarizes the adsorption capacity of different heavy metals using various mesoporous silica based and chitosan based materials along with other materials.

Conclusions and future perspectives

Nanomaterials are widely studied as highly efficient adsorbents for the removal of heavy metals from aqueous solutions and wastewater. They possess advantages such as high capacities, fast kinetics and preferable sorption toward certain heavy metals in water and wastewater in which it has been demonstrated that addition of nanomaterials aides vastly improves both selectivity and efficiency of separations.

Table 4. Adsorption capacity of other absorbents toward heavy metals ions.

Adsorbent	Target contaminant	Adsorption capacity (mg/g)	References
M-MCM-41/PVOHNC	Cd(II)	46.73	Soltani et al. 2018
NH$_2$-MCM-41/NY6NC	Cd(II)	27.59	Dinari et al. 2016
APTS-MCM-41/PMMA	Cd(II)	24.75	Mohammadnezhad et al. 2017
APTS-MCM-41/PS	Cd(II)	10.42	Mohammadnezhad et al. 2017
NH$_2$-MCM-41	Cd(II)	18.25	Heidari et al. 2009
SNHS	Cd(II)	20.82	Najafi et al. 2012
Si-NH$_2$	Cd(II)	31.89	Najafi et al. 2012
NH$_2$-SNHS	Cd(II)	40.73	Najafi et al. 2012
Si-DHB	Cd(II)	3.60	Venkatesh et al. 2004
Si-DHAQ	Cd(II)	8.10	Goswami and Singh 2002
Si-APTS-EDTA	Cd(II)	23.67	Melo et al. 2013
HMS-NH$_2$	Cd(II)	28.10	Machida et al. 2012
HMS-SH	Cd(II)	14.61	Machida et al. 2012
Al(OH)CO$_3$ nanospheres	As(V)	170	Min et al. 2015
MgSiO$_3$ spheres	Pb(II), As(V)	118 mg/g for Pb(II), 68 mg/g for As(V)	Han et al. 2016
SnO$_2$/C without MgSiO$_3$ coating	Pb(II), As(V)	83 mg/g for Pb(II), 55 mg/g for As(V)	Han et al. 2016
MgSiO$_3$-coated SnO$_2$/C	Pb(II), As(V)	185 mg/g for Pb(II), 109 mg/g for As(V)	Han et al. 2016
CCN	Cr(VI)	357.14	Pandey and Mishra 2011
CS-ENM	As(V)	30.8	Min et al. 2015
MICB	As(III), As(V)	35.3 for As(III), 35.7 for As(V)	Wang et al. 2014

SNHS: silica nano hollow sphere; Si: silica; DHB: 3,4-digydroxybenzene; DHAQ: 1,8-dihydroxyanthraquinone; APTS: 3-aminopropyltriethoxysilane; EDTA: ethylenediaminetetraacetic acid; HMS: hexagonal mesoporous silica; CCN: chitosan/organoclay bionanocomposites; CS-ENM: chitosan based electrospun nanofiber membrane; MICB: magnetic nanoparticles impregnated chitosan beads.

In order to promote the practical application of nanomaterials in removal of heavy metal ions from solutions, the current limitations have to be addressed. These include the following: (a) nanomaterials tend to aggregate into large-size particles thereby reducing their adsorption capacity; (b) blocking and fouling problems in subsequent filtration steps exist due to their nano-sizes and finding approaches for better dispersion of nanomaterials need to be further explored; (c) nanoparticles are often not fully recovered leading to secondary contamination; and (d) an efficient, simple and cost effective separation of the exhausted nanomaterials from water/wastewater still remains an interesting and challenging task. It would appear that magnetic-assistance based separations could possibly provide long-term benefits and the design of magnetic separation systems might thus be one new frontier for research. The excessive pressure drop caused by nanomaterials should also be considered when column operation is performed. Fortunately, fabrication of new carbon- and magnetic-based composite adsorbents seems to be an effective approach to address all the above technical problems. However, various issues still need to be solved concerning the development of efficient and cost effective processes to obtain the composite adsorbents. The type of interaction between the heavy metal ions and the supported nanomaterials, physicochemical properties of the composite adsorbents such as hydrophilicity, porosity, charge density, thermal and mechanical stability, the long-term performance of the composite adsorbents, high desorption, regeneration and reusability are important parameters in water treatment that will require intensive work to find solutions for. The potential effects of leached nanomaterials on the environment and nanomaterial leakage and its environmental toxicity also need to be systematically investigated. Finally, very few reports exist on the large scale production and industrial application of membrane fabrication; this process needs to be evaluated for its cost-effectiveness including the supply of nanomaterials. New surgical tools need to be developed to improve on current surgical procedures, and new procedures need to be developed to reduce the invasiveness of surgery, and reduce the complexity of surgical procedures; Also, new procedures need to be developed for nanomaterial incorporation and monitoring of the long-term stability of membranes under practical application conditions.

Reference cited

Addo Ntim, S. and S. Mitra. 2011. Removal of trace arsenic to meet drinking water standards using iron oxide coated multiwall carbon nanotubes. J. Chem. Eng. Data 56: 2077–2083.

Afkhami, A. and R. Norooz-Asl. 2009. Removal, preconcentration and determination of Mo(VI) from water and wastewater samples using maghemite nanoparticles. Colloids and Surf. A Physicochem. Eng. Asp. 346: 52–57.

Afkhami, A., M. Saber-Tehrani and H. Bagheri. 2010. Simultaneous removal of heavy-metal ions in wastewater samples using nano-alumina modified with 2,4-dinitrophenylhydrazine. J. Hazard. Mater. 181: 836–844.

Al Nafiey, A., A. Addad, B. Sieber, G. Chastanet, A. Barras, S. Szunerits and R. Boukherroub. 2017. Reduced graphene oxide decorated with Co_3O_4 nanoparticles (rGO-Co_3O_4) nanocomposite: A reusable catalyst for highly efficient reduction of 4-nitrophenol, and Cr(VI) and dye removal from aqueous solutions. Chem. Eng. J. 322: 375–384.

Ali, A., H. Zafar, M. Zia, I. ul Haq, A.R. Phull, J.S. Ali and A. Hussain. 2016. Synthesis, characterization, applications, and challenges of iron oxide nanoparticles. Nanotechnol. Sci. Appl. 9: 49–67.

Ali, I. 2012. New generation adsorbents for water treatment. Chem. Rev. 112: 5073–5091.

Ali, K., Y. Javed and Y. Jamil. 2017. Size and shape control synthesis of iron oxide-based nanoparticles: current status and future possibility. pp. 39–81. *In*: Sharma, S.K. (ed.). Complex Magnetic Nanostructure, Synthesis, Assembly and Application, Springer International Publishing.

Anitha, K., S. Namsani and J.K. Singh. 2015. Removal of heavy metal ions using a functionalized single-walled carbon nanotube: a molecular dynamics study. J. Phys. Chem. A 119: 8349–8358.

Anku, W.W., S.O.B. Oppong, S.K. Shukla, E.S. Agorku and P.P. Govender. 2016. Palladium-doped–ZrO_2-multiwalled carbon nanotubes nanocomposite: an advanced photocatalyst for water treatment. Appl. Phys. A: Mater Sci. Processing 122: 1–8.

Arjunan, N., C. Singaravelu, J. Kulanthaivel and J. Kandasamy. 2017. A potential photocatalytic, antimicrobial and anticancer activity of chitosan-copper nanocomposite. Inter. J. Biolog. Macromol. 104: 1774–1782.

Arshad, A., J. Iqbal and Q. Mansoor. 2017. NiO-nanoflakes grafted graphene: an excellent photocatalyst and a novel nanomaterial for achieving complete pathogen control. Nanoscale 9: 16321–16328.

Arshad, A., J. Iqbal, M. Siddiq, M. Ali, A. Ali, H. Shabbir, U. Nazeer and M. Saleem. 2017. Solar light triggered catalytic performance of graphene-CuO nanocomposite for waste water treatment. Ceramics Inter. 43: 10654–10660.

Asahi, R. 2001. Visible-light photocatalysis in nitrogen-doped titanium oxides. Science 293: 269–271.

Avouris, P. and C. Dimitrakopoulos. 2012. Graphene: synthesis and applications. Mater. Today 15: 86–97.

Badruddoza, A.Z.M., Z.B.Z. Shawon, W.J.D. Tay, K. Hidajat and M.S. Uddin. 2013. Fe_3O_4/cyclodextrin polymer nanocomposites for selective heavy metals removal from industrial wastewater. Carbohydr. Polym. 91: 322–332.

Barakat, M.A., M.H. Ramadan, M.A. Alghamdi, S.S. Al-Garny, H.L. Woodcock and J.N. Kuhn. 2013. Remediation of Cu(II), Ni(II), and Cr(III) ions from simulated wastewater by dendrimer/titania composites. J. Environ. Manage. 117: 50–57.

Boparai, H.K., M. Joseph and D.M. O'Carroll. 2011. Kinetics and thermodynamics of cadmium ion removal by adsorption onto nano zerovalent iron particles. J. Hazard. Mater. 186: 458–465.

Boyd, R. 2010. Heavy metal pollutants and chemical ecology: exploring new frontiers. J. Chem. Ecology 36: 46–58.

Cantarella, M., R. Sanz, M. Buccheri, L. Romano and V. Privitera. 2016. PMMA/TiO_2 nanotubes composites for photocatalytic removal of organic compounds and bacteria from water. Mater. Sci. Semiconductor Processing 42: 58–61.

Cao, C.Y., P. Li, J. Qu, Z.F. Dou, W.S. Yan, J.F. Zhu, Z.Y. Wu and W.G. Song. 2012. J. Mater. Chem. 22: 19898–19903.

Cao, C.Y., F. Wei, J. Qu and W.G. Song. 2013. Programmed synthesis of magnetic magnesium silicate nanotubes with high adsorption capacities for lead and cadmium ions. Chem. Eur. J. 19: 1558–1562.

Cao, W., Y.R. Ma, W. Zhou and L. Guo. 2015. One-pot hydrothermal synthesis of rGO-Fe_3O_4 hybrid nanocomposite for removal of Pb(II) magnetic separation. Chem. Res. Chin. Univ. 31: 508–513.

Chandra, V., J. Park, Y. Chun, J.W. Lee, I.C. Hwang and K.S. Kim. 2010. Water-dispersible magnetite-reduced graphene oxide composites for arsenic removal. ACS Nano 4: 3979–3986.

Chandraboss, V., J. Kamalakkannan, S. Prabha and S. Senthilvelan. 2015. An efficient removal of methyl violet from aqueous solution by an AC-Bi/ZnO nanocomposite material. RSC Adv. 5: 25857–25869.

Chen, A., G. Zeng, G. Chen, X. Hu, M. Yan, S. Guan, C. Shang, L. Lu, Z. Zou and G. Xie. 2012. Novel thiourca-modified magnetic ion-imprinted chitosan/TiO_2 composite for simultaneous removal of cadmium and 2,4-dichlorophenol. Chem. Eng. J. 191: 85–94.

Chen, C., W. Ma and J. Zhao. 2010. Semiconductor-mediated photodegradation of pollutants under visible-light irradiation. Chem. Soc. Rev. 39: 4206–4219.

Chen, D., W. Li, Y. Wu, Q. Zhu, Z. Lu and G. Du. 2013. Preparation and characterization of chitosan/montmorillonite magnetic microspheres and its application for the removal of Cr(VI). Chem. Eng. J. 221: 8–15.

Chen, H., J. Li, D. Shao, X. Ren and X. Wang. 2012. Poly(acrylic acid) grafted multiwall carbon nanotubes by plasma techniques for Co(II) removal from aqueous solution. Chem. Eng. J. 210: 475–481.

Chen, H., J. Luo, X. Wang, X. Liang, Y. Zhao, C. Yang, M. Baikenov and X. Su. 2018. Synthesis of Al_2O_3/carbon composites from wastewater as superior adsorbents for Pb(II) and Cd(II) removal. Micropor. Mesopor. Mater. 255: 69–75.

Chen, H.Y., K.L. Lv, Y. Du, H.P. Ye and D.Y. Du. 2016. Microwave-assisted rapid synthesis of Fe_2O_3/ACF hybrid for high efficient As(V) removal. J. Alloys Compd. 674: 399–405.

Chen, J.S., Y. Zhang and X.W. Lou. 2011. One-pot synthesis of uniform Fe_3O_4 nanospheres with carbon matrix support for improved lithium storage capabilities. ACS Appl. Mater. Interfaces 3: 3276–3279.

Chen, X., L. Liu, P. Yu and S. Mao. 2011. Increasing solar absorption for photocatalysis with black hydrogenated titanium dioxide nanocrystals. Science 331: 746–750.

Chen, Y., H. Li, Q. Ma, Z. Zhang, J. Wang, G. Wang, Q. Che and P. Yang. 2017. A novel electrospun approach for highly-dispersed carambola-like SnO_2/C composite microparticles with superior photocatalytic performance. Mater. Lett. 202: 17–20.

Chen, Z., Y. Li, M. Guo, F. Xu, P. Wang, Y. Du and P. Na. 2016. One-pot synthesis of Mn-doped TiO_2 grown on graphene and the mechanism for removal of Cr(VI) and Cr(III). J. Hazard. Mater. 310: 188–198.

Choi, J., S. Chan, H. Joo, H. Yang and F. Ko. 2016. Three-dimensional (3D) palladium-zinc oxide nanowire nanofiber as photo-catalyst for water treatment. Water Res. 101: 362–369.

Chong, M., B. Jin, C. Chow and C. Saint. 2010. Recent developments in photocatalytic water treatment technology: A review. Water Res. 44: 2997–3027.

Convertino, A., L. Maiolo, V. Scuderi, A. Di Mauro, M. Scuderi, G. Nicotra, G. Impellizzeri, G. Fortunato and V. Privitera. 2016. A forest of SiO_2 nanowires covered by a TiO_2 thin film for an efficient photocatalytic water treatment. RSC Advances 6: 91121–91126.

Crane, R.A., M. Dickinson, I.C. Popescu and T.B. Scott. 2011. Magnetite and zero-valent iron nanoparticles for the remediation of uranium contaminated environmental water. Water Res. 45: 2931–2942.

Crane, R.A. and T.B. Scott. 2012. Nanoscale zero-valent iron: future prospects for an emerging water treatment technology. J. Hazard. Mater. 211-212: 112–125.

Cui, L., Y. Wang, L. Gao, L. Hu, L. Yan, Q. Wei and B. Du. 2015. EDTA functionalized magnetic graphene oxide for removal of Pb(II), Hg(II) and Cu(II) in water treatment: Adsorption mechanism and separation property. Chem. Eng. J. 281: 1–10.

Daou, T.j., G. Pourroy, S. Bégin-Colin, J.M. Grenèche, C. Ulhaq-Bouillet, P. Legaré, P. Bernhardt, C. Leuvrey and G. Rogez. 2006. Hydrothermal synthesis of monodisperse magnetite nanoparticles. Chem. Mater. 18: 4399–4404.

Deng, Y., L. Wang, W. Yang, S. Fu and A. Elaïssari. 2003. Preparation of magnetic polymeric particles via inverse microemulsion polymerization process. J. Magn. Magn. Mater. 257: 69–78.

Depault, F., M. Cojocaru, F. Fortin, S. Chakrabarti and N. Lemieux. 2006. Genotoxic effects of chromium(VI) and cadmium(II) in human blood lymphocytes using the electron microscopy *in situ* end-labeling (EM-ISEL) assay. Toxicology *In Vitro* 20: 513–518.

Dette, C., M. Pérez-Osorio, C. Kley, P. Punke, C. Patrick, P. Jacobson, F. Giustino, S. Jung and K. Kern. 2014. TiO_2 anatase with a bandgap in the visible region. Nano Lett. 14: 6533–6538.

Dinari, M., G. Mohammadnezhad and R. Soltani. 2016. Fabrication of poly(methyl methacrylate)/silica KIT-6 nanocomposites via *in situ* polymerization approach and their application for removal of Cu^{2+} from aqueous solution. RSC Advances 6: 11419–11429.

Dong, C., R. Eldawud, A. Wagner and C. Dinu. 2016. Hybrid nanocomposites with enhanced visible light photocatalytic ability for next generation of clean energy systems. Appl. Catal. A: Gen. 524: 77–84.

Dong, P., X. Wu, Z. Sun, J. Hu and S. Yang. 2014. Removal performance and the underlying mechanisms of plasma-induced CD/MWCNT/iron oxides towards Ni(II). Chem. Eng. J. 256: 128–136.

Dozzi, M.V., A. Saccomanni and E. Selli. 2012. Cr(VI) photocatalytic reduction: effects of simultaneous organics oxidation and of gold nanoparticles photodeposition on TiO_2. J. Hazard. Mater. 211: 188–195.

Drexler, K. 2000. Engines of creation; Eric Dresler: Los Altos, CA.

Du, J. and C. Jing. 2011. Preparation of Fe_3O_4@Ag SERS substrate and its application in environmental Cr(VI) analysis. J. Colloid Interface Sci. 358: 54–61.

Fan, L., C. Luo, M. Sun and H. Qiu. 2012. Synthesis of graphene oxide decorated with magnetic cyclodextrin for fast chromium removal. J. Mater. Chem. 22: 24577–24583.

Fan, L., C. Luo, M. Sun, X. Li and H. Qiu. 2013. Highly selective adsorption of lead ions by water-dispersible magnetic chitosan/graphene oxide composites. Colloids Surf. B Biointerfaces 103: 523–529.

Fana, H.L., S.F. Zhou, W.Z. Jiao, G.S. Qi and Y.Z. Liu. 2017. Removal of heavy metal ions by magnetic chitosan nanoparticles prepared continuously via high-gravity reactive precipitation method. Carbohydr. Polym. 174: 1192–1200.

Feng, L., M. Cao, X. Ma, Y. Zhu and C. Hu. 2012. Superparamagnetic high-surface-area Fe_3O_4 nanoparticles as adsorbents for arsenic removal. J. Hazard. Mater. 217: 439–446.

Feynman, R.P. 1960. There's Plenty of Room at the Bottom. Eng. Sci. 23: 22–36.

Fu, F. and Q. Wang. 2011. Removal of heavy metal ions from wastewaters: A review. J. Environ. Management 92: 407–418.

Fujishima, A. and K. Honda. 1972. Electrochemical photolysis of water at a semiconductor electrode. Nature 238: 37–38.

Fujishima, A., T.N. Rao and D.A. Tryk. 2000. Titanium dioxide photocatalysis. J. Photochem. Photobio. C: Photochem. Rev. 1: 1–21.

Ge, F., M.M. Li, H. Ye and B.X. Zhao. 2012. Effective removal of heavy metal ions Cd^{2+}, Zn^{2+}, Pb^{2+}, Cu^{2+} from aqueous solution by polymer-modified magnetic nanoparticles. J. Hazard. Mater. 211: 366–372.

Ge, Y., Z. Li, D. Xiao, P. Xiong and N. Ye. 2014. Sulfonated multi-walled carbon nanotubes for the removal of copper(II) from aqueous solutions. J. Ind. Eng. Chem. 20: 1765–1771.

Gjipalaj, J. and I. Alessandri. 2017. Easy recovery, mechanical stability, enhanced adsorption capacity and recyclability of alginate-based TiO_2 macrobead photocatalysts for water treatment. J. Environ. Chem. Eng. 5: 1763–1770.

Goswami, A. and A.K. Singh. 2002. 1,8-Dihydroxyanthraquinone anchored on silica gel: synthesis and application as solid phase extractant for lead(II), zinc(II) and cadmium(II) prior to their determination by flame atomic absorption spectrometry. Talanta 58: 669–678.

Grover, V.A., J. Hu, K.E. Engates and H.J. Shipley. 2012. Adsorption and desorption of bivalent metals to hematite nanoparticles. Environ. Toxicol. Chem. 31: 86–92.

Gu, C., S. Xiong, Z. Zhong, Y. Wang and W. Xing. 2017. A promising carbon fiber-based photocatalyst with hierarchical structure for dye degradation. RSC Advances 7: 22234–22242.

Gu, H., S. Rapole, Y. Huang, D. Cao, Z. Luo, S. Wei and Z. Guo. 2013. Synergistic interactions between multi-walled carbon nanotubes and toxic hexavalent chromium. J. Mater. Chem. A 1: 2011–2021.

Guo, X., B. Du, Q. Wei, J. Yang, L. Hu, L. Yan and W. Xu. 2014. Synthesis of aminofunctionalized magnetic graphenes composite material and its application to remove Cr(VI), Pb(II), Hg(II), Cd(II), and Ni(II) from contaminated water. J. Hazard. Mater. 278: 211–220.

Gupta, A., M. Yunus and N. Sankararamakrishnan. 2012. Zerovalent iron encapsulated chitosan nanospheres—A novel adsorbent for the removal of total inorganic Arsenic from aqueous systems. Chemosphere 86: 150–155.

Gupta, V., O. Moradi, I. Tyagi, S. Agarwal, H. Sadegh, R. Shahryari-Ghoshekandi, A. Makhlouf, M. Goodarzi and A. Garshasbi. 2015. Study on the removal of heavy metal ions from industry waste by carbon nanotubes: Effect of the surface modification: a review. Critical Rev. Environ. Sci. Tech. 46: 93–118.

Gupta, V.K., I. Tyagi, H. Sadegh, R. Shahryari-Ghoshekand, A.S.H. Makhlouf and B. Maazinejad. 2015. Nanoparticles as adsorbent; a positive approach for removal of noxious metal ions: a review. Sci. Technol. Dev. 34: 195–214.

Han, T., X. Zhang, X. Fu and J. Liu. 2016. Three-dimensional $MgSiO_3$-coated SnO_2/C nanostructures for efficient adsorption of heavy metal ions from aqueous solution. RSC Adv. 6: 73412–73420.

Hao, L., H. Song, L. Zhang, X. Wan, Y. Tang and Y. Lv. 2012. SiO_2/graphene composite for highly selective adsorption of Pb(II) ion. J. Colloid Interface Sci. 369: 381–387.

Hayati, B., A. Maleki, F. Najafi, H. Daraei, F. Gharibi and G. McKay. 2017. Super high removal capacities of heavy metals (Pb^{2+} and Cu^{2+}) using CNT dendrimer. J. Hazard. Mater. 336: 146–157.

He, Y.Q., N.N. Zhang and X.D. Wang. 2011. Adsorption of graphene oxide/chitosan porous materials for metal ions. Chin. Chem. Lett. 22: 859–862.

Heidari, A., H. Younesi and Z. Mehraban. 2009. Removal of Ni(II), Cd(II), and Pb(II) from a ternary aqueous solution by amino functionalized mesoporous and nano mesoporous silica. Chem. Eng. J. 153: 70–79.

Hintsho, N., L. Petrik, A. Nechaev, S. Titinchi and P. Ndungu. 2014. Photo-catalytic activity of titanium dioxide carbon nanotube nano-composites modified with silver and palladium nanoparticles. Appl. Catal. B: Environ. 156-157: 273–283.

Hou, W. and S. Cronin. 2012. A review of surface plasmon resonance-enhanced photocatalysis. Adv. Funct. Mater. 23: 1612–1619.

Hu, X.J., Y.G. Liu, H. Wang, A.W. Chen, G.M. Zeng, S.M. Liu, Y.M. Guo, X. Hu, T.T. Li, Y.Q. Wang, L. Zhou and S.H. Liu. 2013. Removal of Cu(II) ions from aqueous solution using sulfonated magnetic graphene oxide composite. Sep. Purif. Technol. 108: 189–195.

Hu, X.J., Y.G. Liu, G.M. Zeng, S.H. You, H. Wang, X. Hu, Y.M. Guo, X.F. Tan and F.Y. Guo. 2014. Effects of background electrolytes and ionic strength on enrichment of Cd(II) ions with magnetic graphene oxide-supported sulfanilic acid. J. Colloid Interface Sci. 435: 138–144.

Hua, M., S. Zhang, B. Pan, W. Zhang, L. Lv and Q. Zhang. 2012. Heavy metal removal from water/ wastewater by nanosized metal oxides: A review. J. Hazard. Mater. 211-212: 317–331.

Huang, Z., X. Zheng, W. Lv, M. Wang, Q. Yang and F. Kang. 2011. Adsorption of lead(II) ions from aqueous solution on low-temperature exfoliated graphene nanosheets. Langmuir 27: 7558–7562.

Hur, J., J. Shin, J. Yoo and Y.S. Seo. 2015. Competitive adsorption of metals onto magnetic graphene oxide: comparison with other carbonaceous adsorbents. Sci. World J. 11.

Ihsanullah, F. Al-Khaldi, B. Abu-Sharkh, A. Abulkibash, M. Qureshi, T. Laoui and M. Atieh. 2015. Effect of acid modification on adsorption of hexavalent chromium (Cr(VI)) from aqueous solution by activated carbon and carbon nanotubes. Desalination Water Treatment 57: 7232–7244.

Ihsanullah, F. Al-Khaldi, B. Abusharkh, M. Khaled, M. Atieh, M. Nasser, T. laoui, T. Saleh, S. Agarwal, I. Tyagi and V.K. Gupta. 2015. Adsorptive removal of cadmium(II) ions from liquid phase using acid modified carbon-based adsorbents. J. Mol. Liq. 204: 255–263.

Islama, M.S., W.S. Choi, B. Nam, C. Yoon and H.J. Lee. 2017. Needle-like iron oxide@$CaCO_3$ adsorbents for ultrafast removal of anionic and cationic heavy metal ions. Chem. Eng. J. 307: 208–219.

Jamei, M.R., M.R. Khosravi and B. Anvaripour. 2014. A novel ultrasound assisted method in synthesis of nZVI particles. Ultrason. Sonochem. 21: 226–233.

Ji, L.L., W. Chen, L. Duan and D.Q. Zhu. 2009. Mechanisms for strong adsorption of tetracycline to carbon nanotubes: a comparative study using activated carbon and graphite as adsorbents. Environ. Sci. Technol. 43: 2322–2327.

Jiang, L., P. Liu and S. Zhao. 2015. Magnetic ATP/FA/Poly(AA-co-AM) ternary nanocomposite microgel as selective adsorbent for removal of heavy metals from wastewater. Colloids and Surf. A Physicochem. Eng. Asp. 470: 31–38.

Jiang, Y.J., X.Y. Yu, T. Luo, Y. Jia, J.H. Liu and X.J. Huang. 2013. γ-Fe_2O_3 nanoparticles encapsulated millimeter-sized magnetic chitosan beads for removal of Cr(VI) from water: Thermodynamics, kinetics, regeneration, and uptake mechanisms. J. Chem. Eng. Data 58: 3142–3149.

Joshi, K.M. and V.S. Shrivastava. 2011. Photocatalytic degradation of Chromium(VI) from wastewater using nanomaterials like TiO_2, ZnO, and CdS. Appl. Nanosci. 1: 147–155.

Jourshabani, M., Z. Shariatinia and A. Badiei. 2017. Facile one-pot synthesis of cerium oxide/sulfur-doped graphitic carbon nitride (g-C_3N_4) as efficient nanophotocatalysts under visible light irradiation. J. Colloid Interface Sci. 507: 59–73.

Kamat, P.V. 2002. Photophysical, photochemical and photocatalytic aspects of metal nanoparticles. J. Phys. Chem. B 106: 7729–7744.

Kanel, S.R., J.M. Greneche and H. Choi. 2006. Arsenic(V) removal from groundwater using nano scale zero-valent iron as a colloidal reactive barrier material. Environ. Sci. Technol. 40: 2045–2050.

Kaur, A. and U. Gupta. 2009. A review on applications of nanoparticles for the preconcentration of environmental pollutants. J. Mater. Chem. 19: 8279–8289.

Khan, S. 2002. Efficient photochemical water splitting by a chemically modified n-TiO_2. Science 297: 2243–2245.

Khin, M., A. Nair, V. Babu, R. Murugan and S. Ramakrishna. 2012. A review on nanomaterials for environmental remediation. Energy Environ. Sci. 5: 8075.

Kim, G. and W. Choi. 2010. Charge-transfer surface complex of EDTA-TiO_2 and its effect on photocatalysis under visible light. Appl. Catal. B: Environ. 100: 77–83.

Kim, M.K., K.S. Sundaram, G. AnanthaIyengar and K.P. Lee. 2015. A novel chitosan functional gel included with multiwall carbon nanotube and substituted polyaniline as adsorbent for efficient removal of chromium ion. Chem. Eng. J. 267: 51–64.

Kim, S., M. Kim, S.H. Hwang and S.K. Lim. 2012. Enhancement of photocatalytic activity of titania-titanate nanotubes by surface modification. Appl. Catal. B: Environ. 123: 391–397.

Kosaa, S.A., G. Al-Zhrani and M.A. Salam. 2012. Removal of heavy metals from aqueous solutions by multi-walled carbon nanotubes modified with 8-hydroxyquinoline. Chem. Eng. J. 181-182: 159–168.

Ku, Y. and I. Jung. 2001. Photocatalytic reduction of Cr(VI) in aqueous solution by UV irradiation with presence of titanium dioxide. Water Res. 35: 135–142.

Kumar, S., R.R. Nair, P.B. Pillai, S.N. Gupta, M.A.R. Iyengar and A.K. Sood. 2014. Graphene oxide–$MnFe_2O_4$ magnetic nanohybrids for efficient removal of lead and arsenic from water. ACS Appl. Mater. Interfaces 6: 17426–17436.

Kumaresan, S., K. Vallalperuman and S. Sathishkumar. 2017. A novel one-step synthesis of Ag-doped ZnO nanoparticles for high performance photo-catalytic applications. J. Mater. Sci. Mater. Electronics 28: 5872–5879.

Kurniawan, T., G. Chan, W. Lo and S. Babel. 2006. Physico-chemical treatment techniques for wastewater laden with heavy metals. Chem. Eng. J. 118: 83–98.

Lai, K.C.K. and I.M.C. Lo. 2008. Removal of chromium(VI) by acidwashed zero-valent iron under various groundwater geochemistry conditions. Environ. Sci. Technol. 42: 1238–1244.

Laurent, S., D. Forge, M. Port, A. Roch, C. Robic, L. Vander Elst and R.N. Muller. 2008. Magnetic iron oxide nanoparticles: Synthesis, stabilization, vectorization, physico-chemical characterizations and biological applications. Chem. Rev. 108: 2064–2110.

Lee, Y.C. and J.W. Yang. 2012. Self-assembled flower-like TiO_2 on exfoliated graphite oxide for heavy metal removal. J. Ind. Eng. Chem. 18: 1178–1185.

Li, G., Z. Zhao, J. Liu and G. Jiang. 2011. Effective heavy metal removal from aqueous systems by thiol functionalized magnetic mesoporous silica. J. Hazard. Mater. 192: 277–283.

Li, J., S. Chen, G. Sheng, J. Hu, X. Tan and X. Wang. 2011. Effect of surfactants on Pb(II) adsorption from aqueous solutions using oxidized multiwall carbon nanotubes. Chem. Eng. J. 166: 551–558.

Li, J., S. Zhang, C. Chen, G. Zhao, X. Yang, J. Li and X. Wang. 2012. Removal of Cu(II) and fulvic acid by graphene oxide nanosheets decorated with Fe_3O_4 nanoparticles. ACS Appl. Mater. Interfaces 4: 4991–5000.

Li, L., L. Fan, M. Sun, H. Qiu, X. Li, H. Duan and C. Luo. 2013. Adsorbent for chromium removal based on graphene oxide functionalized with magnetic cyclodextrin–chitosan. Colloids Surf. B 107: 76–83.

Li, L., H. Duan, X. Wang and C. Luo. 2014. Adsorption property of Cr(VI) on magnetic mesoporous titanium dioxide-graphene oxide core-shell microspheres. New. J. Chem. 38: 6008–6016.

Li, L., C. Luo, X. Li, H. Duan and X. Wang. 2014. Preparation of magnetic ionic liquid/chitosan/graphene oxide composite and application for water treatment. Int. J. Biol. Macromol. 66: 172–178.

Li, N., L.D. Zhang, Y.Z. Chen, M. Fang, J.X. Zhang and H.M. Wang. 2012. Highly efficient, irreversible and selective ion exchange property of layered titanate nanostructures. Adv. Funct. Mater. 22: 835–841.

Li, N.A., L.D. Zhang, Y.Z. Chen, Y. Tian and H.M. Wang. 2011. Adsorption behavior of Cu(II) onto titanate nanofibers prepared by alkali treatment. J. Hazard. Mater. 189: 265–272.

Li, S.L., W. Wang, W.L. Yan and W.X. Zhang. 2014. Nanoscale zero-valent iron (nZVI) for the treatment of concentrated Cu(II) wastewater: a field demonstration. Environ. Sci. Proc. Imp. 16: 524–533.

Li, X., H. Zhou, W. Wu, S. Wei, Y. Xu and Y. Kuang. 2015. Studies of heavy metal ion adsorption on chitosan/sulfydryl-functionalized graphene oxide composites. J. Colloid Interface Sci. 448: 389–397.

Li, Y., F. Liu, B. Xia, Q. Du, P. Zhang, D. Wang, Z. Wang and Y. Xia. 2010. Removal of copper from aqueous solution by carbon nanotube/calcium alginate composites. J. Hazard. Mater. 177: 876–880.

Li, Y.H., S. Wang, J. Wei, X. Zhang, C. Xu, Z. Luan, D. Wu and B. Wei. 2002. Lead adsorption on carbon nanotubes. Chem. Phys. Lett. 357: 263–266.

Li, Y.H., J. Ding, Z. Luan, Z. Di, Y. Zhu, C. Xu, D. Wu and B. Wei. 2003. Competitive adsorption of Pb^{2+}, Cu^{2+} and Cd^{2+} ions from aqueous solutions by multi walled carbon nanotubes. Carbon 41: 2787–2792.

Lingamdinne, L., J. Koduru, R. Jyothi, Y. Chang and J. Yang. 2015. Factors affect on bioremediation of Co(II) and Pb(II) onto Lonicera japonica flowers powder. Desalination Water Treatment 57: 13066–13080.

Lingamdinne, L.P., J.R. Koduru, Y.L. Choi, Y.Y. Chang and J.K. Yang. 2016. Studies on removal of Pb(II) and Cr(III) using graphene oxide based inverse spinel nickel ferrite nano-composite as sorbent. Hydrometallurgy 165: 64–72.

Lingamdinne, L.P., J.K. Yang, Y.Y. Chang and J.R. Koduru. 2016. Low-cost magnetized Lonicera japonica flower biomass for the sorption removal of heavy metals. Hydrometallurgy 165: 81–89.

Lingamdinne, L.P., Y.-Y. Chang, J.-K. Yang, J. Singh, E.-H. Choi, M. Shiratani, J.R. Koduru and P. Attri. 2017. Biogenic reductive preparation of magnetic inverse spinel iron oxide nanoparticles for the adsorption removal of heavy metals. Chem. Eng. J. 307: 74–84.

Linic, S., P. Christopher and D. Ingram. 2011. Plasmonic-metal nanostructures for efficient conversion of solar to chemical energy. Nature Mater. 10: 911–921.

Litter, M.I. 1999. Heterogeneous photocatalysis: transition metal ions in photocatalytic systems. Appl. Catal. B 23: 89–114.

Liu, L., S.X. Liu, Q.P. Zhang, C. Li, C.L. Bao, X.T. Liu and P.F. Xiao. 2013. Adsorption of Au(III), Pd(II), and Pt(IV) from aqueous solution on to graphene oxide. J. Chem. Eng. Data 58: 209–216.

Liu, M., T. Wen, X. Wu, C. Chen, J. Hu, J. Li and X. Wang. 2013. Synthesis of porous Fe_3O_4 hollow microspheres/graphene oxide composite for Cr(VI) removal. Dalton Trans. 42: 14710–14717.

Liu, M., R. Inde, M. Nishikawa, X. Qiu, D. Atarashi, E. Sakai, Y. Nosaka, K. Hashimoto and M. Miyauchi. 2014. Enhanced photoactivity with nanocluster-grafted titanium dioxide photocatalysts. ACS Nano 8: 7229–7238.

Liu, S., L. Fu, Y. Liu, L. Meng, Y. Dong, Y. Li and M. Ma. 2016. Cu/C or Cu_2O/C composites: Selective synthesis, characterization, and applications in water treatment. Sci. Adv. Mater. 8: 2045–2053.

Liu, W., T. Wang, A.G.L. Borthwick, Y. Wang, X. Yin, X. Li and J. Ni. 2013. Adsorption of Pb^{2+}, Cd^{2+}, Cu^{2+} and Cr^{3+} onto titanate nanotubes: Competition and effect of inorganic ions. Sci. Total Environ. 456-457: 171–180.

Liu, W., J. Ni and X. Yin. 2014. Synergy of photocatalysis and adsorption for simultaneous removal of Cr(VI) and Cr(III) with TiO_2 and titanate nanotubes. Water Res. 53: 12–25.

Liu, W., W. Sun, Y. Han, M. Ahmad and J. Ni. 2014. Adsorption of Cu(II) and Cd(II) on titanate nanomaterials synthesized via hydrothermal method under different NaOH concentrations: Role of sodium content. Colloids Surf A Physicochem. Eng. Asp. 452: 138–147.

Lo, I.M.C., C.S.C. Lam and K.C.K. Lai. 2006. Hardness and carbonate effects on the reactivity of zero-valent iron for Cr(VI) removal. Water Res. 40: 595–605.

Lu, A.H., S.J. Zhong, J. Chen, J.X. Shi, J.L. Tang and X.Y. Lu. 2006. Removal of Cr(VI) and Cr(III) from aqueous solutions and industrial wastewaters by natural clino-pyrrhotite. Environ. Sci. Technol. 40: 3064–3069.

Lu, C. and H. Chiu. 2006. Adsorption of zinc(II) from water with purified carbon nanotubes. Chem. Eng. Sci. 61: 1138–1145.

Lu, C. and C. Liu. 2006. Removal of nickel(II) from aqueous solution by carbon nanotubes. J. Chem. Technol. Biotechnol. 81: 1932–1940.

Lu, C., C. Liu and G.P. Rao. 2008. Comparisons of sorbent cost for the removal of Ni^{2+} from aqueous solution by carbon nanotubes and granular activated carbon. J. Hazard. Mater. 151: 239–246.

Lu, C. and H. Chiu. 2008. Chemical modification of multiwalled carbon nanotubes for sorption of Zn^{2+} from aqueous solution. Chem. Eng. J. 139: 462–468.

Lu, C.S., H. Chiu and C.T. Liu. 2006. Removal of zinc(II) from aqueous solution by purified carbon nanotubes: kinetics and equilibrium studies. Ind. Eng. Chem. Res. 45: 2850–2855.

Lv, S., X. Chen, Y. Ye, S. Yin, J. Cheng and M. Xia. 2009. Ricehull/$MnFe_2O_4$ composite: preparation, characterization and its rapid microwave-assisted COD removal for organic wastewater. J. Hazard. Mater. 171: 634–639.

Ma, H., Y. Zhang, Q. Hu, D. Yang, Z. Yu and M. Zhai. 2012. Chemical reduction and removal of Cr(VI) from acidic aqueous solution by ethylenediamine-reduced graphene oxide. J. Mater. Chem. 22: 5914–5916.

Ma, Q., H. Zhang, R. Guo, Y. Cui, X. Deng, X. Cheng, M. Xie, Q. Cheng and B. Li. 2017. A novel strategy to fabricate plasmonic Ag/AgBr nano-particle and its enhanced visible photocatalytic performance and mechanism for degradation of acetaminophen. J. Taiwan Instit. Chem. Eng. 80: 176–183.

Ma, X., Y. Wang, M. Gao, H. Xu and G. Li. 2010. A novel strategy to prepare ZnO/PbS heterostructured functional nanocomposite utilizing the surface adsorption property of ZnO nanosheets. Catal. Today 158: 459–463.

Machida, M., B. Fotoohi, Y. Amamo, T. Ohba, H. Kanoh and L. Mercier. 2012. Cadmium(II) adsorption using functional mesoporous silica and activated carbon. J. Hazard. Mater. 221: 220–227.

Madadrang, C.J., H.Y. Kim, G.H. Gao, N. Wang, J. Zhu, H. Feng, M. Gorring, M.L. Kasner and S.F. Hou. 2012. Adsorption behavior of EDTA-graphene oxide for Pb(II) removal. ACS Appl. Mater. Interfaces 4: 1186–1193.

Mahfouz, M.G., A.A. Galhoum, N.A. Gomaa, S.S. Abdel-Rehem, A.A. Atia, T. Vincent and E. Guibal. 2015. Uranium extraction using magnetic nano-based particles of diethylenetriamine-functionalized chitosan: Equilibrium and kinetic studies. Chem. Eng. J. 262: 198–209.

Mahmood, A., S. Mahmood Ramay, Y. Al-Zaghayer, S. Atiq and S. Ansari. 2015. Thermal/plasma treatment effect on photocatalytic degradation of aqueous solution of methylene blue using Au-doped Fe/TiO$_2$ photocatalyst. Desalination Water Treatment 57: 5183–5192.

Mahmoud, M.E., M.S. Abdelwaha and E.M. Fathallah. 2013. Design of novel nano-sorbents based on nano-magnetic iron oxide-bound-nano-silicon oxide-immobilized-triethylenetetramine for implementation in water treatment of heavy metals. Chem. Eng. J. 223: 318–327.

Mai, N., J. Bae, I. Kim, S. Park, G. Lee, J. Kim, D. Lee, H. Son, Y. Lee and J. Hur. 2017. A recyclable, recoverable, and reformable hydrogel-based smart photocatalyst. Environ. Sci. Nano 4: 955–966.

Maity, D. and D.C. Agrawal. 2007. Synthesis of iron oxide nanoparticles under oxidizing environment and their stabilization in aqueous and non-aqueous media. J. Magn. Magn. Mater. 308: 46–55.

Mallakpour, S. and E. Khadem. 2016. Carbon nanotube–metal oxide nanocomposites: fabrication, properties, and applications. Chem. Eng. J. 302: 344–367.

Melo, D.Q., V.O. Neto, J.T. Oliveira, A.L. Barros, E.C. Gomes, G.S. Raulino, E. Longuinotti and R.F. Nascimento. 2013. Adsorption equilibria of Cu^{2+}, Zn^{2+}, and Cd^{2+} on EDTA-functionalized silica spheres. J. Chem. Eng. Data 58: 798–806.

Meng, Y., D. Chen, Y. Sun, D. Jiao, D. Zeng and Z. Liu. 2015. Adsorption of Cu^{2+} ions using chitosan-modified magnetic Mn ferrite nanoparticles synthesized by microwave-assisted hydrothermal method. Appl. Surf. Sci. 324: 745–750.

Min, L.L., Z.H. Yuan, L.B. Zhong, Q. Liu, R.X. Wu and Y.M. Zheng. 2015. Preparation of chitosan based electrospun nanofiber membrane and its adsorptive removal of arsenate from aqueous solution. Chem. Eng. J. 267: 132–141.

Mishra, A.K. and S. Ramaprabhu. 2010. Magnetite decorated multiwalled carbon nanotube based supercapacitor for arsenic removal and desalination of seawater. J. Phys. Chem. C 114: 2583–2590.

Mishra, A.K. and S. Ramaprabhu. 2011. Functionalized graphene sheets for arsenic removal and desalination of sea water. Desalination 282: 39–45.

Mishra, A.K. and S. Ramaprabhu. 2012. The role of functionalized multiwalled carbon nanotubes based super capacitor for arsenic removal and desalination of sea water. J. Exp. Nanosci. 7: 85–97.

Mizutani, N., T. Iwasaki, S. Watano, T. Yanagida, H. Tanaka and T. Kawai. 2008. Effect of ferrous/ferric ions molar ratio on reaction mechanism for hydrothermal synthesis of magnetite nanoparticles. Bull. Mater. Sci. 31: 713–717.

Mohammadnezhad, G., S. Abad, R. Soltani and M. Dinari. 2017. Study on thermal, mechanical and adsorption properties of amine-functionalized MCM-41/PMMA and MCM-41/PMMA nanocomposites prepared by ultrasonic irradiation. Ultrason. Sonochem. 39: 765–773.

Mu, R.X., Z.Y. Xu, L.Y. Li, Y. Shao, H.Q. Wan and S.R. Zheng. 2010. On the photocatalytic properties of elongated TiO$_2$ nanoparticles for phenol degradation and Cr(VI) reduction. J. Hazard. Mater. 176: 495–502.

Mubarak, N., J. Sahu, E. Abdullah and N. Jayakumar. 2013. Removal of heavy metals from wastewater using carbon nanotubes. Sep. Purif. Rev. 43: 311–338.

Nagarajah, R., K.T. Wong, G. Lee, K.H. Chu, Y. Yoon, N.C. Kimb and M. Jang. 2017. Synthesis of a unique nanostructured magnesium oxide coated magnetite cluster composite and its application for the removal of selected heavy metals. Sep. Purif. Technol. 174: 290–300.

Najafi, M., Y. Yousefi and A. Rafati. 2012. Synthesis, characterization and adsorption studies of several heavy metal ions on amino-functionalized silica nanohollow sphere and silica gel. Sep. Purif. Technol. 85: 193–205.

Nandi, D., T. Basu, S. Debnath, A.K. Ghosh, A. Ghosh and U.C. De. 2013. Mechanistic insight for the sorption of Cd(II) and Cu(II) from aqueous solution on magnetic Mn-doped Fe(III) oxide nanoparticle implanted graphene. J. Chem. Eng. Data 58: 2809–2818.

National Primary Drinking Water Regulation. 2016. US, https://www.epa.gov/ground-water-and-drinking-water/table-regulated-drinking-water-contaminants#Inorganic.

Nejaei, Arezoo, Rahmanifar, Bahar, Maleki, Aghil Esmaeilzadeh, Gazani and Zhila Nasouri. 2017. Collagen-based CuO nanoparticles: a new adsorbent for removal of cypermethrin from water, Fresenius Environ. Bull. 26(4): 2739–2746.

Neyens, E. and J. Baeyens. 2003. A review of classic Fenton's peroxidation as an advanced oxidation technique. J. Hazard. Mater. 98: 33–50.

O'Carroll, D., B. Sleep, M. Krol, H. Boparai and C. Kocur. 2013. Nanoscale zero valent iron and bimetallic particles for contaminated site remediation. Adv. Water Resour. 51: 104–122.

Ollis, D. 1993. Comparative aspects of advanced oxidation processes, Chapter 2. pp. 18–34. *In*: Tedder, D. and F. Pohland (eds.). Emerging Technologies in Hazardous Waste Management III. ACS Symposium Series, Washington, DC, Vol. 518.

Ou, Q., L. Zhou, S. Zhao, H. Geng, J. Hao, Y. Xu, H. Chen and X. Chen. 2012. Self-templated synthesis of bifunctional Fe_3O_4@$MgSiO_3$ magnetic sub-microspheres for toxic metal ions removal. Chemical Engineering Journal 180: 121–127.

Pan, B., D.H. Lin, H. Mashayekhi and B.S. Xing. 2008. Adsorption and hysteresis of bisphenol A and 17 alpha-ethinyl estradiol on carbon nanomaterials. Environ. Sci. Technol. 42: 5480–5485.

Pan, B. and B.S. Xing. 2008. Adsorption mechanisms of organic chemicals on carbon nanotubes. Environ. Sci. Technol. 42: 9005–9013.

Pandey, S. and S.B. Mishra. 2011. Organic–inorganic hybrid of chitosan/organoclay bionanocomposites for hexavalent chromium uptake. J. Colloid Interface Sci. 361: 509–520.

Park, J. 2011. Synthesis of TiO_2-based photocatalyst and their environmental applications; Washington University: Saint Louis, Mo.

Park, S., J.H. An, I.W. Jung, R.D. Piner, S.J. An, X.S. Li, A. Velamakanni and R.S. Ruoff. 2009. Colloidal suspensions of highly reduced graphene oxide in a wide variety of organic solvents. Nano Lett. 9: 1593–1597.

Peng, H., J. Ouyang and Y. Peng. 2017. A simple approach for the synthesis of bi-functional Fe_3O_4@WO_{3-x} core–shell nanoparticles with magnetic-microwave to heat responsive properties. Inorg. Chem. Commun. 84: 138–143.

Prakash, K., P. Kumar, P. Latha, K. Stalin Durai, R. Shanmugam and S. Karuthapandian. 2017. Dry synthesis of water lily flower like SrO_2/g-C_3N_4 nanohybrids for the visible light induced superior photocatalytic activity. Mater. Res. Bull. 93: 112–122.

Qiu, R., D. Zhang, Z. Diao, X. Huang, C. He, J.L. Morel and Y. Xiong. 2012. Visible light induced photocatalytic reduction of Cr(VI) over polymer-sensitized TiO_2 and its synergism with phenol oxidation. Water Res. 46: 2299–2306.

Rahimi-Nasrabadi, M., S. Pourmortazavi, M. Aghazadeh, M. Ganjali, M. Karimi and P. Norouzi. 2017. Application of Taguchi robust design to the optimization of the synthesis of holmium carbonate and oxide nanoparticles and exploring their photocatalyst behaviors for water treatment. Journal of Materials Science: Mater Electronics 28: 11383–11392.

Rahman, M. and M. Islam. 2009. Effects of pH on isotherms modeling for Cu(II) ions adsorption using maple wood sawdust. Chem. Eng. J. 149: 273–280.

Ramos, M.A.V., W. Yan, X.Q. Li, B.E. Koel and W.X. Zhang. 2009. Simultaneous oxidation and reduction of arsenic by zero-valent iron nanoparticles: understanding the significance of the core-shell structure. J. Phys. Chem. C 113: 14591–14594.

Rani, M., U. Shanker and A. Chaurasia. 2017. Catalytic potential of laccase immobilized on transition metal oxides nanomaterials: Degradation of alizarin red S dye. J. Environ. Chem. Eng. 5: 2730–2739.

Rao, G., C. Lu and F. Su. 2007. Sorption of divalent metal ions from aqueous solution by carbon nanotubes: a review. Sep. Purif. Technol. 58: 224–231.

Razzaz, A., S. Ghorban, L. Hosayni, M. Irani and M. Aliabadi. 2016. Chitosan nanofibers functionalized by TiO_2 nanoparticles for the removal of heavy metal ions. J. Taiwan Inst. Chem. Eng. 58: 333–343.

Ren, J., Y. Chai, Q. Liu, L. Zhang and W. Dai. 2017. Intercorrelated Ag_3PO_4 nanoparticles decorated with graphic carbon nitride: Enhanced stability and photocatalytic activities for water treatment. Appl. Surface Sci. 403: 177–186.

Ren, X., J. Li, X. Tan and X. Wang. 2013. Comparative study of graphene oxide, activated carbon and carbon nanotubes as adsorbents for copper decontamination. Dalton Trans. 42: 5266–5274.

Ren, Y., H.A. Abbood, F. He, H. Peng and K. Huang. 2013. Magnetic EDTA-modified chitosan/SiO_2/Fe_3O_4 adsorbent: Preparation, characterization, and application in heavy metal adsorption. Chem. Eng. J. 226: 300–311.

Renuka, L., K. Anantharaju, Y. Vidya, H. Nagaswarupa, S. Prashantha, S. Sharma, H. Nagabhushana and G. Darshan. 2017. A simple combustion method for the synthesis of multi-functional ZrO_2/CuO nanocomposites: Excellent performance as sunlight photocatalysts and enhanced latent fingerprint detection. Appl. Catal. B: Environ. 210: 97–115.

Robel, I., V. Subramanian, M. Kuno and P. Kamat. 2006. Quantum dot solar cells harvesting light energy with CdSe nanocrystals molecularly linked to mesoscopic TiO_2 films. J. Am. Chem. Soc. 128: 2385–2393.

Sajjad, S., S.A.K. Leghari, F. Chen and J. Zhang. 2010. Bismuth-doped ordered mesoporous TiO_2: visible-light catalyst for simultaneous degradation of phenol and chromium. Chem. Eur. J. 16: 13795–13804.

Salehi, E., S. Madaeni, L. Rajabi, V. Vatanpour, A. Derakhshan, S. Zinadini, S. Ghorabi and H. Ahmadi Monfared. 2012. Novel chitosan/poly(vinyl) alcohol thin adsorptive membranes modified with amino functionalized multi-walled carbon nanotubes for Cu(II) removal from water: Preparation, characterization, adsorption kinetics and thermodynamics. Sep. Purif. Technol. 89: 309–319.

Salehi, E., S.S. Madaeni, L. Rajabi, A.A. Derakhshan, S. Daraei and V. Vatanpour. 2013. Static and dynamic adsorption of copper ions on chitosan/polyvinyl alcohol thin adsorptive membranes: Combined effect of polyethylene glycol and aminated multi-walled carbon nanotubes. Chem. Eng. J. 215-216: 791–801.

Santhosh, C., V. Velmurugan, G. Jacob, S. Jeong, A. Grace and A. Bhatnagar. 2016. Role of nanomaterials in water treatment applications: A review. Chem. Eng. J. 306: 1116–1137.

Savage, N. and M. Diallo. 2005. Nanomaterials and water purification: opportunities and challenges. J. Nanoparticle Res. 7: 331–342.

Schwarzer, H.C. and W. Peukert. 2004. Tailoring particle size through nanoparticle precipitation. Chem. Eng. Commun. 191: 580–606.

Scuderi, V., G. Impellizzeri, L. Romano, M. Scuderi, M. Brundo, K. Bergum, M. Zimbone, R. Sanz, M. Buccheri, F. Simone, G. Nicotra, B. Svensson, M.G. Grimaldi and V. Privitera. 2014. An enhanced photocatalytic response of nanometric TiO_2 wrapping of Au nanoparticles for eco-friendly water applications. Nanoscale 6: 11189–11195.

Selvi, K. 2001. Removal of Cr(VI) from aqueous solution by adsorption onto activated carbon. Bioresource Tech. 80: 87–89.

Shadbad, M., A. Mohebbi and A. Soltani. 2011. Mercury(II) removal from aqueous solutions by adsorption on multi-walled carbon nanotubes. Korean J. Chem. Eng. 28: 1029–1034.

Shanmugam, M., A. Alsalme, A. Alghamdi and R. Jayavel. 2016. *In situ* microwave synthesis of graphene–TiO_2 nanocomposites with enhanced photocatalytic properties for the degradation of organic pollutants. J. Photochem. Photobio. B: Biology 163: 216–223.

Shao, D., J. Hu and X. Wang. 2010. Plasma induced grafting multiwalled carbon nanotube with chitosan and its application for removal of UO_2^{2+}, Cu^{2+}, and Pb^{2+} from aqueous solutions. Plasma Processes Polym. 7: 977–985.

Shin, K., J. Hong and J. Jang. 2011. Heavy metal ion adsorption behavior in nitrogen-doped magnetic carbon nanoparticles: Isotherms and kinetic study. J. Hazard. Mater. 190: 36–44.

Sitko, R., M. Musielak, B. Zawisza, E. Talik and A. Gagor. 2016. Graphene oxide/cellulose membranes in adsorption of divalent metal ions. RSC Adv. 6: 96595–96605.

Sobhana, L., M. Sarakha, V. Prevot and P. Fardim. 2016. Layered double hydroxides decorated with Au-Pd nanoparticles to photodegradate Orange II from water. Appl. Clay Sci. 134: 120–127.

Soltani, R., M. Dinari and G. Mohammadnezhad. 2018. Ultrasonic-assisted synthesis of novel nanocomposite of poly(vinylalcohol) and amino-modified MCM-41: A green adsorbent for Cd(II) removal. Ultrason. Sonochem. 40: 533–542.

Song, H.J., L.C. Zhang, C.L. He, Y. Qu, Y.F. Tian and Y. Lv. 2011. Graphene sheets decorated with SnO_2 nanoparticles: *in situ* synthesis and highly efficient materials for cataluminescence gas sensors. J. Mater. Chem. 21: 5972–5977.

Sreeprasad, T.S., S.M. Maliyekkal, K.P. Lisha and T. Pradeep. 2011. Reduced graphene oxide-metal/metal oxide composites: facile synthesis and application in water purification. J. Hazard. Mater. 186: 921–931.

Stafiej, A. and K. Pyrzynska. 2008. Solid phase extraction of metal ions using carbon nanotubes. Microchem. J. 89: 29–33.

Su, Q., B. Pan, S. Wan, W. Zhang and L. Lv. 2010. Use of hydrous manganese dioxide as a potential sorbent for selective removal of lead, cadmium, and zinc ions from water. J. Colloid Interface Sci. 349: 607–612.

Sun, S. and H. Zeng. 2002. Size-controlled synthesis of magnetite nanoparticles. J. Am. Chem. Soc. 124: 8204–8205.

Tawabini, B., S. Al-Khaldi, M. Atieh and M. Khaled. 2010. Removal of mercury from water by multi-walled carbon nanotubes. Water Sci. Tech. 61: 591–598.

Tawabini, B., S. Al-Khaldi, M. Khaled and M. Atieh. 2011. Removal of arsenic from water by iron oxide nanoparticles impregnated on carbon nanotubes. J. Environ. Sci. Health, Part A 46: 215–223.

Teja, A.S. and P.Y. Koh. 2009. Synthesis, properties, and applications of magnetic iron oxide nanoparticles. Prog. Cryst. Growth & Charact. Mater. 55: 22–45.

Testa, J.J., M.A. Grela and M.I. Litter. 2001. Experimental evidence in favor of an initial one-electron-transfer process in the heterogeneous photocatalytic reduction of chromium(VI) over TiO_2. Langmuir 17: 3515–3517.

Testa, J.J., M.A. Grela and M.I. Litter. 2004. Heterogeneous photocatalytic reduction on chromium(VI) over TiO_2 particles in the presence of oxalate: involvement of Cr(V) species. Environ. Sci. Technol. 38: 1589–1594.

Thangavel, S., S. Thangavel, N. Raghavan, R. Alagu and G. Venugopal. 2017. Efficient visible-light photocatalytic and enhanced photocorrosion inhibition of Ag_2WO_4 decorated MoS_2 nanosheets. J. Phys. Chem. Solids 110: 266–273.

The Millennium Development Goals Report. 2008. United Nations, http://www.un.org/millenniumgoals/2008highlevel/pdf/newsroom/mdg%20reports/MDG_Report_2008_ENGLISH.pdf.

The United Nations World Water Development Report. 2016. http://unesdoc.unesco.org/images/0024/002439/243938e.pdf.

Thinh, N.N., P.T. Hanh, L.T.T. Ha, L.N. Anh, T.V. Hoang, V.D. Hoang, L.H. Dang, N.V. Khoi and T.D. Lam. 2013. Magnetic chitosan nanoparticles for removal of Cr(VI) from aqueous solution. Mat. Sci. Eng. C-Mater. 33: 1214–1218.

Tian, Y., L. Huang, X. Zhou and C. Wu. 2012. Electroreduction of hexavalent chromium using a polypyrrole-modified electrode under potentiostatic and potentiodynamic conditions. J. Hazard. Mater. 225: 15–20.

Tiraferri, A., K.L. Chen, R. Sethi and M. Elimelech. 2008. Reduced aggregation and sedimentation of zero-valent iron nanoparticles in the presence of guar gum. J. Colloid Interface Sci. 324: 71–79.

Tofighy, M.A. and T. Mohammadi. 2011. Adsorption of divalent heavy metal ions from water using carbon nanotube sheets. J. Hazard. Mater. 185: 140–147.

Tosco, T., M. Petrangeli Papini, C. Cruz Viggi and R. Sethi. 2014. Nanoscale zerovalent iron particles for groundwater remediation: a review. J. Clean. Prod. 77: 10–21.

Tu, Y.J., S.C. Lo and C.F. You. 2015. Selective and fast recovery of neodymium from seawater by magnetic iron oxide Fe_3O_4. Chem. Eng. J. 262: 966–972.

Umukoro, E., M. Peleyeju, J. Ngila and O. Arotiba. 2016. Photocatalytic degradation of acid blue 74 in water using Ag–Ag_2O–Zno nanostuctures anchored on graphene oxide. Solid State Sci. 51: 66–73.

United Nations Educational, Scientific and Cultural Organization, Water for sustainable world. 2015. https://www.unesco-ihe.org/sites/default/files.

Upadhyayula, V., S. Deng, M. Mitchell and G. Smith. 2009. Application of carbon nanotube technology for removal of contaminants in drinking water: A review. Sci. Total Environ. 408: 1–13.

Vellaichamy, S. and K. Palanivelu. 2011. Preconcentration and separation of copper, nickel and zinc in aqueous samples by flame atomic absorption spectrometry after column solid-phase extraction onto MWCNTs impregnated with D2EHPA-TOPO mixture. J. Hazard. Mater. 185: 1131–1139.

Venkatesh, G., A. Singh and B. Venkataramani. 2004. Silica gel loaded with o-dihydroxybenzene: Design, metal sorption equilibrium studies and application to metal enrichment prior to determination by flame atomic absorption spectrometry. Microchimica Acta 144: 233–241.

Venkateswarlu, S. and M.Y. Yoon. 2015. Core–shell ferromagnetic nanorod based on amine polymer composite (Fe_3O_4@DAPF) for fast removal of Pb(II) from aqueous solutions. ACS Appl. Mater. Interfaces 7: 25362–25372.

Vilela, D., J. Parmar, Y. Zeng, Y. Zhao and S. Sánchez. 2016. Graphene-based microbots for toxic heavy metal removal and recovery from water. Nano Lett. 16: 2860–2866.

Vojoudi, H., A. Badiei, S. Bahar, G.M. Ziarani, F. Faridbod and M.R. Ganjali. 2017. A new nano-sorbent for fast and efficient removal of heavy metals from aqueous solutions based on modification of magnetic mesoporous silica nanospheres. J. Magn. Magn. Mater. 441: 193–203.

Vuković, G., A. Marinković, M. Čolić, M. Ristić, R. Aleksić, A. Perić-Grujić and P. Uskoković. 2010. Removal of cadmium from aqueous solutions by oxidized and ethylenediamine-functionalized multi-walled carbon nanotubes. Chem. Eng. J. 157: 238–248.

Vuković, G.D., A.D. Marinković, S.D. Skapin, M.Đ. Ristić, R. Aleksić, A.A. Perić-Grujić, S. Petar and P.S. Uskoković. 2011. Removal of lead from water by amino modified multi-walled carbon nanotubes. Chem. Eng. J. 173: 855–865.

Waalkes, M. 2000. Cadmium carcinogenesis in review. J. Inorg. Biochem. 79: 241–244.

Wang, H., X.Z. Yuan, Y. Wu, H.J. Huang, G.M. Zeng, Y. Liu, X.L. Wang, N.B. Lin and Y. Qi. 2013. Adsorption characteristics and behaviors of graphene oxide for Zn(II) removal from aqueous solution. Appl. Surf. Sci. 279: 432–440.

Wang, J., W. Xu, L. Chen, X. Huang and J. Liu. 2014. Preparation and evaluation of magnetic nanoparticles impregnated chitosan beads for arsenic removal from water. Chem. Eng. J. 251: 25–34.

Wang, J., L. Peng, F. Cao, B. Su and H. Shi. 2016. A Fe_3O_4-SiO_2-TiO_2 core-shell nanoparticle: Preparation and photocatalytic properties. Inorg. Nano-Metal Chem. 47: 396–400.

Wang, L., Q. Zhao, J. Hou, J. Yan, F. Zhang, J. Zhao, H. Ding, Y. Li and L. Ding. 2015. One-step solvothermal synthesis of magnetic Fe_3O_4–graphite composite for Fenton-like degradation of levofloxacin. J. Environ. Sci. Health, Part A 51: 52–62.

Wang, N., L. Zhu, K. Deng, Y. She, Y. Yu and H. Tang. 2010. Visible light photocatalytic reduction of Cr(VI) on TiO_2 *in situ* modified with small molecular weight organic acids. Appl. Catal. B: Environ. 95: 400–407.

Wang, Q., J. Li, C. Chen, X. Ren, J. Hu and X. Wang. 2011. Removal of cobalt from aqueous solution by magnetic multiwalled carbon nanotube/iron oxide composites. Chem. Eng. J. 174: 126–133.

Wang, Q., X. Shi, J. Xu, J.C. Crittenden, E. Liu, Y. Zhang and Y. Cong. 2016. Highly enhanced photocatalytic reduction of Cr(VI) on AgI/TiO_2 under visible light irradiation: Influence of calcination temperature. J. Hazard. Mater. 307: 213–220.

Wang, T., W. Liu, N. Xu and J. Ni. 2013. Adsorption and desorption of Cd(II) onto titanate nanotubes and efficient regeneration of tubular structures. J. Hazard. Mater. 250-251: 379–386.

Wang, Y., S. Lin and R. Juang. 2003. Removal of heavy metal ions from aqueous solutions using various low-cost adsorbents. J. Hazard. Mater. 102: 291–302.

Wang, Y., S. Liang, B. Chen, F. Guo, S. Yu and Y. Tang. 2013. Synergistic removal of Pb(II), Cd(II) and humic acid by Fe_3O_4@mesoporous silica-graphene oxide composites. PLoS ONE 8: e65634.

Wang, Z., Y. Du, Y. Liu, B. Zou, J. Xiao and J. Ma. 2016. Degradation of organic pollutants by $NiFe_2O_4$/peroxymonosulfate: efficiency, influential factors and catalytic mechanism. RSC Advances 6: 11040–11048.

Wang, Z., J. Xu, Y. Hu, H. Zhao, J. Zhou, Y. Liu, Z. Lou and X. Xu. 2016. Functional nanomaterials: Study on aqueous Hg(II) adsorption by magnetic Fe_3O_4@SiO_2-SH nanoparticles. J. Taiwan Inst. Chem. Eng. 60: 394–402.

Wei, Y., G. Chao, F.L. Meng, H.H. Li, L. Wang, J.H. Liu and X.J. Huang. 2012. SnO_2/Reduced graphene oxide nanocomposite for the simultaneous electrochemical detection of cadmium(II), lead(II), copper(II), and mercury(II): an interesting favorable mutual interference. J. Phys. Chem. C 116: 1034–1041.

Wu, W., C.Z. Jiang and V.A.L. Roy. 2016. Designed synthesis and surface engineering strategies of magnetic iron oxide nanoparticles for biomedical applications. Nanoscale 8: 19421–19474.

Xiong, L., C. Chen, Q. Chen and J. Ni. 2011. Adsorption of Pb(II) and Cd(II) from aqueous solutions using titanate nanotubes prepared via hydrothermal method. J. Hazard. Mater. 189: 741–748.

Xu, P., G.M. Zeng, D.L. Huang, C.L. Feng, S. Hu, M.H. Zhao, C. Lai, Z. Wei, C. Huang, G.X. Xie and Z.F. Liu. 2012. Use of iron oxide nanomaterials in wastewater treatment: a review. Sci. Total Environ. 424: 1–10.

Yan, W., H.L. Lien, B.E. Koel and W.X. Zhang. 2013. Iron nanoparticles for environmental clean-up: recent developments and future outlook. Environ. Sci.: Process. Impacts 15: 63–77.

Yang, K., L. Zhu and B. Xing. 2006. Adsorption of polycyclic aromatic hydrocarbons by carbon nanomaterials. Environ. Sci. Technol. 40: 1855–1861.

Yang, K. and B.S. Xing. 2010. Adsorption of organic compounds by carbon nanomaterials in aqueous phase: Polanyi theory and its application. Chem. Rev. 110: 5989–6008.

Yang, S., J. Hu, C. Chen, D. Shao and X. Wang. 2011. Few-layered graphene oxide nanosheets as superior sorbents for heavy metal ion pollution management. Environ. Sci. Technol. 45: 3621–3627.

Yang, S., P. Wu, M. Chen, Z. Huang, W. Li, N. Zhu and Y. Ji. 2016. Enhanced photo-degradation of bisphenol a under simulated solar light irradiation by Zn–Ti mixed metal oxides loaded on graphene from aqueous media. RSC Adv. 6: 26495–26504.

Yang, Y., Y. Xie, L. Pang, M. Li, X. Song, J. Wen and H. Zhao. 2013. Preparation of reduced graphene oxide/poly(acrylamide) nanocomposite and its adsorption of Pb(II) and methylene blue. Langmuir 29: 10727–10736.

Yantasee, W., C.L. Warner, T. Sangvanich, R.S. Addleman, T.G. Carter, R.J. Wiacek, G.E. Fryxell, C. Timchalk and M.G. Warner. 2007. Removal of heavy metals from aqueous systems with thiol functionalized superparamagnetic nanoparticles. Environ. Sci. Technol. 41: 5114–5119.

Yavuz, E., S. Tokalıoglu, H. Sxahan and S. Patat. 2013. A graphene/Co$_3$O$_4$ nanocomposite as a new adsorbent for solid phase extraction of Pb(II), Cu(II) and Fe(III) ions in various samples. RSC Adv. 3: 24650–24657.

Ye, Y., Y. Feng, H. Bruning, D. Yntema and H. Rijnaarts. 2017. Photocatalytic degradation of metoprolol by TiO$_2$ nanotube arrays and UV-LED: Effects of catalyst properties, operational parameters, commonly present water constituents, and photo-induced reactive species. Appl. Catal. B: Environ. 220: 171–181.

Yosefi, L. and M. Haghighi. 2018. Fabrication of nanostructured flowerlike p-BiOI/p-NiO heterostructure and its efficient photocatalytic performance in water treatment under visible-light irradiation. Appl. Catal. B: Environ. 220: 367–378.

Yu, H., H. Irie and K. Hashimoto. 2010. Conduction band energy level control of titanium dioxide: Toward an efficient visible-light-sensitive photocatalyst. J. Am. Chem. Soc. 132: 6898–6899.

Yu, J., X. Zhao, L. Yu, F. Jiao, J. Jiang and X. Chen. 2013. Removal, recovery and enrichment of metals from aqueous solutions using carbon nanotubes. J. Radioanalytical Nuclear Chem. 299: 1155–1163.

Yu, R.-F., F.H. Chi, W.P. Cheng and J.C. Chang. 2014. Application of pH, ORP, and DO monitoring to evaluate chromium(VI) removal from wastewater by the nanoscale zerovalent iron (nZVI) process. Chem. Eng. J. 255: 568–576.

Zeng, G., M. Chen and Z. Zeng. 2013. Risks of neonicotinoid pesticides. Science 340: 1403–1403.

Zeng, H., X. Liu, T. Wei, X. Li, T. Liu, X. Min, Q. Zhu, X. Zhao and J. Li. 2017. Boosting visible light photo-/Fenton-catalytic synergetic activity of BiOIO$_3$ by coupling with Fe$_2$O$_3$. RSC Advances 7: 23787–23792.

Zhang, A., W. Wang, D. Pei and H. Yu. 2016. Degradation of refractory pollutants under solar light irradiation by a robust and self-protected ZnO/CdS/TiO$_2$ hybrid photocatalyst. Water Res. 92: 78–86.

Zhang, A., Y. He, T. Lin, N. Huang, Q. Xu and J. Feng. 2017. A simple strategy to refine Cu$_2$O photocatalytic capacity for refractory pollutants removal: Roles of oxygen reduction and Fe(II) chemistry. J. Hazard. Mater. 330: 9–17.

Zhang, G., G. Kim and W. Choi. 2014. Visible light driven photocatalysis mediated via ligand-to-metal charge transfer (LMCT): an alternative approach to solar activation of titania. Energy Environ. Sci. 7: 954–966.

Zhang, J., X. Wang, P. Xia, X. Wang, J. Huang, J. Chen, B. Louangsouphom and J. Zhao. 2016. Enhanced sunlight photocatalytic activity and recycled Ag–N co-doped TiO$_2$ supported by expanded graphite C/C composites for degradation of organic pollutants. Res. Chem. Intermed. 42: 5541–5557.

Zhang, K., H. Li, X. Xu and H. Yu. 2018. Synthesis of reduced graphene oxide/NiO nanocomposites for the removal of Cr(VI) from aqueous water by adsorption. Microporous Mesoporous Mater. 255: 7–14.

Zhang, L., R. He and H.C. Gu. 2006. Oleic acid coating on the monodisperse magnetite nanoparticles. Appl. Surf. Sci. 253: 2611–2617.

Zhang, N., H. Qiu, Y. Si, W. Wang and J. Gao. 2011. Fabrication of highly porous biodegradable monoliths strengthened by graphene oxide and their adsorption of metal ions. Carbon 49: 827–837.

Zhang, Q., X. Quan, H. Wang, S. Chen, Y. Su and Z. Li. 2017. Constructing a visible-light-driven photocatalytic membrane by g-C$_3$N$_4$ quantum dots and TiO$_2$ nanotube array for enhanced water treatment. Scientific Reports, 7.

Zhang, X., L. Chen, T. Yuan, H. Huang, Z. Sui, R. Du, X. Li, Y. Lu and Q. Li. 2014. Dendrimer-linked, renewable and magnetic carbon nanotube aerogels. Mater. Horiz. 1: 232–236.

Zhang, Y., H. Ma, J. Peng, M. Zhai and Z. Yu. 2013. Cr(VI) removal from aqueous solution using chemically reduced and functionalized graphene oxide. J. Mater. Sci. 48: 1883–1889.

Zhang, Y., L. Yan, W. Xu, X. Guo, L. Cui, L. Gao, Q. Wei and B. Du. 2014. Adsorption of Pb(II) and Hg(II) from aqueous solution using magnetic $CoFe_2O_4$-reduced graphene oxide. J. Mol. Liq. 191: 177–182.

Zhang, Y., T. Yan, L. Yan, X. Guo, L. Cui, Q. Wei and B. Du. 2014. Preparation of novel cobalt ferrite/ chitosan grafted with graphene composite as effective adsorbents for mercury ions. J. Mol. Liq. 198: 381–387.

Zhang, Y., B. Wu, H. Xu, H. Liu, M. Wang, Y. He and B. Pan. 2016. Nanomaterials-enabled water and wastewater treatment. NanoImpact 3-4: 22–39.

Zhang, Y.J., H.J. Chi, W.H. Zhang, Y.Y. Sun, Q. Liang, Y. Gu and R.Y. Jing. 2014. Highly efficient adsorption of copper ions by a PVP-reduced graphene oxide based on a new adsorptions mechanism. Nano-Micro Lett. 6: 80–87.

Zhao, G., X. Ren, X. Gao, X. Tan, J. Li, C. Chen, Y. Huang and X. Wang. 2011. Removal of Pb(II) ions from aqueous solutions on few layered graphene oxide nanosheets. Dalton Trans. 40: 10945–10952.

Zhao, J., J. Liu, N. Li, W. Wang, J.N.Z. Zhao and F. Cui. 2016. Highly efficient removal of bivalent heavy metals from aqueous systems by magnetic porous Fe_3O_4-MnO_2: adsorption behavior and process study. Chem. Eng. J. 304: 737–746.

Zhao, Y.L. and J.F.R. Stoddart. 2009. Noncovalent functionalization of single-walled carbon nanotubes. Acc. Chem. Res. 42: 1161–1171.

Zhou, L., H.P. Deng, J.L. Wan, J. Shi and T. Su. 2013. A solvothermal method to produce RGO-Fe_3O_4 hybrid composite for fast chromium removal from aqueous solution. Appl. Surf. Sci. 283: 1024–1031.

Zhou, Y.T., H.L. Nie, C. Branford-White, Z.Y. He and L.M. Zhu. 2009. Removal of Cu^{2+} from aqueous solution by chitosan-coated magnetic nanoparticles modified with α-ketoglutaric acid. J. Colloid Interface Sci. 330: 29–37.

Zhu, M., L. Zhu, J. Wang, T. Yue, R. Li and Z. Li. 2017. Adsorption of Cd(II) and Pb(II) by *in situ* oxidized Fe_3O_4 membrane grafted on 316 L porous stainless steel filter tube and its potential application for drinking water treatment. J. Environ. Manage. 196: 127–136.

Zhu, W.P., S.P. Sun, J. Gao, F.J. Fu and T.S. Chung. 2014. Dual-layer polybenzimidazole /polyethersulfone (PBI/PES) nanofiltration (NF) hollow fiber membranes for heavy metals removal from wastewater. J. Membrane Sci. 456: 117–127.

Zhu, Y.W., S. Murali, W.W. Cai, X.S. Li, J.W. Suk, J.R. Potts and R.S. Ruoff. 2010. Graphene and graphene oxide: synthesis, properties, and applications. Adv. Mater. 22: 3906–3924.

Zimbone, M., G. Cacciato, M. Boutinguiza, V. Privitera and M. Grimaldi. 2017. Laser irradiation in water for the novel, scalable synthesis of black TiOx photocatalyst for environmental remediation. Beilstein J. Nanotech. 8: 196–202.

Chapter 8

Nanocatalysts Based in Zeolites for Environmental Applications

Isabel Correia Neves,[1,2,*] *António M. Fonseca*[1,2] and
Pier Parpot[1,2]

INTRODUCTION

Over the past years, zeolites have always found several new applications in the different fields. In the environmental applications, zeolites continue to be a very interesting material in the areas such as catalysis, separation and ion exchange, due to their unique physiochemical properties (Millini and Bellussi 2017; Sebastián et al. 2010; Serrano et al. 2017; Snyder et al. 2018).

The modifications of the zeolites by the insertion of metal complexes or metal species aims to prepare heterogeneous catalysts that perform important tasks ranging from selective oxidation to reduction reactions applied in the environmental field, especially in the treatment of the liquid or gas effluents (Kuźniarska-Biernacka et al. 2016; Begum et al. 2016; Centi and Perathoner 2010; Freitas et al. 2015; Hamid et al. 2017; Snyder et al. 2018; Soares et al. 2015c; Zeng et al. 2017) or transformation of chemical compounds into high value added products (Ammar et al. 2017; Dai et al. 2017; Danish et al. 2017; Ramírez-Garza et al. 2018; Sharifi et al. 2015).

Zeolites are very interesting porous materials with frameworks which contain silicon (Si), aluminium (Al) and oxygen (O) linked to form nano-channels and cages of regular dimensions resulting in different structures. The peculiar physicochemical properties, such as cation exchange capacity, high specific surface area, shape selectivity, high adsorption capacity, thermal stability and catalytic activity, privilege the use of zeolites in the environmental field (Centi and Perathoner 2010; Snyder

[1] CQUM, Centre of Chemistry, Chemistry Department, University of Minho, Campus de Gualtar, 4710-057 Braga, Portugal.
[2] CEB—Centre of Biological Engineering, University of Minho, 4710-057 Braga, Portugal.
* Corresponding author: ineves@quimica.uminho.pt

et al. 2018). Thanks to these properties, these materials find enormous industrial applications. The most important applications are those in the field of detergents (water softening, ion-exchange and purification), as catalysts (for the petroleum refining and petrochemical industry) and as adsorbents (gas separation and purification processes) with Western Europe, United States and China being the bigger consumers (Bellussi et al. 2010; Snyder et al. 2018).

The catalytic properties of the zeolites are improved by the introduction of the metal species into the framework. In the last years, the bimetallic catalysts based on zeolites have attracted considerable attention because they lead to the stable and selective heterogeneous catalysts with higher activities than those found for monometallic ones or even for the parent zeolites. This chapter briefly highlights the new trends in the preparation of the nanocatalysts based on zeolites for environmental applications. Some examples of catalytic processes which use bimetallic zeolite nanocatalysts in the environmental field are also included.

Bimetallic nanocatalysts based on zeolites

Recently, several works were carried out for the introduction of two distinct metal ions into the framework of the zeolites using ion exchange (Freitas et al. 2015; Soares et al. 2015c; Zeng et al. 2017) or impregnation (Hamid et al. 2016; Hamid et al. 2017) methods, in order to obtain stable heterogeneous catalysts. These currents methods are not new and are usually used in the preparation of the heterogeneous catalysts (Campanati et al. 2003; Otto et al. 2016).

Thanks to the high cation exchange capacity of the zeolites depending on Si/Al ratio, more than one metal can be introduced into the zeolite structure by the ion exchange method (Snyder et al. 2018). Zeolites with low Si/Al ratios shows selectivity for small and highly hydrated cations; however, zeolites with high Si/Al ratio give better results with less hydrated larger cations (Dyer 2005).

The ion exchange method can be schematically represented as follows by two steps (Ferreira et al. 2015; Zeng et al. 2017). In the first step, a monometallic nanocatalyst is produced:

$$ZEO\ (Z) + M_1^{n+}\ (aq) \rightarrow M_1\text{-}ZEO\ (Z) + n\ Na^+\ (aq) \tag{1}$$

In the second step, the bimetallic nanocatalyst is obtained:

$$M_1\text{-}ZEO\ (Z) + M_2^{n+}\ (aq) \rightarrow M_1M_2\text{-}ZEO\ (Z) + n\ Na^+\ (aq) \tag{2}$$

The locations of the metals in the bulk structure will determine the catalytic properties of the bimetallic nanocatalysts based on zeolites; however, the intrinsic characteristics of the zeolite also enhances the catalytic properties of the nanocatalyst. In this method, the amount of the metals introduced is restricted to the capacity of cation exchange of the zeolite structure and higher dispersion of metals is achieved since the reaction occurs on the whole available surface of the structure (Dyer 2005).

For the impregnation method, a given volume of both metals is added to the zeolite support and the solvent is removed. Depending of the solution volume, two approaches could be carried out: wet impregnation and incipient wetness impregnation

(Campanati et al. 2003). The metals are impregnated on the outer surface of the zeolite and the substrates have direct contact with higher amount of metals than those found for than ion-exchange method, since they are not limited to the ion exchange capacity of the zeolite. Despite the method used for the preparation of the bimetallic nanocatalysts, the zeolite support has a direct effect on the modifications of the electronic properties of the metal species introduced or even on the participation of the catalytic reaction (Chaplin et al. 2012; Freitas et al. 2015; Snyder et al. 2018).

Modification of nanocatalysts based on zeolites with two metals is very promising considering that the use of these bimetallic catalysts for the treatment of the liquid or gas phase effluents has shown better results for the catalytic reduction of the contaminants than those found for the monometallic catalysts.

The metal catalysed reduction with hydrogen (H_2) as a reducing agent could be a very promising approach for the removal of some organic pollutants like oxyanions, N-nitrosamines, halogenated alkanes, alkenes and aromatics (Chaplin et al. 2012). Concerning the group oxyanions, the main contaminants to be removed from drinking water are bromate, perchlorate and nitrate anions, with a special concern focused on bromates and nitrates.

Bromate (BrO_3^-) is a stable, water-soluble inorganic contaminant with very low volatility and it has been classified as possibly carcinogenic to humans by the World Health Organization (WHO) and the United States Environmental Protection Agency (EPA). This ion is produced from the ozonation or chlorination of bromide-containing water (Freitàs et al. 2015), while the nitrate (NO_3^-) ion, which is harmful to the mammalian organisms, is accumulated in water due to the excessive agricultural fertilizers and insufficient treatment of domestic and industrial effluents (Jung et al. 2014; Hamid et al. 2017; Vorlop and Prüsse 1999). Nitrate is transformed into ammonium (NH_4^+) and partially into nitrite ion (NO_2^-), which is responsible for the blue baby syndrome, and is also a precursor to the carcinogenic nitrous amine (Pintar 2003) by the action of living organisms. The agencies mentioned above established the maximum limits for nitrates in drinking water as 50 (WHO) and 25 mg/L (EPA) (Barrabés et al. 2006).

For these pollutants, the presence of noble metal in the bimetallic catalysts is crucial for the catalytic process, since it can activate hydrogen (H_2) and catalyse reductive transformation of a number of priority drinking water contaminants. So, this synergistic effect due to the presence of both the promoter and the noble metal enhance their catalytic properties and the selectivity of the reaction towards less desired toxic products (Chaplin et al. 2012). In this context, the modifications of the zeolite structures by the introduction of the bimetallic pair, Pd-M or M-Pd (M = non-noble metal), by ion-exchange or impregnation methods, lead to the catalytic reduction of these water contaminants (Fig. 1).

Nitrate reduction over metal modified catalysts is extensively studied and the first reported work on the topic was by Vorlop and Tacke (1989). Since then, several studies have been published using different supports (Horold et al. 1993; Vorlop and Prüsse 1999; Chaplin et al. 2012), one of the more employed being carbon materials (Mikami et al. 2003; Soares et al. 2009a; Soares et al. 2010b; Soares et al. 2011c). In this reaction, the presence of the promoter metal enhances the reduction of NO_3^- into

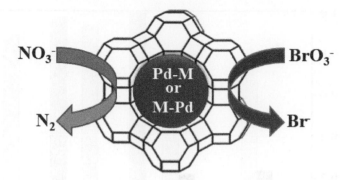

Fig. 1. Schematic transformation of bromates and nitrates by bimetallic nanocatalysts based on zeolites (adapted from Chaplin et al. 2012).

NO_2^- while the noble metal allows the reduction of the NO_2^- ions into N_2/NH_4^+ (Soares et al. 2011c).

Recently, zeolite structures have been explored as supports for this reaction and the Cu-Pd pair modified zeolite seems to be the more efficient. Several studies (Fateminia and Falamaki 2013; Nakamura et al. 2005; Nakamura et al. 2006; Hamid et al. 2016; Hamid et al. 2017; Soares et al. 2015d; Zeng et al. 2017) evaluated a variety of the zeolite structures including clinoptilolite, mordonite, ferrite, ZSM5 and Y for nitrate reduction. Soares and coworkers (2015d) show that the bimetallic nanocatalysts are more active than the monometallic nanocatalysts for this reaction and the order of metals introduced into the zeolite NaY influence nitrate reduction performance, and especially nitrogen (N_2) selectivity. The authors also studied different combinations of metal pairs (CuAg and CuZn) without the noble metal and they concluded that the presence of this metal is important for achieving high selectivity towards nitrogen (Fig. 2).

The CuAg and CuZn pairs show similar nitrate conversions in comparison with the monometallic nanocatalysts, the worst one being CuZn-Y. The bimetallic nanocatalysts using the noble metal achieved 100% of nitrate conversion with 94% and 89% selectivity to nitrogen for CuPd-Y and PdCu-Y, respectively. These results were obtained with a low amount of metals (0.64 wt % of Cu and 1.80 wt % of Pd for CuPd-Y, and 0.84 wt % of Cu and 1.60 wt % of Pd for PdCu-Y) introduced in NaY by the ion exchange method followed by the reduction step. These good results were obtained thanks to the presence of the noble metal on the accessible sites of the zeolite. The structure of the zeolite NaY is characterized by the hexagonal prism, sodalite cages and supercages, which offer different sites for the exchangeable cations (Dyer 2005; Gu et al. 2000; Seo et al. 2012; Snyder et al. 2018). In the bimetallic CuPd-Y nanocatalyst, copper and palladium ions occupy sodalite cages and supercages, respectively, with the last one more available for nitrate and hydrogen adsorption (Soares et al. 2015d).

Between the catalysts studied by the Nakamura group (Nakamura et al. 2005a; Nakamura et al. 2006b), the best one for this reaction was CuPd-mordonite prepared by the ion-exchange method with 1.2 wt % of Cu and 0.5 wt % of Pd. The authors

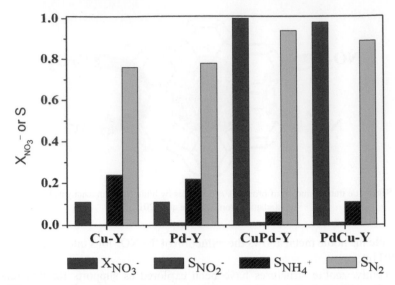

Fig. 2. Nitrate conversions ($X_{NO_3^-}$) and nitrite, ammonium and nitrogen selectivities ($S_{NO_2^-}$, $S_{NH_4^+}$, S_{N_2}) of mono and bimetallic nanocatalysts based on NaY after 5 h of reaction (unpublished results; Soares et al. 2015d).

concluded that the selectivity of the bimetallic catalysts is dependent on the type of zeolite and the Cu/Pd ratio.

Likewise, Zeng and coworkers (2017) studied the nitrate reduction on bimetallic NaY nanocatalysts prepared by the ion exchange method using iron (Fe), copper (Cu) and manganese (Mn) at different solution pHs (3.0 to 9.0). The authors observed that the prepared bimetallic catalysts were not sensitive to the initial solution pH, even at a high pH of 9.0 and they provide the nitrate reduction above 94%. However, in order to achieve these results, very high amounts of Cu (20 wt %) and Fe (41 wt %) were exchanged. These results prove that the presence of a noble metal allows the use of smaller amounts of metal with high selectivity for this reaction (Soares et al. 2015d).

Similarly, the presence of palladium increases the catalytic activity of the nanocatalysts prepared by the impregnation method, despite the significant amounts of metals used for this purpose. Two nanozeolite structures, nanoZSM5 (Hamid et al. 2016) and nanoBEA (Hamid et al. 2017) were used as supports in order to impregnate several metal pairs. For nanoZSM5, the bimetallic catalysts were prepared with different promoter metals (Sn, Cu, Ag, Ni) and noble metals (Pd, Pt, Au), the concentration range being respectively, 0–3.4 wt % and 0–2.8 wt %.

The authors found that all bimetallic nanocatalysts prepared with both zeolite structures achieved complete nitrate removal with significant selectivity to nitrogen. For the bimetallic nanocatalysts based on nanoZSM5, the SnPd (1.0 wt % of Sn and 1.6 wt % of Pd) pair provides the highest selectivity towards N$_2$ (> 81%), with a negligible leaching of the metals even after five successive cycles (Hamid et al. 2016). Also, for the bimetallic nanocatalysts prepared using nanoBEA with copper (Cu), tin (Sn) and Indium (In), as promoter metals, and palladium (Pd) as a noble metal, in the same conditions, the best results were achieved with the CuPd-nanoBeta catalyst (Hamid et al. 2017).

Since, the WHO and EPA agencies regulate the maximum authorized concentration of bromate in water as 10 μg/L, the reduction of bromates with hydrogen has been an important issue for an effective treatment of water. Unlike nitrate reduction, few studies were dedicated to this reaction. The Pereira group extensively studied this reaction using heterogeneous catalysts based on carbon materials (Restivo et al. 2015a; Restivo et al. 2015b; Restivo et al. 2017c).

Like in the nitrate reduction, the palladium plays an important role in this reaction and the heterogeneous catalysts using this noble metal are very active for bromate reduction into bromide (Restivo et al. 2017c). For both the reduction reactions of nitrates or bromates, the presence of palladium in bimetallic catalysts allows researchers to dissociate the hydrogen and reduce the catalyst metal-oxo species generated upon oxygen atom transfer from the oxyanion to the catalyst metal decreasing the oxidation state of the contaminant (Chaplin et al. 2012).

For the bromate reduction reaction, the bimetallic nanocatalysts also show important catalytic activity since the bromate in the presence of hydrogen is reduced into bromide with high selectivity. The bromide ion and water were obtained both by the direct reaction of bromate with hydrogen in solution and by reduction with hydrogen on the metal surface after the adsorption step. At the end of bromate reduction, the hydrogen reduces the metal species in order to make them available again for participating in the bromate reduction (Restivo et al. 2015a; Restivo et al. 2015b; Restivo et al. 2017c).

There are very few examples using zeolites as support for this reaction, with only ZSM5 (Freitas et al. 2015) and Y (Soares et al. 2016e) zeolites. The authors used these structures as supports for preparing heterogeneous catalysts using different metals, copper (Cu), palladium (Pd), rhodium (Rh) and thorium (Th), by ion-exchange and impregnation methods. The presence of the metals in both zeolite structures enhances the catalytic activity for the reduction reaction. The bimetallic nanocatalysts show higher activities than those found for the monometallic ones, the reaction being faster in the presence of noble metal. The complete bromate reduction into bromide was achieved with both zeolite structures after 10 min of reaction (Freitas et al. 2015; Soares et al. 2016e).

Soares et al. (2016e) mentioned that among the bimetallic nanocatalysts modified by ion exchange and impregnation methods, those prepared by the ion exchange method provided the best catalytic results due to the better dispersion of metals throughout the structure. In the incipient wetness impregnation method, the metals impregnated on the zeolite structure provoke an obstruction of the pores as it was confirmed by the decrease of the surface area (Soares et al. 2016e).

The redox properties and the stability of the bimetallic nanocatalysts were studied by cyclic voltammetry using Carbon Toray (CT) as an electrode support (Soares et al. 2015d). The bimetallic nanocatalysts modified electrodes were prepared according to a published method (Fonseca et al. 2009). The cyclic voltammograms of the bimetallic catalysts (CuPd-Y and PdCu-Y) in 0.10 M NaCl⁺ acetate buffer (pH 4) medium are displayed in Fig. 3. Successive cyclic voltammograms of the

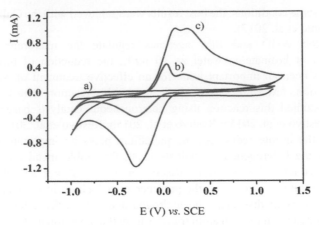

Fig. 3. Cyclic voltammograms of (a) NaY/CT, (b) CuPd-Y/CT and (c) PdCu-Y/CT modified electrodes in 0.10 M NaCl⁺ acetate buffer (pH 4) medium (Soares et al. 2015d).

prepared electrodes in this medium demonstrated good reproducibility indicating that the nanocatalyst modified electrodes are mechanically and chemically stable (Soares et al. 2015d).

The voltammetric study carried out with mono metallic catalysts separately allows researchers to conclude that the bimetallic nanocatalyst shows similar redox processes corresponding to the potential values observed for the monometallic ones. These results confirm the presence of both metals species stabilized by the zeolite structure and show that cyclic voltammetry put in evidence distinctions between the redox process of different metals and consequently allows the characterization of the prepared heterogeneous catalysts, proving their readiness to participate in the catalytic activity (Soares et al. 2015d). These studies confirm the potential of these bimetallic nanocatalysts for application in the electrochemical reduction for water treatment, since the electrochemical technologies were already used for that (Chen 2004). Wang and coworkers (Wang et al. 2007; Wang and Huang 2008) show that the electrocatalytic reduction of perchlorate in dilute aqueous solutions over monometallic nanocatalysts using a catalytic membrane as a support may be a very promising approach.

The support also has an important effect on the catalytic activities of these bimetallic nanocatalysts. In both reactions, using zeolite structures, the Cu-Pd pair was found as the most promising among the bimetallic nanocatalysts with a remarkable activity for the reduction of bromate into bromide (Freitas et al. 2015; Soares et al. 2016e) and for nitrate reduction in the presence of hydrogen (Soares et al. 2015d). For the same reactions, but using carbon materials as support, it was concluded that the Pd-Cu pair provides the best results for nitrate reduction (Soares et al. 2009a) and for bromate reduction (Restivo et al. 2015a).

Other examples of application of the bimetallic nanocatalysts based on zeolites in the environmental field were mentioned in the literature. Danish et al. (2017) show that a composite based on iron-nickel bimetallic ultra-fine nanoparticles supported on a natural zeolite has been demonstrated as an efficient heterogeneous catalyst for the degradation of trichloroethene as a model compound for ground water contaminants.

Another benefit of the bimetallic nanocatalyst is the increment of the active sites due to the high dispersion of Fe-Ni particles on the natural zeolite structure which enhances the catalytic efficiency of the catalyst. Also, the introduction of Pd-Ni in the HZSM-5 zeolite as a bimetallic catalyst improved the production of aromatics and hydrocarbon components of the biojet fuel derived from two biomass resources by the dehydroaromatization of the terpene limonene and the hydrodeoxygenation of stearic acid (Zhang and Zhao 2016).

The catalytic conversion of 1,2-dichloroethane over bimetallic Cu-Ni loaded BEA zeolites was achieved by using a beta structure with two different Si/Al ratios of 17 and 1500. These bimetallic catalysts were produced by the conventional wet impregnation method. Both bimetallic catalysts showed excellent results for the complete hydrodechlorination of 1,2-dichloroethane into ethylene. The catalytic conversion of 1,2-dichloroethane depends mainly on the nature of bimetallic Cu-Ni nanoparticles, their dispersion and the acidity of zeolite support (Srebowata et al. 2014).

The catalytic performances of the bimetallic nanocatalysts using ZSM5 and zinc (Zn) and cobalt (Co) for thermal decomposition of hexamethylene-1,6-dicarbamate (HDC) into hexamethylene-1,6-diisocyanate (HDI) were evaluated and the total conversion higher selectivity towards HDI was found for Zn-Co/ZSM5 with different concentrations (Ammar et al. 2017).

For the removal of greenhouses gases, an active bimetallic nanocatalyst based on a nickel (7.0 wt %) and cobalt (5.0 wt %) pair supported on zeolite Y was developed by the sonochemical method for bio gas reforming in order to obtain syngas with high added value (Sharifi et al. 2015).

Zinc (Zn) and nickel (Ni) bimetallic oxides supported on HZSM-5 were synthetized by a simple ultrasonic impregnation method as active bimetallic catalysts for the aromatization of methanol. The catalytic performance results showed that the Zn-Ni bimetallic catalyst provides higher conversion of methanol for a longer time and a lower average rate of coke formation compared to the monometallic catalysts (Jia et al. 2017).

Begum et al. 2016, studied the catalytic activity of several bimetallic nanocatalysts based on ZSM5 (MFe-ZSM5; M = Ce, Pr, Nd, and Sm) for the selective reduction of NO by NH_3 in comparison with the monometallic catalysts, M-ZSM5, using the same metals and concluded that the bimetallic nanocatalysts are more active than the monometallic ones. Likewise, bimetallic catalysts with copper-silver supported on mordenite zeolite were prepared for NO reduction (Ramírez-Garza et al. 2018). The bimetallic catalysts exhibit higher catalytic activity and the improvements in catalytic activity due to the presence of silver can be explained with a strong Cu-Ag interaction, where silver may be acting as a bridge promoting the copper redox cycle. The catalysts are hydrothermally stable since they were exposed to water vapour stream at high temperature during several hours without a detected change in their structure (Ramírez-Garza et al. 2018).

Bimetallic catalysts with Fe-Cu oxides were encapsulated in a hollow silicalite zeolite for phenol degradation. The particle size of encapsulated Fe_2O_3-CuO is larger than that of zeolite micropores which help to prevent the leaching of metal oxide and thus greatly improves the catalyst reusability (Dai et al. 2017).

Recently, the presence of two metal centres in complexes led to improved catalytic activities and their encapsulation in the available space of the zeolite structures allowed researchers to obtain stable bimetallic nanocatalysts. So, the preparation of copper(II) and iron(II) pyrrolyl-azine complexes encapsulated in the NaY zeolite was successfully achieved and these novel heterogeneous nanocatalysts were tested for phenol oxidation (Kuźniarska-Biernacka et al. 2016a). The pyrrolyl-azine complexes are formed with the metal:ligand 2:1 molar ratio and their structures were further confirmed by DFT calculations (Fig. 4).

These metal pyrrolyl-azine complexes encapsulated in the zeolite NaY exhibited increased activity for the phenol oxidation, in the presence of *t*BuOOH as an oxygen source, under mild conditions, compared to the homogeneous systems. The heterogeneous nanocatalysts were achieved by two different methodologies reported in the literature (Kuźniarska-Biernacka et al. 2013b; Kuźniarska-Biernacka et al. 2012c). In this context, the heterogeneous nanocatalysts were prepared by two steps: introduction of metals by ion exchanged into zeolite followed by *in situ* complexation

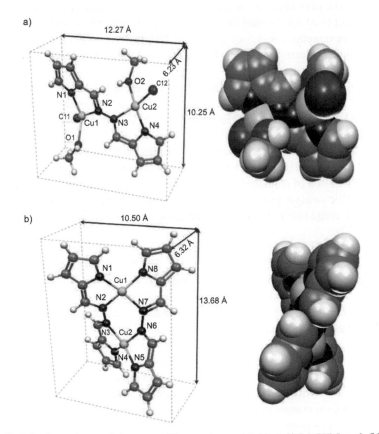

Fig. 4. Optimized structures of the copper(II) complexes: (a) [Cu$_2$LCl$_2$(MeOH)$_2$] and (b) [Cu$_2$L$_2$] determined by DFT calculations at the M062X/6-311++G(d,p) level of theory, where L is pyrrolyl-azine derivative. The boxes indicate the molecular dimensions, size and volume, are estimated from the electronic isodensity surfaces of 0.005 e/ao3 (adapted from Kuźniarska-Biernacka et al. 2016a).

with the pyrrolyl-azine derivative (L, ligand). The prepared catalysts showed high conversion for the phenol oxidation under optimized reaction conditions with 100% selectivity towards catechol.

Acknowledgments

The experimental results and insight of this topic could never have been achieved without the help of a large number of post-graduate and graduate students and without sponsoring from the FCT (Foundation for Science and Technology, Portugal) programs from the Portuguese Government and also the financial programs supported by the Northern Portugal Regional Operational Programme (NORTE 2020), under the Portugal 2020 Partnership Agreement, through the European Regional Development Fund (FEDER). In particular, the authors are grateful to Manuel F.R. Pereira, Olívia S.G.P. Soares, José J.M. Órfão, Iwona Kuźniarska-Biernacka, Kristof Biernacki and Alexandre L. Magalhães.

Reference cited

Ammar, M., Y. Cao, P. He, L.-G. Wang, J.-Q. Chen and H.-Q. Li. 2017. Zn-Co bimetallic supported ZSM-5 catalyst for phosgene-free synthesis of hexamethylene-1,6-diisocyanate by thermal decomposition of hexamethylene-1,6-dicarbamate. Chin. Chem. Lett. 28: 1583–1589.

Barrabés, N., J. Just, A. Dafinov, F. Medina, J.L.G. Fierro, J.E. Sueiras, P. Salagre and Y. Cesteros. 2006. Catalytic reduction of nitrate on Pt-Cu and Pd-Cu on active carbon using continuous reactor. The effect of copper nanoparticles. Appl. Catal. B. 62: 77–85.

Begum, S.H., Ch.-T. Hung, Y.-T. Chen, Sh.-J. Huang, P.-H. Wu, X. Han and Sh.-B. Liu. 2016. Acidity-activity correlation over bimetallic iron-based ZSM-5 catalysts during selective catalytic reduction of NO by NH$_3$. J. Mol. Catal. A Chem. 423: 423–432.

Bellussi, G., A. Carati and R. Millini. 2010. Industrial potential of zeolites. pp. 449–492. *In*: Čejka, J., A. Corma and S. Zones (eds.). Zeolites and Catalysis: Synthesis, Reactions and Applications. Wiley-VCH, USA.

Campanati, M., G. Fornasari and A. Vaccari. 2003. Fundamentals in the preparation of heterogeneous catalysts. Catal. Today 77: 299–314.

Centi, G. and S. Perathoner. 2010. Environmental catalysis over zeolites. pp. 745–774. *In*: Čejka, J., A. Corma and S. Zones (eds.). Zeolites and Catalysis: Synthesis, Reactions and Applications. Wiley-VCH, USA.

Chaplin, B.P., M. Reinhard, W.F. Schneider, Ch. Schuth, J.R. Shapley, T.J. Strathmann and Ch.J. Werth. 2012. Critical review of Pd-based catalytic treatment of priority contaminants in water. Environ. Sci. Technol. 46: 3655–3670.

Chen, G. 2004. Electrochemical technologies in wastewater treatment. Sep. Purif. Technol. 38: 11–41.

Dai, Ch., A. Zhang, L. Luo, X. Zhang, M. Liu, J. Wang, X. Guo and Ch. Song. 2017. Hollow zeolite-encapsulated Fe-Cu bimetallic catalysts for phenol degradation. Catal. Today 297: 335–343.

Danish, M., X. Gu, Sh. Lu, M.L. Brusseau, A. Ahmad, M. Naqvi, U. Farooq, W.Q. Zaman, X. Fu and Z. Miao. 2017. An efficient catalytic degradation of trichloroethene in a percarbonate system catalyzed by ultra-fine heterogeneous zeolite supported zero valent iron-nickel bimetallic composite. Appl. Catal. A 531: 177–186.

Dyer, A. 2005. Biomass conversion over zeolite catalysts. pp. 181–204. *In*: Čejka, J. and H. van Bekkum (eds.). Zeolites and Ordered Mesoporous Materials: Progress and Prospects. Studies in Surface Science and Catalysis 157, Elsevier, The Netherlands.

Fateminia, F.S. and C. Falamaki. 2013. Zero valent nano-sized iron/clinoptilolite modified with zero valent copper for reductive nitrate removal. Process Saf. Environ. Prot. 91: 304–310.

Ferreira, L., C. Almeida-Aguiar, P. Parpot, A.M. Fonseca and I.C. Neves. 2015. Preparation and assessment of antimicrobial properties of bimetallic materials based on NaY zeolite. RSC Adv. 5: 37188–37195.

Fonseca, A.M., S. Goncalves, P. Parpot and I.C. Neves. 2009. Host–guest chemistry of the (N,N'-diarylacetamidine)rhodium(III) complex in zeolite Y. Phys. Chem. Chem. Phys. 11: 6308–6314.

Freitas, C.M.A.S., O.S.G.P. Soares, J.J.M. Órfão, A.M. Fonseca, M.F.R. Pereira and I.C. Neves. 2015. Highly efficient reduction of bromate to bromide over mono and bimetallic ZSM5 catalysts. Green Chem. 17: 4247–4254.

Gu, B., L. Wang, S. Wang, D. Zhao, V.H. Rotberg and R.C. Ewing. 2000. The effect of H irradiation on the Cs-ion exchange capacity of zeolite-NaY. J. Mater. Chem. 10: 2610–2616.

Hamid, Sh., M.A. Kumar and W. Lee. 2016. Highly reactive and selective Sn-Pd bimetallic catalyst supported by nanocrystalline ZSM-5 for aqueous nitrate reduction. Appl. Catal. B. 187: 37–46.

Hamid, Sh., M.A. Kumar, J.-In Han, H. Kimb and W. Lee. 2017. Nitrate reduction on the surface of bimetallic catalysts supported by nano-crystalline beta-zeolite (NBeta). Green Chem. 19: 853–866.

Horold, S., K.D. Vorlop, T. Tacke and M. Sell. 1993. Development of catalysts for a selective nitrate and nitrite removal from drinking water. Catal. Today 17: 21–30.

Jia, Y., J. Wang, K. Zhang, W. Feng, Sh. Liu, Ch. Ding and P. Liu. 2017. Promoted effect of zinc–nickel bimetallic oxides supported on HZSM-5 catalysts in aromatization of methanol. J. Energy Chem. 26: 540–548.

Jung, S., S. Bae and W. Lee. 2014. Development of Pd-Cu/hematite catalyst for selective nitrate reduction. Environ. Sci. Technol. 48: 9651–9658.

Kuźniarska-Biernacka, I., O. Rodrigues, M.A. Carvalho, I.C. Neves and A.M. Fonseca. 2012c. Encapsulation of manganese(III) complex in NaY nanoporosity for heterogeneous catalysis. Appl. Organometal. Chem. 26: 44–49.

Kuźniarska-Biernacka, I., O. Rodrigues, M.A. Carvalho, P. Parpot, K. Biernacki, A.L. Magalhães, A.M. Fonseca and I.C. Neves. 2013b. Electrochemical and catalytic studies of a manganese(III) complex with a tetradentate Schiff-base ligand encapsulated in NaY zeolite. Eur. J. Inorg. Chem. 2768–2776.

Kuźniarska-Biernacka, I., M.M.M. Raposo, R. Batista, P. Parpot, K. Biernacki, A.L. Magalhães, A.M. Fonseca and I.C. Neves. 2016a. Highly efficient heterogeneous catalysts for phenol oxidation: Binuclear pyrrolyl-azine metal complexes encapsulated in NaY zeolite. Microporous Mesoporous Mater. 227: 272–280.

Mikami, I., Y. Sakamoto, Y. Yoshinaga and T. Okuhara. 2003. Kinetic and adsorption studies on the hydrogenation of nitrate and nitrite in water using Pd-Cu on active carbon support. Appl. Catal. B 44: 79–86.

Millini, R. and G. Bellussi. 2017. Zeolite science and perspectives. pp. 1–36. *In*: Čejka, J., R.E. Morris and P. Nachtigall (eds.). Zeolites in Catalysis: Properties and Applications. Royal Society of Chemistry, UK.

Nakamura, K., Y. Yoshida, I. Mikami and T. Okuhara. 2005a. Cu-Pd/-zeolites as highly selective catalysts for the hydrogenation of nitrate with hydrogen to harmless products. Chem. Lett. 34: 678–679.

Nakamura, K., Y. Yoshida, I. Mikami and T. Okuhara. 2006b. Selective hydrogenation of nitrate in water over Cu-Pd/mordenite. Appl. Catal. B 65: 31–36.

Otto, T., J.M. Ramallo-López, L.J. Giovanetti, F.G. Requejo, S.I. Zones and E. Iglesia. 2016. Synthesis of stable monodisperse AuPd, AuPt, and PdPt bimetallic clusters encapsulated within LTA-zeolites. J. Catal. 342: 125–137.

Pintar, A. 2003. Catalytic processes for the purification of drinking water and industrial effluents. Catal. Today 77: 451–465.

Ramírez-Garza, R.E., I. Rodríguez-Iznaga, A. Simakov, M.H. Farías and F.F. Castillón-Barraza. 2018. Cu-Ag/mordenite catalysts for NO reduction: Effect of silver on catalytic activity and hydrothermal stability. Mater. Res. Bull. 97: 369–378.

Restivo, J., O.S.G.P. Soares, J.J.M. Órfão and M.F.R. Pereira. 2015a. Metal assessment for the catalytic reduction of bromate in water under hydrogen. Chem. Eng. J. 263: 119–126.

Restivo, J., O.S.G.P. Soares, J.J.M. Órfão and M.F.R. Pereira. 2015b. Bimetallic activated carbon supported catalysts for the hydrogen reduction of bromate in water. Catal. Today 249: 213–219.

Restivo, J., O.S.G.P. Soares, J.J.M. Órfão and M.F.R. Pereira. 2017c. Catalytic reduction of bromate over monometallic catalysts on different powder and structured supports. Chem. Eng. J. 309: 197–205.

Sebastián, V., C. Casado and J. Coronas. 2010. Zeolite science and perspectives. pp. 389–410. *In*: Čejka, J., A. Corma and S. Zones (eds.). Zeolites and Catalysis: Synthesis, Reactions and Applications, Wiley-VCH, USA.

Seo, S.M., W.T. Lim and K. Seff. 2012. Crystallographic verification that copper(II) coordinates to four of the oxygen atoms of zeolite 6-rings. two single-crystal structures of fully dehydrated, largely Cu^{2+}-exchanged zeolite Y (FAU, Si/Al = 1.56). J. Phys. Chem. C 116: 963–974.

Serrano, D.P., J.A. Melero, J.M. Coronado, P. Pizarro and G. Morales. 2017. Biomass conversion over zeolite catalysts. pp. 441–480. *In*: Čejka, J., R.E. Morris and P. Nachtigall (eds.). Zeolites in Catalysis: Properties and Applications, Royal Society of Chemistry, UK.

Sharifi, M., M. Haghighi and M. Abdollahifar. 2015. Sono-dispersion of bimetallic Ni-Co over zeolite Y used in conversion of greenhouse gases CH_4/CO_2 to high valued syngas. J. Nat. Gas. Sci. Eng. 23: 547–558.

Snyder, B.E.R., M.L. Bols, R.A. Schoonheydt, B.F. Sels and E.I. Solomon. 2018. Iron and copper active sites in zeolites and their correlation to metalloenzymes. Chem. Rev. in press. DOI: 10.1021/acs. chemrev.7b00344.

Soares, O., J.J.M. Órfão and M.F.R. Pereira. 2009a. Bimetallic catalysts supported on activated carbon for the nitrate reduction in water: Optimization of catalysts composition. Appl. Catal. B 91: 441–448.

Soares, O.S.G.P., J.J.M. Órfão, J. Ruiz-Martínez, J. Silvestre-Albero, A. Sepúlveda-Escribano and M.F.R. Pereira. 2010b. Pd-Cu/AC and Pt-Cu/AC catalysts for nitrate reduction with hydrogen: Influence of calcination and reduction temperatures. Chem. Eng. J. 165: 78–88.

Soares, O.S.G.P., J.J.M. Órfão and M.F.R. Pereira. 2011c. Nitrate reduction with hydrogen in the presence of physical mixtures with mono and bimetallic catalysts and ions in solution. Appl. Catal. B 102: 424–432.

Soares, O.S.G.P., L. Marques, C.M.A.S. Freitas, A.M. Fonseca, P. Parpot, J.J.M. Órfão, M.F.R. Pereira and I.C. Neves. 2015d. Mono and bimetallic NaY catalysts with high performance in nitrate reduction in water. Chem. Eng. J. 281: 411–417.

Soares, O.S.G.P., C.M.A.S. Freitas, A.M. Fonseca, J.J.M. Órfão, M.F.R. Pereira and I.C. Neves. 2016e. Bromate reduction in water promoted by metal catalysts prepared over faujasite zeolite. Chem. Eng. J. 291: 199–205.

Srebowata, A., R. Baran, S. Casale, I.I. Kamínska, D. Łomot, D. Lisovytskiy and S. Dzwigaj. 2014. Catalytic conversion of 1,2-dichloroethane over bimetallic Cu-Ni loaded BEA zeolites. Appl. Catal. B 152-153: 317–327.

Vorlop, K.D. and T. Tacke. 1989. 1st steps towards noble-metal catalyzed removal of nitrate and nitrite from drinking-water. Chem. Ing. Tech. 61: 836–837.

Vorlop, K.D. and U. Prüsse. 1999. Catalytically removing nitrate from water. pp. 369. *In*: Janssen, F.J.J.G. and R.A. van Santen (eds.). Environmental Catalysis. Imperial College Press. London, UK.

Wang, D.M., C.P. Huang, J.G. Chen, H.Y. Lin and S.I. Shah. 2007. Reduction of perchlorate in dilute aqueous solutions over monometallic nano-catalysts: Exemplified by tin. Sep. Purif. Technol. 58: 129–137.

Wang, D.M. and C.P. Huang. 2008. Electrodialytically assisted catalytic reduction (EDACR) of perchlorate in dilute aqueous solutions. Sep. Purif. Technol. 59: 333–341.

Zeng, Y., H. Walker and Q. Zhu. 2017. Reduction of nitrate by NaY zeolite supported Fe, Cu/Fe and Mn/Fe nanoparticles. J. Hazard. Mater. 324: 605–616.

Zhang, J. and Ch. Zhao. 2016. Development of a bimetallic Pd-Ni/HZSM-5 catalyst for the tandem limonene dehydrogenation and fatty acid deoxygenation to alkanes and arenes for use as biojet fuel. ACS Catal. 6(7): 4512–4525.

Chapter 9

Bioelectrocatalysis
Basic Knowledge and Applications

Cristina Gutiérrez-Sánchez

INTRODUCTION

Catalysis occupies an important place in chemistry, directing in three directions: heterogeneous, homogeneous and enzymatic systems. Nevertheless, all of them have the same goal, that is, the improvement of catalytic performance.

Since the end of the 1980s, and with the development of nanosciences, nanocatalysis has clearly emerged as a domain at the interface between homogeneous and heterogeneous catalysis, offering unique solutions to answer the demanding conditions for catalyst improvement. The main aim is to develop well defined catalysts, which may include metal nanoparticles and a nanomaterial as support. These nanocatalysts should be able to display the ensuing benefits of homogenous and heterogeneous systems, namely high efficiency and selectivity, stability and easy catalyst recovery or recycling. Specific reactivity can be anticipated due to nanodimensions, which can afford specific properties that cannot be achieved with regular bulk materials.

Biocatalysis, also known as enzymatic catalysis or biotransformation, can also be included in this approach. Enzymes are proteins that catalyze chemical reactions in living organisms. They are high molecular weight compounds made up principally of chains of amino acids linked together by peptide bonds, which contain one or more catalytic sites, forming a complex globular structure of nanometric diameter (Fig. 1).

A common way of classifying biocatalysts is based on their main catalytic activity. This allows distributing them in six large groups such as oxidoreductases (EC 1), transferases (EC 2), hydrolases (EC 3), lyases (EC 4), isomerases (EC 5) and

Instituto de Catálisis y Petroleoquímica, CSIC, c/ Marie Curie 2, L10, 28049 Madrid, Spain.
Email: cgutierrez@icp.csic.es

Fig. 1. Representation of the secondary structure of the enzyme BOxidase composed of β-sheets, α-helices and random coils.

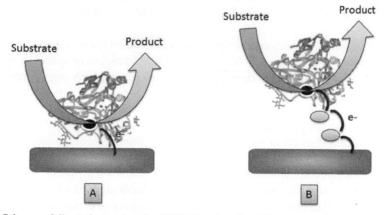

Fig. 2. Schemes of direct electron transfer (DET, A) and mediated electron transfer (MET, B) between the redox centre of an enzyme (black circle) and an electrode for the oxidation of a substrate. The grey circles represent a redox mediator.

ligases (EC 6), where the acronym EC represents the Enzyme Commission. There is a growing interest in the discovery of new biocatalytic activities.

Bioelectrocatalysis is a term used to indicate electrochemical processes where biological systems (oxidoreductases, EC 1), are involved in the acceleration of reactions. Direct or mediated (a redox mediator is needed to shuttle electrons) bioelectrocatalysis allows the interaction of the active centre of the enzyme with the electrode exchanging electrons, that is, the surface of the electrode acts as substrate of the enzyme (Fig. 2). These redox enzymes catalyse a large panoply of oxidation and reduction reactions and have been subject to multiple studies.

The activity of redox enzymes can be controlled and monitored by electrochemical methods, thus, the application of these proteins into bionanoelectronics seems to be ideal.

The use of enzymes, either isolated ones or whole biological cells that contain them, as catalysts presents several advantages over traditional metal catalysts. First, the enzymes are very effective catalysts, which provide higher reaction rates at lower concentrations than those obtained with chemical catalysts. Second, enzymes work at mild temperature and pH conditions, as well as in aqueous environments, allowing green chemistry processes. Thirdly, enzymes offer a high specificity, as they may be chemospecific, regiospecific, diastereospecific and enantiospecific towards their substrates and products. In spite of these clear advantages, the use of these biocatalysts has not been generalized in the industry, mainly due to their limited stability since they are proteins that may be denaturalized under non-natural conditions, thus losing their catalytic activity. Moreover, an additional difficulty is the enzyme separation from the substrates and products in the reaction medium, which prevents reusing them.

Immobilization methods

Enzyme immobilization has overcome the disadvantages described above and has allowed for many biocatalyst industrial processes that are economically profitable today.

Immobilization of enzymes is a process that in the 1970s was defined as that in which the enzyme is confined or localized in a defined region of space, to give rise to insoluble forms that retain their catalytic activity and which can be repeatedly reused. Subsequently, this definition has been extended to those processes in which the degree of freedom of movement of enzymes, organelles or cells is restricted completely or partially by their binding to a support.

This process restricts or reduces to a greater or lesser extent the conformational mobility of the enzymes by their binding to a support, obtaining insoluble forms that retain their catalytic activity and can be reused.

In general, immobilization methods are often classified in two broad categories: physical retention and chemical bonding. The adsorption and entrapment of enzymes in porous substrates or the confinement of enzymes in semipermeable membranes are the main methods of immobilization by physical retention, while the covalent binding of enzymes to supports and cross-linking are the more prominent methods by chemical bonding (Fig. 3) (Mateo 2007; Torres-Salas et al. 2011).

Among other factors, the choice of immobilization method must take into account the conditions of the reaction, the type of reactor to be used and the type of substrate that has to be processed.

Generally, an increase in the stability of the enzymes after their immobilization is observed, which is mainly due to a conformational stabilization of the enzyme due to the existence of enzyme-support bonds. The tertiary structure of the enzyme acquires greater rigidity and becomes more resistant to thermal or chemical deactivation. In this way, intermolecular aggregation is avoided by keeping molecules of the enzyme retained in a specific region. These types of advantages are obtained only in those methods involving covalent bonds.

The covalent attachment of an enzyme is perhaps the most interesting method of immobilization from the industrial point of view.

Figure 4 shows the operational stability of the metaloenzyme hydrogenase immobilized in two different ways on graphite electrodes. On one hand, the catalytic current was very low when the enzyme was adsorbed directly on the functionalized surface and after six days, there was seldom any catalytic current of H_2 oxidation. On

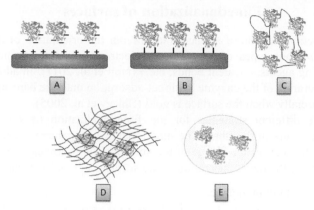

Fig. 3. Immobilization of enzymes by chemical interactions: (A) Adsorption on the support by electrostatic interactions; (B) Covalent bonds to the support; (C) Crosslinking of enzyme molecules. Enzyme immobilization by physical interactions; (D) Entrapment of enzyme molecules in a solid and porous matrix; (E) Encapsulation of enzymes by the membrane.

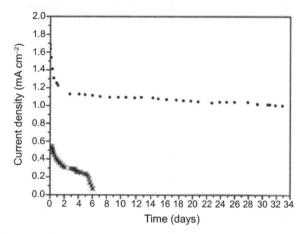

Fig. 4. Chronoamperogram of H_2 oxidation at −520 mV with MWCNT electrodes modified with covalently immobilized hydrogenase (•) and with adsorbed hydrogenase (*). Measurement conditions were 100 mM phosphate buffer, pH 7.0, 40°C, 2500 rpm electrode rotation rate, bubbling of the solution with 1 atm H_2. Reproduced from Ref. (Alonso-Lomillo et al. 2007) with permission from the American Society. Copyright©2007 American Chemical Society.

the other hand, the electrode with hydrogenase immobilized by covalent bonds was very stable. After an initial loss of the catalytic current, the electrode maintained a stable electrocatalytic activity after a month of continuous operation. The instability of the first electrode could be due to the electrostatic interactions of the enzyme that are not strong enough for immobilization of the hydrogenase during operation conditions (Alonso-Lomillo et al. 2007).

Covalent bonding of enzymes to supports requires the reaction of certain amino acids exposed at the enzyme surface with chemical groups of a pre-functionalized support.

Functionalization of surfaces

Adequate functionalization of surfaces may favour the orientation of immobilized enzymes by various types of interactions (electrostatic, hydrophobic, hydrogen bonds, chelating agents, covalent bonds, etc.) (Noll et al. 2011). In addition, it can prevent denaturation of the enzyme by direct adsorption onto the bare surface of the electrode, especially when the surface is gold (Shleev et al. 2005).

There are different strategies for the functionalization of surfaces, which depend on the nature of the same. The functional groups generated can change the hydrophobic character or the charge of the surface. The following are the methods generally used for the functionalization of the surfaces in bioelectrochemistry:

• Self-assembled thiol monolayers:

The formation of self-absorbed monolayers (SAM) of thiols is one of the simplest methods for the functionalization of metal surfaces. This molecular ordering occurs at the interface between a solid surface and an active substance in solution. The chemisorption of thiols on gold occurs spontaneously, in which the elimination of H_2 occurs. Thiol monolayers generally exhibit optimum organization, compaction and stability as a result of Au-S bonds formation. In spite of this, it has been found that in the monolayers there may be certain defects of self-organization, especially when the carbon chain is very short and has some charged functional group (Troughton et al. 1988).

The process of chemisorption on a solid surface is influenced, among other factors, by the chemical structure of the receptor and can be described by two fundamental stages: (1) Physical process of adsorption, which is a fairly fast stage. (2) Formation of the chemical bond and molecule orientation, which leads to the formation of the monolayer and is usually a slower stage.

The thiol SAMs are not stable at negative potentials because they undergo reductive thiol desorption from the surface. At sufficiently negative potentials, the bond formed between sulfur and gold is broken, and the thiol is released into the solution (Walczak et al. 1991).

• The most common methods for the electrochemical modification of surfaces are:

Modification by aromatic diazonium salts reduction:

This modification allows the formation of monolayers or multilayers of aromatic compounds on the surface of the electrodes. In the 1990s, Delamar et al. (1992) modified for the first time carbon electrodes with aromatic derivatives of diazonium salts. For this, they reduced the diazonium salt electrochemically, forming a covalent C-C bond with the electrode (Allongue et al. 1997). The electrochemical modification is a simple and fast method that allows modifying the surface with a great variety of monolayers by covalent union, which offers great stability. It also allows controlling the formation of a monolayer or multilayers by the redox potential and the application time. Carbon surfaces can also be modified with diazonium salts spontaneously, without the need to apply any potential (Adenier et al. 2005).

Some aromatic diazonium salts may be purchased commercially; if not available, they can be prepared *in situ* by the reduction of a primary aromatic amine to an azo group. The application of a reductive electric potential ensures that the azo group is released as a molecule of N_2 and an aryl radical is generated, which will later react with the surface of the electrode forming a C-C bond.

The theoretical coating for a compact monolayer of aromatic rings is 1.35×10^{-9} mol/cm^2 (Pinson and Podvorica 2005), but it has been demonstrated by techniques such as XPS, Raman spectroscopy or AFM, that the experimental coatings obtained are greater. This is due to the formation of multilayers, as the electrochemically generated aryl radicals can react with aryls already anchored to the surface through nucleophilic substitutions (Brooksby and Downard 2004).

Flat metal surfaces, such as gold, and even nanostructured surfaces such as gold nanoparticles, can also be modified by this method (Lyskawa and Bélanger 2006; Liu et al. 2007), generating covalent Au-C bonds (Laurentius et al. 2011). Gold surfaces can also be modified spontaneously, without the need to apply a potential (Podvorica et al. 2009). These organic films generated on the gold have greater mechanical and thermal stability than those obtained by SAM formation of thiols (Kullapere et al. 2010). In addition, they resist prolonged sonication and reflux times in acetonitrile, whereas monolayers formed by thiols are not so stable under such conditions. This demonstrates the stability of the gold-modified diazonium derivatives compared to the analogous thiol monolayers. However, by this method, layers three times thicker than those formed by their analogous thiol derivative are obtained (Shewchuk and McDermott 2009). By the reduction of diazonium salts, modified gold surfaces can be obtained with finer layers by decreasing the number of cycles during electrodeposition.

Other electrochemical methods of surface modification are:

Oxidation of amines, arylacetates and oxidation of alcohols.

Redox polymers: They contain redox species (organometallic complexes or organic compounds covalent bound to the polymer backbone chain by flexible chains) that can exchange electrons by a hopping mechanism between the surface of an electrode and the active centre of the enzyme. The advantage of using redox polymers in bioanalytical strategies over direct electron transfer based on adsorbed enzymes lies in the possibility to electrochemically wire multiple layers of the biorecognition element with the electrode surface, since a large number of enzymes do not take part in the electrochemical reaction.

Works have been published where immobilization of the enzyme was performed by entrapment with polymers modified with osmium complexes (Suraniti 2013; Lopez 2017; Tapia et al. 2017), ruthenium complexes (Reuillard et al. 2014) or mixed monolayers (Groppi et al. 2016).

It has been also demonstrated that the use of redox polymers protects the enzyme from external damage, such as O_2 inhibition of immobilized hydrogenase (Plumeré et al. 2014).

Pyrene derivative: Another strategy is based on using 1-pyrenehexanoic acid (PHA) and its derivatives adsorbed via π-π stacking onto a hierarchical carbon structure (Chen 2001; Giroud and Minteer 2013; Giroud et al. 2017).

Although many immobilization techniques have been applied to numerous enzymes, it is recognized that there is no valid universal method. However, from all the available information the most suitable technique to immobilize an enzyme intended for a particular application can be selected as per the requirement.

Immobilization supports

The support chosen for immobilization of the redox enzymes may be metallic and non-metallic materials capable of conducting electricity and are used as a working electrode. It must also have a mechanical resistance that suits the operating conditions, and be easily separable from the liquid medium so that it can be reused.

Nowadays, the use of nanostructured surfaces have proliferated because they offer a large surface area, allowing researchers to immobilize a large amount of enzyme while maintaining a very small electrode size and a high signal transduction (Katz and Willner 2004). A nanostructured electrode, strictly speaking, refers not only to a surface containing conductive nanoelements, such as carbon nanotubes (Monsalve et al. 2016), carbon nanofibres (Gutiérrez-Sánchez et al. 2012a), graphene (Di Bari et al. 2016) or nanoparticles such as gold nanoparticles (Monsalve et al. 2015b); it also should indicate a nanoscale control of the structural organization and chemical composition of the surface.

Nanoparticles of different shape and material have been shown to have different activity and selectivity as well as stability in biocatalytic reactions. Gold nanoparticles (Dagys et al. 2017), nanotriangles (Zhu et al. 2016), nanostarts (Hsiangkuo et al. 2012), nanocones (Toma et al. 2017), nanoporous (Salaj-Kosla et al. 2012) and nanorods (Di Bari 2016; Alagiri et al. 2017) are some of the nanoparticles employed.

Much work in the field has focused on the elucidation of the effects of nanoparticle size on catalytic behaviour. As early as 1966, Boudart asked fundamental questions about the underlying relationship between particle size and catalysis, such as how catalyst activity is affected by size in the regime between atoms and bulk, whether some minimum bulk-like lattice is required for normal catalytic behaviour and whether an intermediate ideal size exists for which catalytic activity is maximized (Boudart et al. 1966). Somorjai's group has studied this issue extensively, concluding that there is a tremendous variation in the relationships between size and activity depending on the choices of catalyst and reaction (Che and Bennett 1989).

Several authors have studied the relationship between the size of nanostructured elements, in particular gold nanoparticles, and enzymes. Shleev et al. 2014b use the enzyme *Mv*BOx and gold nanoparticles as a nanoelement conductor with a diameter higher than the diameter of the enzyme. Their results demonstrated no relationship between the bioelectrocatalytic parameters and the nanoparticle diameter (Pankratov et al. 2014b). However, the results obtained in this study cannot be extrapolated to other nano-bio-modified surfaces. Reducing the particle size of magnetic nanoparticles down to 18 nm can increase the activity retention of the conjugated lactase enzyme. The authors suggested that the increase of retained activity when decreasing nanoparticle size could be due to less interactions of the enzyme with the surface of the particle, as they are reduced with the increasing curvature (Talbert and Goddard 2013).

The efficient electronic coupling between ThLc and gold nanoparticles depends not only on the adequate orientation of the immobilized enzyme molecules, but also on the size of the AuNPs, obtaining best results when the AuNPs have similar dimensions to those of ThLc, approximately 5 nm diameter (Gutiérrez-Sánchez et al. 2012b). In addition, using the same strategy of immobilization and the same enzyme but varying the morphology of the nanoconductor, using in this case, gold nanorods. The electrocatalytic response obtained was different, possibly due to the large dimension of the nanostructure (Di Bari et al. 2016).

It is of particular interest to study the influence of size in the case of nanoparticles with a diameter smaller than the size of the enzyme that can promote the transfer of electrons between the enzyme and the surface of the electrode. In this way, it can act as an electronic bridge facilitated by the coordination of its metal centre site to the ligands in the gold cluster (Abad et al. 2009).

Quantification of immobilized biocatalysts

Immobilized enzymes on different kind of surfaces show a loss in electrocatalytic activity with time, which limits their applicability to technological applications such as heterogeneous biocatalysis, biosensors and biofuel cells and more recently, in supercapacitors and semiconductors. However, the amount of enzyme that effectively participates in the catalytic current is unknown, as it is difficult to elucidate the causes of its decrease. A possible cause could be the release of the enzyme from the support, but also the change in orientation or integrity of the structural conformation of the enzyme over time. The applied potential or environmental conditions may explain the evolution of the signal. One possible way of calculating the amount of immobilized enzyme is to detect non-catalytic electron transfer redox processes between the redox site/sites of the enzyme and the electrode. The integration of the obtained signal would allow us to determine the amount of enzyme that is oriented on the surface of the electrode for direct electron transfer. However, detection of non-catalytic signals that can be attributed to metal enzyme centres is difficult. It is probably due to the small amount of enzyme immobilized on the surface of the electrode, since when the enzyme coatings are below $3-4$ pmol/cm^2, it is very difficult to detect non-catalytic redox signals from the background current (Sosna et al. 2010); thus, they have rarely been detected (Lalaoui et al. 2015).

Another way is to couple other characterization methods with electrochemistry is to obtain quantitative information. One example is the use of quartz crystal microbalance (E-QCM) to calculate enzyme coatings (Singh et al. 2013). It has been demonstrated that Surface Plasmon Resonance (SPR) coupled with electrochemistry allows researchers to obtain information about adsorbed species at the electrochemical interface (Wang et al. 2000). Adsorption experiments with the enzyme Box over gold electrode were studied by SPR and demonstrated that there are no desorption processes of the enzyme, since the SPR response is very stable with time. The loss of activity may be associated with changes in the structure of the protein (Gutierrez-Sanchez et al. 2016). The immobilization support is also extremely important, since the immobilization of certain enzymes on highly ordered

Au (111) causes its inactivation. It is probably due to the enzyme flattening on the metal surface (Pankratov et al. 2014b). Knowledge of kinetic mechanisms for immobilized enzymes will help in the development of new generations of enzymes by improving the rate limiting step (Stines-Chaumeil et al. 2017).

Researchers continue to work on discovering the causes of inactivation of the catalytic activity of enzymes to improve their operational stability, and thus improve any kind of device with commercial applications.

Multiple applications/biotechnology

Advances in biocatalysis have been possible due to the development of new technologies such as bioinformatics, high-throughput screening, directed evolution, protein engineering and other techniques, such as enzyme immobilization.

Below is a quick overview of the enzymatic processes currently used in many sectors to reduce the chemical load by eliminating from industrial production aggressive and toxic substances, or more simply, pollutants (Choi et al. 2015):

- Detergent industry: enzymatic degradation of proteins, starch and fat stains in laundry. Use of lipolytic enzymes in dishwashing substances. Use of enzymes as surfactants.

- Textile industry: stone wash of jeans, ecological whitening, enzymatic scouring of cotton fabrics, enzymatic degumming of the silk.

- Starch industry: enzymatic production of dextrose, fructose and special syrups for pastry, confectionery and soft drink industries.

- Brewing industry: enzymatic degradation of starch, proteins and glucans from the mixture of cereals used in brewing.

- Pastry and baking industry: enzymatic modification of carbon hydrate and proteins from the cereal to improve bread properties.

- Wine and juice industry: enzymatic degradation of fruit pectin in the processing of juices and wines.

- Alcohol industry: degradation of carbohydrates into sugars and subsequent fermentation to alcohol.

- Food and additive industry: improvement of the nutritional and functional properties of animal and vegetable proteins, conversion of lactose from milk in sugars that are better digested and production of cheese aromas, among others.

- Animal feed industry: enzymatic hydrolysis of matter from slaughterhouses to obtain flours of high nutritional value destined for animal feed.

- Cosmetic industry: biotechnological production of collagen and other application products in beauty creams.

- Paper industry: ecological bleaching of paper pulp, enzymatic control of the viscosity of the plasters with starch.

- Tanning industry: preparation of the skin and removal of hair and fat.
- Oil and fat industry: enzymatic hydrolysis of fats and lecithin and synthesis of esters.
- Fine chemical industry: synthesis of organic substances.

In the field of bioelectrochemistry, the effective immobilization of metalloenzymes on electrodes has made possible their electrochemical study and the application of bioanodes and biocathodes in fuel cells or biosensors (Nogala et al. 2006).

The bio-fuel cells (BFC) are fuel cells that operate at low temperatures and instead of a metal catalyst (usually platinum), they have a biological catalyst. BFC are not limited to the use of hydrogen or methanol as fuels; electricity can be obtained through organic substrates, provided that the biocatalyst catalyses its oxidation. BFC can be classified into those that use living cells such as bacteria and algae, and those that use catalysts extracted from cells, such as enzymes and more recently mitochondria as biological catalysts (Moehlenbrock et al. 2010).

The use of cells as biocatalysts has been of great interest since the last decade because of their applications in the transformation of biomass into electricity (Zhao et al. 2009), and now-a-days microbial fuel cells are currently studied for their use in waste disposal by anaerobic digestion with the consequent generation of electricity (Santoro et al. 2017), or to provide energy to sensors in marine environments (Girguis et al. 2010).

Enzymatic fuel cells (Fig. 5A) may consist entirely of enzymatic catalysts either at the anode or the cathode, or have a hybrid constitution where only one of the two electrodes is based on enzymatic catalysis, while the other is based on processing the electrocatalytic characteristics of traditional fuel cells.

The first BFC developed to function under *in vivo* conditions appeared in the 1960s, based on glucose oxidation and O_2 reduction (Yahiro et al. 1964). However, due to its low operational stability and power, it was not applicable, whereas in the 1970s, battery-based technology emerged for chemical energy conversion to electricity. Currently, the development of implanted BFC focuses on miniaturizing

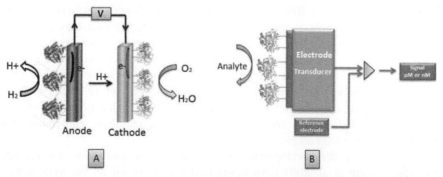

Fig. 5. (A) Model of a BFC without separating membrane. Anode constituted by hydrogenase and cathode by laccase. Both enzymes are covalently immobilized on the electrodes. (B) Representation of an electrochemical and enzimatic biosensor.

the device by the use of nanostructured materials and removing the separating membrane between the electrodes (Shleev 2017). Glucose BFC appears more promising for biomedical applications, as they produce electrical energy from glucose and dioxygen, two substrates present in physiological fluids. Heller and collaborators 2003 developed a BFC composed of carbon fibres containing redox osmium polymers where they immobilized GOx in the anode and BOx in the cathode. The mini-electronic device was implanted in a grape with dimensions of 7 μm diameter and 2 cm length (Mano et al. 2003). The operating voltage obtained was eight times higher (0.52 vs. 0.06 V) and the power density 12 times higher (4.3 vs. 0.35 μW/mm^2) than those previously reported. However, the developed BFCs could not be implanted in animals, mainly because the enzymes on which they are based require a low pH or are inhibited by chloride or urate anions, present in the extra cell fluid. Moreover, the BFC required the presence of toxic redox mediators. Cosnier et al. 2010 developed composite graphite discs containing glucose oxidase and ubiquinone at the anode, polyphenol oxidase and quinone at the cathode using dialysis bags implanted in rats. In this device, polyphenol oxidase reduces dioxygen in water at pH 7 and in the presence of chloride ions and urates at physiological concentrations. A peak specific power of 24.4 μW/mL was achieved and the redox mediators were mechanically confined, thus improving its biocompatibility (Cinquin et al. 2010). Other organisms with an internal cavity rich in glucose, analogous to blood in a human have been used until now for studying implantable BFC, such as in snails, clams or lobsters, producing electric power densities of 30 μW/cm^2 (Halámková et al. 2012), 40 μW/cm^2 (Halámková et al. 2012) and 640 μW/cm^2 (MacVittie et al. 2013), respectively. These values are insufficient to run electronics such as a clock or pacemaker, for instance, but this was remedied by connecting several lobsters and clamping "live batteries" in series, or by using a charge pump and a DC-DC converter circuit. A breakthrough has been in the development of the first BFC that operates in human blood in homeostatic conditions, composed of CDh in the anode and BOx in the cathode that generated sustained electricity, which was enough to power an e-ink screen (Pankratov et al. 2016b).

The development of flexible materials has opened the door to the design of foldable, flexible and adhesive electronic biodevices (Wen-chang and Steve Lien-Chung 2007). Among the BFC, *ex vivo* (semi-implantable) devices have also been developed. One example is the design of an electronic contact lens composed by transparent materials that can operate in physiological fluids such as tears (Pankratov et al. 2016a). Fabrication of BFC adhesives to the epidermis (epidermial electronics) (Kim et al. 2011) used materials such as stretchable graphene (Yun et al. 2017) or buckypaper (Yu et al. 2017) among others, which would measure various electrophysiological and motion signals of humans with applications sensing electrophysiological signals, frequency/postural pressure, body temperature and large-scale strain among others.

In addition to the development of BFC in *ex vivo* conditions for medical purposes, Vincent et al. 2007 have developed a BFC based on the hydrogenase Ralstonia metallidurans on the anode and laccase Trametes Versicolor on the cathode. Three cells connected in series were needed to provide enough power to supply a wristwatch for 24 hours using hydrogen as fuel (Vincent et al. 2007). BFC formed by

3D electrodes constructed from compacted porous carbon loaded with hydrogenase (anode) and bilirubin oxidase (cathode) was able to illuminate a small house that contained five red LED and a miniature clock (Xu and Armstrong 2015).

Another potential application of fuel BFC focuses on providing power to portable electronic devices (Gellett 2010; Falk 2014; Monsalve et al. 2015a). The maximum power density obtained to date is 275 $\mu W/cm^2$ in 5 mM glucose in PBS, providing enough power to allow the wireless transmission of a signal to a data logger. However, the maximum power density in human blood and unstimulated human saliva is 73 and 6 $\mu W/cm^2$ for the same fuel cell configuration, respectively (Ó Conghaile et al. 2016).

Despite the interesting applications of BFC, some aspects must be investigated and intensively developed before their commercial application. Factors such as cell voltage, power, catalytic current density and above all, operational stability should be greatly improved. Furthermore, the BFC should operate efficiently at ambient temperature, at least for implanted devices.

Another field of application of the enzymatic electrodes is that of the biosensors. Biosensors conjugate the specificity of the biological reactions with the great sensitivity that the last advances have conferred to the instrumental techniques. A biosensor is a device in which a biological element acts as a recognition element that is connected or integrated within an appropriate signal transducer. The biosensors are based on a variety of transduction principles such as electrochemical, optical, masssentive, piezoelectric or thermal. In Fig. 5B, a biosensor is schematically shown.

The biological element is composed of a bioactive layer of molecules responsible for the biological recognition of the analyte coupled to a transducer, capable of converting the biological recognition response into a quantifiable process-able signal. The most commonly used biological elements are enzymes (enzyme biosensors), antibodies (immunosensors), nucleic acids (DNA chips) and microorganisms.

Biosensors have many advantages compared to conventional analysis techniques, such as high specificity provided by the biological element, high sensitivity, short analysis times and the possibility of reusing the biological component.

The first biosensor developed was due to Clark and Lyons in the 1960s, which immobilized glucose oxidase on a Pt electrode to determine glucose. It revealed the possibility of enzyme immobilization on an electrochemical detector, resulting in enzyme-based amperometric biosensors that allow researchers to detect electroactives species. Since then, the advances in this field have been directed towards the development of new techniques of immobilization and stabilization of biomolecules, and toward the coupling and use of more sensitive new transduction media.

The new trend in the development of enzymatic biosensors is also the use of nanostructured surfaces, in particular, different carbon nanomaterials such as graphene nanosheets, carbon nanotubes and gold nanoparticles have mostly been employed (Martínez-Periñán et al. 2016).

The glucose biosensor is the most widely used amperometric biosensor on the market and is present in various shapes and sizes. It has been widely studied over the last few decades since the monitoring of blood glucose has been established as a tool for diabetes control. Glucose oxidase and glucose-1-dehydrogenase are the standard enzymes used for the majority of amperometric biosensors (Yoo and Lee 2010).

The enzyme laccase is a very versatile enzyme because it catalyses the reduction of O_2 to water by several organic compounds. For example, ortho and para-diphenols, aminophenols, polyphenols and lignins. The laccase is also being able to reduce some inorganic compounds. Its immobilization on gold electrodes (Casero et al. 2013) or incorporated into a carbon paste has allowed the design of biosensors for the determination of phenolic content and antioxidant power of honey samples (de Oliveira Neto 2017), the determination of phenolics for environmental analysis (Gil et al. 2009), the determination of phenolic micropollutants in surface waters or for evaluating the antioxidant capacity of wines. It has also allowed the design of an O_2 biosensor that can be applied to acidic biofluids or other acidic solutions (Gutiérrez-Sánchez et al. 2015). Due to growing interest in sensing O_2 in living systems and organisms, another O_2 biosensor was designed employing instead Box as an enzyme, showing a detection limit for oxygen of 6 ± 1 µM with a linear range of 6–300 µM, in a buffer that mimics the content and the composition of human physiological fluids (Pita et al. 2013). Other examples of redox enzymes used for the development of enzymatic amperometric biosensors are peroxidase (Muthurasu and Ganesh 2014), glutamate oxidase (Soldatkina et al. 2017) or the co-mobilization of several enzymes in the same biosensor (Conzuelo 2010; Ziller et al. 2017).

Research in the design of biosensors should focus on finding better alternatives to the development of low-cost portables biosensors and on the improvement of the biosensor's selectivity and detection range. The fields of application of enzymatic amperometric biosensors are numerous, mainly the food industry, the medicine sector and environmental sector.

Biotechnology is considered to be one of the options that can be considered for the implementation of new technologies that avoid potentially polluting raw materials. It aims for the application of good environmental practices, integrated into the product redesign, favouring reduction in the origin of waste streams, thus allowing cleaner production. These methods fulfil the principles of green chemistry, that is, using chemical processes that contribute to the improvement of the environment and sustainability, by avoiding the use of harmful chemicals.

The field of biocatalysis continues to grow as chemists strive to find more efficient processes by sustainable chemistry, while reducing costs. The use of enzymes can often reduces the costs of chemical synthesis and these biocatalysts can be produced on a large scale by biotechnological engineering. Furthermore, applied biocatalysis could be placed in a priority position in the business, political and academic agendas, because it has enormous potential benefits for the economy, as well as for the society and the environment.

The future of biocatalysis is promising and we can foresee an increasing number of applications of enzymes in biotechnology.

Acknowledgments

The author thanks Dr. Antonio López De Lacey for critical reading, language corrections and fruitful discussions. This work was supported financially by the MINECO/FEDER project CTQ2015-71290-R.

Reference cited

Abad, J.M., M. Gass, A. Bleloch and D.J. Schiffrin. 2009. Direct electron transfer to a metalloenzyme redox center coordinated to a monolayer-protected cluster. J. Am. Chem. Soc. 131(29): 10229–10236.

Adenier, A., E. Cabet-Deliry, A. Chaussé, S. Griveau, F. Mercier, J. Pinson and C. Vautrin-Ul. 2005. Grafting of nitrophenyl groups on carbon and metallic surfaces without electrochemical induction. Chem. of Mater. 17(3): 491–501.

Alagiri, M., P. Rameshkumar and A. Pandikumar. 2017. Gold nanorod-based electrochemical sensing of small biomolecules: A review. Microchim. Acta. 184(9): 3069–3092.

Alonso-Lomillo, M.A., O. Rüdiger, A. Maroto-Valiente, M. Velez, I. Rodríguez-Ramos, F.J. Muñoz, V.M. Fernández and A.L. De Lacey. 2007. Hydrogenase-coated carbon nanotubes for efficient H_2 oxidation. Nano Lett. 7(6): 1603–1608.

Allongue, P., M. Delamar, B. Desbat, O. Fagebaume, R. Hitmi, J. Pinson and J.-M. Savéant. 1997. Covalent modification of carbon surfaces by aryl radicals generated from the electrochemical reduction of diazonium salts. J. Am. Chem. Soc. 119(1): 201–207.

Boudart, M., A. Aldag, J.E. Benson, N.A. Dougharty and C. Girvin Harkins. 1966. On the specific activity of platinum catalysts. Journal of Catalysis 6(1): 92–99.

Brooksby, P.A. and A.J. Downard. 2004. Electrochemical and atomic force microscopy study of carbon surface modification via diazonium reduction in aqueous and acetonitrile solutions. Langmuir 20(12): 5038–5045.

Casero, E., M.D. Petit-Domínguez, L. Vázquez, I. Ramírez-Asperilla, A.M. Parra-Alfambra, F. Pariente and E. Lorenzo. 2013. Laccase biosensors based on different enzyme immobilization strategies for phenolic compounds determination. Talanta 115(Supplement C): 401–408.

Che, M. and C.O. Bennett. 1989. The influence of particle size on the catalytic properties of supported metals. pp. 55–172. *In*: Eley, D.D., H. Pines and P.B. Weisz (eds.). Advances in Catalysis. Academic Press, 36.

Chen, R.J., Y. Zhang, D. Wang and H. Dai. 2001. Noncovalent sidewall functionalization of single-walled carbon nanotubes for protein immobilization. J. Am. Chem. Soc. 123(16): 3838–3839.

Choi, J.-M., S.-S. Han and H.-S. Kim. 2015. Industrial applications of enzyme biocatalysis: Current status and future aspects. Biotechnology Advances 33(7): 1443–1454.

Cinquin, P., C. Gondran, F. Giroud, S. Mazabrard, A. Pellissier, F. Boucher, J.-P. Alcaraz, K. Gorgy, F. Lenouvel, S. Mathé, P. Porcu and S. Cosnier. 2010. A glucose biofuel cell implanted in rats. PLOS ONE 5(5): e10476.

Conzuelo, F., M. Gamella, S. Campuzano, M.A. Ruiz, A.J. Reviejo and J.M. Pingarrón. 2010. An integrated amperometric biosensor for the determination of lactose in milk and dairy products. J. Agric. Food Chem. 58(12): 7141–7148.

Dagys, M., A. Laurynenas, D. Ratautas, J. Kulys, R. Vidziunaite, M. Talaikis, G. Niaura, L. Marcinkeviciene, R. Meskys and S. Shleev. 2017. Oxygen electroreduction catalysed by laccase wired to gold nanoparticles via the trinuclear copper cluster. Energy Environ. Sci. 10(2): 498–502.

de Oliveira Neto, J.R. 2017. Electroanalysis and laccase-based biosensor on the determination of phenolic content and antioxidant power of honey samples. Food Chemistry 237: pp. 1118-1123-2017 v.1237.

Delamar, M., R. Hitmi, J. Pinson and J.M. Saveant. 1992. Covalent modification of carbon surfaces by grafting of functionalized aryl radicals produced from electrochemical reduction of diazonium salts. J. Am. Chem. Soc. 114(14): 5883–5884.

Di Bari, C., A. Goñi-Urtiaga, M. Pita, S. Shleev, M.D. Toscano, R. Sainz and A.L. De Lacey. 2016. Fabrication of high surface area graphene electrodes with high performance towards enzymatic oxygen reduction. Electrochim. Acta 191(Supplement C): 500–509.

Di Bari, C., S. Shleev, A.L. De Lacey and M. Pita. 2016. Laccase-modified gold nanorods for electrocatalytic reduction of oxygen. Bioelectrochemistry 107(Supplement C): 30–36.

Falk, M., M. Alcalde, P.N. Bartlett, A.L. De Lacey, L. Gorton, C. Gutierrez-Sanchez, R. Haddad, J. Kilburn, D. Leech, R. Ludwig, E. Magner, D.M. Mate, P.Ó. Conghaile, R. Ortiz, M. Pita, S. Pöller, T. Ruzgas, U. Salaj-Kosla, W. Schuhmann, F. Sebelius, M. Shao, L. Stoica, C. Sygmund, J. Tilly, M.D. Toscano, J. Vivekananthan, E. Wright and S. Shleev. 2014. Self-powered wireless carbohydrate/oxygen sensitive biodevice based on radio signal transmission. PLOS ONE 9(10): e109104.

Gellett, W., M. Kesmez, J. Schumacher, N. Akers and S.D. Minteer. 2010. Biofuel cells for portable power. Electroanalysis 22(7-8): 727–731.

Gil, E.S., L. Muller, M.F. Santiago and T.A. Garcia. 2009. Biossensor a base de extrato bruto de lacase (pycnoporus sanguineus) para análise ambiental de compostos fenólicos. Portugaliae Electrochimica Acta 27: 215–225.

Girguis, P.R., M.E. Nielsen and I. Figueroa. 2010. Harnessing energy from marine productivity using bioelectrochemical systems. Current Opinion in Biotechnology 21(3): 252–258.

Giroud, F. and S.D. Minteer. 2013. Anthracene-modified pyrenes immobilized on carbon nanotubes for direct electroreduction of O_2 by laccase. Electrochemistry Communications 34(Supplement C): 157–160.

Giroud, F., K. Sawada, M. Taya and S. Cosnier. 2017. 5,5-Dithiobis(2-nitrobenzoic acid) pyrene derivative-carbon nanotube electrodes for NADH electrooxidation and oriented immobilization of multicopper oxidases for the development of glucose/O_2 biofuel cells. Biosens. Bioelectron. 87(Supplement C): 957–963.

Groppi, J., P.N. Bartlett and J.D. Kilburn. 2016. Toward the control of the creation of mixed monolayers on glassy carbon surfaces by amine oxidation. Chemistry—A European Journal 22(3): 1030–1036.

Gutiérrez-Sánchez, C., W. Jia, Y. Beyl, M. Pita, W. Schuhmann, A.L. De Lacey and L. Stoica. 2012a. Enhanced direct electron transfer between laccase and hierarchical carbon microfibers/carbon nanotubes composite electrodes. Comparison of three enzyme immobilization methods. Electrochimica Acta 82(Supplement C): 218–223.

Gutiérrez-Sánchez, C., M. Pita, C. Vaz-Domínguez, S. Shleev and A.L. De Lacey. 2012b. Gold nanoparticles as electronic bridges for laccase-based biocathodes. J. Am. Chem. Soc. 134(41): 17212–17220.

Gutiérrez-Sánchez, C., S. Shleev, A.L. De Lacey and M. Pita. 2015. Third-generation oxygen amperometric biosensor based on Trametes hirsuta laccase covalently bound to graphite electrode. Chemical Papers 69(1): 237–240.

Gutiérrez-Sánchez, C., A. Ciaccafava, P.Y. Blanchard, K. Monsalve, M.T. Giudici-Orticoni, S. Lecomte and E. Lojou. 2016. Efficiency of enzymatic O_2 reduction by myrothecium verrucaria bilirubin oxidase probed by surface plasmon resonance, pmirras, and electrochemistry. ACS Catal. 6(8): 5482–5492.

Halámková, L., J. Halámek, V. Bocharova, A. Szczupak, L. Alfonta and E. Katz. 2012. Implanted biofuel cell operating in a living snail. J. Am. Chem. Soc. 134(11): 5040–5043.

Hsiangkuo, Y., G.K. Christopher, H. Hanjun, M.W. Christy, A.G. Gerald and V.-D. Tuan. 2012. Gold nanostars: surfactant-free synthesis, 3D modelling, and two-photon photoluminescence imaging. Nanotechnology 23(7): 075102.

Katz, E. and I. Willner. 2004. Biomolecule-functionalized carbon nanotubes: applications in nanobioelectronics. ChemPhysChem. 5(8): 1084–1104.

Kim, D.-H., N. Lu, R. Ma, Y.-S. Kim, R.-H. Kim, S. Wang, J. Wu, S.M. Won, H. Tao, A. Islam, K.J. Yu, T.-i. Kim, R. Chowdhury, M. Ying, L. Xu, M. Li, H.-J. Chung, H. Keum, M. McCormick, P. Liu, Y.-W. Zhang, F.G. Omenetto, Y. Huang, T. Coleman and J.A. Rogers. 2011. Epidermal electronics. Science 333(6044): 838–843.

Kullapere, M., J. Kozlova, L. Matisen, V. Sammelselg, H.A. Menezes, G. Maia, D.J. Schiffrin and K. Tammeveski. 2010. Electrochemical properties of aryl-modified gold electrodes. Journal of Electroanalytical Chemistry 641(1): 90–98.

Lalaoui, N., A. Le Goff, M. Holzinger and S. Cosnier. 2015. Fully oriented bilirubin oxidase on porphyrin-functionalized carbon nanotube electrodes for electrocatalytic oxygen reduction. Chemistry—A European Journal 21(47): 16868–16873.

Laurentius, L., S.R. Stoyanov, S. Gusarov, A. Kovalenko, R. Du, G.P. Lopinski and M.T. McDermott. 2011. Diazonium-derived aryl films on gold nanoparticles: evidence for a carbon–gold covalent bond. ACS Nano 5(5): 4219–4227.

Liu, G., T. Böcking and J.J. Gooding. 2007. Diazonium salts: Stable monolayers on gold electrodes for sensing applications. Journal of Electroanalytical Chemistry 600(2): 335–344.

Lopez, F., S. Ma, R. Ludwig, W. Schuhmann and A. Ruff. 2017. A polymer multilayer based amperometric biosensor for the detection of lactose in the presence of high concentrations of glucose. Electroanalysis 29(1): 154–161.

Lyskawa, J. and D. Bélanger. 2006. Direct modification of a gold electrode with aminophenyl groups by electrochemical reduction of *in situ* generated aminophenyl monodiazonium cations. Chem. of Mater. 18(20): 4755–4763.

MacVittie, K., J. Halamek, L. Halamkova, M. Southcott, W.D. Jemison, R. Lobel and E. Katz. 2013. From "cyborg" lobsters to a pacemaker powered by implantable biofuel cells. Energy Environ. Sci. 6(1): 81–86.

Mano, N., F. Mao and A. Heller. 2003. Characteristics of a miniature compartment-less glucose–O_2 biofuel cell and its operation in a living plant. J. Am. Chem. Soc. 125(21): 6588–6594.

Martínez-Periñán, E., M. Revenga-Parra, M. Gennari, F. Pariente, R. Mas-Ballesté, F. Zamora and E. Lorenzo. 2016. Insulin sensor based on nanoparticle-decorated multiwalled carbon nanotubes modified electrodes. Sens. Actuators B: Chemical 222(Supplement C): 331–338.

Mateo, C., J.M. Palomo, G. Fernandez-Lorente, J.M. Guisan and R. Fernandez-Lafuente. 2007. Improvement of enzyme activity, stability and selectivity via immobilization techniques. Enzyme and Microbial Technology 40(6): 1451–1463.

Moehlenbrock, M.J., T.K. Toby, A. Waheed and S.D. Minteer. 2010. Metabolon catalyzed pyruvate/air biofuel cell. J. Am. Chem. Soc. 132(18): 6288–6289.

Monsalve, K., I. Mazurenko, N. Lalaoui, A. Le Goff, M. Holzinger, P. Infossi, S. Nitsche, J.Y. Lojou, M.T. Giudici-Orticoni, S. Cosnier and E. Lojou. 2015a. A H_2/O_2 enzymatic fuel cell as a sustainable power for a wireless device. Electrochemistry Commun. 60(Supplement C): 216–220.

Monsalve, K., M. Roger, C. Gutiérrez-Sánchez, M. Ilbert, S. Nitsche, D. Byrne-Kodjabachian, V. Marchi and E. Lojou. 2015b. Hydrogen bioelectrooxidation on gold nanoparticle-based electrodes modified by Aquifex aeolicus hydrogenase: Application to hydrogen/oxygen enzymatic biofuel cells. Bioelectrochemistry 106(Part A): 47–55.

Monsalve, K., I. Mazurenko, C. Gutiérrez-Sánchez, M. Ilbert, P. Infossi, S. Frielingsdorf, M.T. Giudici-Orticoni, O. Lenz and E. Lojou. 2016. Impact of carbon nanotube surface chemistry on hydrogen oxidation by membrane-bound oxygen-tolerant hydrogenases. ChemElectroChem. 3(12): 2179–2188.

Muthurasu, A. and V. Ganesh. 2014. Horseradish peroxidase enzyme immobilized graphene quantum dots as electrochemical biosensors. Appl. Biochem. and Biotechnol. 174(3): 945–959.

Nogala, W., E. Rozniecka, I. Zawisza, J. Rogalski and M. Opallo. 2006. Immobilization of ABTS—laccase system in silicate based electrode for biolectrocatalytic reduction of dioxygen. Electrochemistry Commun. 8(12): 1850–1854.

Noll, T. and G. Noll. 2011. Strategies for "wiring" redox-active proteins to electrodes and applications in biosensors, biofuel cells, and nanotechnology. Chem. Soc. Rev. 40(7): 3564–3576.

Ó Conghaile, P., M. Falk, D. MacAodha, M.E. Yakovleva, C. Gonaus, C.K. Peterbauer, L. Gorton, S. Shleev and D. Leech. 2016. Fully enzymatic membraneless glucose|oxygen fuel cell that provides 0.275 ma cm^{-2} in 5 mm glucose, operates in human physiological solutions, and powers transmission of sensing data. Anal. Chem. 88(4): 2156–2163.

Pankratov, D., J. Sotres, A. Barrantes, T. Arnebrant and S. Shleev. 2014a. Interfacial behavior and activity of laccase and bilirubin oxidase on bare gold surfaces. Langmuir 30(10): 2943–2951.

Pankratov, D., E. González-Arribas, Z. Blum and S. Shleev. 2016a. Tear based bioelectronics. Electroanalysis 28(6): 1250–1266.

Pankratov, D., L. Ohlsson, P. Gudmundsson, S. Halak, L. Ljunggren, Z. Blum and S. Shleev. 2016b. *Ex vivo* electric power generation in human blood using an enzymatic fuel cell in a vein replica. RSC Advances 6(74): 70215–70220.

Pankratov, D.V., Y.S. Zeifman, A.V. Dudareva, G.K. Pankratova, M.E. Khlupova, Y.M. Parunova, D.N. Zajtsev, N.F. Bashirova, V.O. Popov and S.V. Shleev. 2014b. Impact of surface modification with gold nanoparticles on the bioelectrocatalytic parameters of immobilized bilirubin oxidase. Acta Naturae 6(1): 102–106.

Pinson, J. and F. Podvorica. 2005. Attachment of organic layers to conductive or semiconductive surfaces by reduction of diazonium salts. Chem. Soc. Rev. 34(5): 429–439.

Pita, M., C. Gutiérrez-Sánchez, M.D. Toscano, S. Shleev and A.L. De Lacey. 2013. Oxygen biosensor based on bilirubin oxidase immobilized on a nanostructured gold electrode. Bioelectrochemistry 94(Supplement C): 69–74.

Plumeré, N., O. Rüdiger, A.A. Oughli, R. Williams, J. Vivekananthan, S. Pöller, W. Schuhmann and W. Lubitz. 2014. A redox hydrogel protects hydrogenase from high-potential deactivation and oxygen damage. Nat. Chem. 6(9): 822–827.

Podvorica, F.I., F. Kanoufi, J. Pinson and C. Combellas. 2009. Spontaneous grafting of diazoates on metals. Electrochimica Acta 54(8): 2164–2170.

Reuillard, B., A. Le Goff and S. Cosnier. 2014. Polypyrrolic bipyridine bis(phenantrolinequinone) Ru(II) complex/carbon nanotube composites for nad-dependent enzyme immobilization and wiring. Anal. Chem. 86(9): 4409–4415.

Salaj-Kosla, U., S. Pöller, Y. Beyl, M.D. Scanlon, S. Beloshapkin, S. Shleev, W. Schuhmann and E. Magner. 2012. Direct electron transfer of bilirubin oxidase (Myrothecium verrucaria) at an unmodified nanoporous gold biocathode. Electrochemistry Commun. 16(1): 92–95.

Santoro, C., C. Arbizzani, B. Erable and I. Ieropoulos. 2017. Microbial fuel cells: From fundamentals to applications. A review. Journal of Power Sources 356(Supplement C): 225–244.

Shewchuk, D.M. and M.T. McDermott. 2009. Comparison of diazonium salt derived and thiol derived nitrobenzene layers on gold. Langmuir 25(8): 4556–4563.

Shleev, S., J. Tkac, A. Christenson, T. Ruzgas, A.I. Yaropolov, J.W. Whittaker and L. Gorton. 2005. Direct electron transfer between copper-containing proteins and electrodes. Biosens. Bioelectron. 20(12): 2517–2554.

Shleev, S. 2017. Quo vadis, implanted fuel cell? ChemPlusChem. 82(4): 522–539.

Singh, K., T. McArdle, P.R. Sullivan and C.F. Blanford. 2013. Sources of activity loss in the fuel cell enzyme bilirubin oxidase. Energy Environ. Sci. 6(8): 2460–2464.

Soldatkina, O.V., O.O. Soldatkin, B.O. Kasap, D.Y. Kucherenko, I.S. Kucherenko, B.A. Kurc and S.V. Dzyadevych. 2017. A novel amperometric glutamate biosensor based on glutamate oxidase adsorbed on silicalite. Nanoscale Research Letters 12(1): 260.

Sosna, M., J.-M. Chretien, J.D. Kilburn and P.N. Bartlett. 2010. Monolayer anthracene and anthraquinone modified electrodes as platforms for Trametes hirsuta laccase immobilisation. Phys. Chem. Chem. Phys. 12(34): 10018–10026.

Stines-Chaumeil, C., E. Roussarie and N. Mano. 2017. The nature of the rate-limiting step of blue multicopper oxidases: Homogeneous studies versus heterogeneous. Biochimie Open 4(Supplement C): 36–40.

Suraniti, E., S. Tsujimura, F. Durand and N. Mano. 2013. Thermophilic biocathode with bilirubin oxidase from Bacillus pumilus. Electrochemistry Commun. 26(Supplement C): 41–44.

Talbert, J.N. and J.M. Goddard. 2013. Influence of nanoparticle diameter on conjugated enzyme activity. Food and Bioproducts Processing 91(4): 693–699.

Tapia, C., R.D. Milton, G. Pankratova, S.D. Minteer, H.-E. Åkerlund, D. Leech, A.L. De Lacey, M. Pita and L. Gorton. 2017. Wiring of photosystem and hydrogenase on an electrode for photoelectrochemical H₂ production by using redox polymers for relatively positive onset potential. ChemElectroChem. 4(1): 90–95.

Toma, M., A. Belu, D. Mayer and A. Offenhäusser. 2017. Flexible gold nanocone array surfaces as a tool for regulating neuronal behavior. Small 13(24): 1700629–n/a.

Torres-Salas, P., A. del Monte-Martinez, B. Cutino-Avila, B. Rodriguez-Colinas, M. Alcalde, A.O. Ballesteros and F.J. Plou. 2011. Immobilized biocatalysts: novel approaches and tools for binding enzymes to supports. Adv. Mater. 23(44): 5275–5282.

Troughton, E.B., C.D. Bain, G.M. Whitesides, R.G. Nuzzo, D.L. Allara and M.D. Porter. 1988. Monolayer films prepared by the spontaneous self-assembly of symmetrical and unsymmetrical dialkyl sulfides from solution onto gold substrates: structure, properties, and reactivity of constituent functional groups. Langmuir 4(2): 365–385.

Vincent, K.A., A. Parkin and F.A. Armstrong. 2007. Investigating and exploiting the electrocatalytic properties of hydrogenases. Chem. Rev. 107(10): 4366–4413.

Walczak, M.M., D.D. Popenoe, R.S. Deinhammer, B.D. Lamp, C. Chung and M.D. Porter. 1991. Reductive desorption of alkanethiolate monolayers at gold: a measure of surface coverage. Langmuir 7(11): 2687–2693.

Wang, S., S. Boussaad, S. Wong and N.J. Tao. 2000. High-sensitivity stark spectroscopy obtained by surface plasmon resonance measurement. Anal Chem. 72(17): 4003–4008.

Wen-chang, L. and H. Steve Lien-Chung. 2007. A novel liquid thermal polymerization resist for nanoimprint lithography with low shrinkage and high flowability. Nanotechnology 18(6): 065303.

Xu, L. and F.A. Armstrong. 2015. Pushing the limits for enzyme-based membrane-less hydrogen fuel cells—achieving useful power and stability. RSC Advances 5(5): 3649–3656.

Yahiro, A.T., S.M. Lee and D.O. Kimble. 1964. Bioelectrochemistry: I. Enzyme utilizing bio-fuel cell studies. Biochim. Biophys. Acta (BBA)—Specialized Section on Biophysical Subjects 88(2): 375–383.

Yoo, E.-H. and S.-Y. Lee. 2010. Glucose biosensors: An overview of use in clinical practice. Sensors 10(5): 4558.

Yu, Y., J. Zhai, Y. Xia and S. Dong. 2017. Single wearable sensing energy device based on photoelectric biofuel cells for simultaneous analysis of perspiration and illuminance. Nanoscale 9(33): 11846–11850.

Yun, Y.J., J. Ju, J.H. Lee, S.-H. Moon, S.-J. Park, Y.H. Kim, W.G. Hong, D.H. Ha, H. Jang, G.H. Lee, H.-M. Chung, J. Choi, S.W. Nam, S.-H. Lee and Y. Jun. 2017. Highly elastic graphene-based electronics toward electronic skin. Adv. Funct. Mater. 27(33): 1701513–n/a.

Zhao, F., R.C.T. Slade and J.R. Varcoe. 2009. Techniques for the study and development of microbial fuel cells: an electrochemical perspective. Chem. Soc. Rev. 38(7): 1926–1939.

Zhu, J., J.-F. Wang, J.-J. Li and J.-W. Zhao. 2016. Specific detection of carcinoembryonic antigen based on fluorescence quenching of Au-Ag core-shell nanotriangle probe. Sens. Actuators B: Chemical 233: 214–222.

Ziller, C., J. Lin, P. Knittel, L. Friedrich, C. Andronescu, S. Pöller, W. Schuhmann and C. Kranz. 2017. Poly(benzoxazine) as an immobilization matrix for miniaturized atp and glucose biosensors. ChemElectroChem. 4(4): 864–871.

Chapter 10

Gold Loaded on Niobium, Zinc and Cerium Oxides
Synthesis, Characterization and Catalytic Application

Maria Ziolek,* Izabela Sobczak and *Lukasz Wolski*

INTRODUCTION

The "well-established" dogma of gold inactivity was broken by Haruta's reports in the late 1980s (Haruta et al. 1987; 1989), showing that gold, when adequately prepared in the form of nanometer-sized particles, exhibits enormous catalytic activity in oxidation processes. It is one of the most pronounced illustrations of the influence of nanostructure on catalytic properties. In particular, a very high catalytic activity of gold loaded on some metal oxides in low temperature CO oxidation has been demonstrated. For some applications, gold properties are much superior to those of the platinum group metals because the bonding strength of adsorbates on Au defective sites is moderate and still weaker than that on Pd and Pt (Scirè and Liotta 2012). Since the first of Haruta's works, the area of catalysis with the use of gold nanoparticles has experienced continuous growth. The beginning of the 21st century was a 'golden' time which brought a huge number of publications devoted to different methods of gold catalysts preparation, different supports for gold loading and different oxidation processes in which gold catalysts were tested. The knowledge on the gold catalysts has been greatly expanded and brought new insight into the effect of various properties of gold particles, apart from their size, on the effectiveness of reactions catalyzed by gold species. Depending on the reaction

Faculty of Chemistry, Adam Mickiewicz University, Umultowska 89b, 61-614 Poznań, Poland.
  Emails: sobiza@amu.edu.pl; wolski.lukasz@amu.edu.pl
* Corresponding author: ziolek@amu.edu.pl

type, metallic gold, cationic gold or negatively charged gold particles have been proposed as active centres. The formation of different gold species strongly depends on the nature of the support. The chemical composition of the support determines the chemical interaction with gold which in turn influences the gold electronic state. Moreover, the support has been reported to participate in the control of the amount of gold anchored to the surface as well as the size and the shape of gold particles (Scirè and Liotta 2012). If reducible metal oxides are applied as supports, the anion vacancies can be easily created and have been suggested as the sites of oxygen adsorption and activation. In general, the performance of the supported gold catalysts is highly dependent on the nature of the support, the size of gold crystallites and the electronic state of gold species.

This chapter presents a thorough analysis of the use and effects of three metal oxides: niobium, zinc and cerium as supports for gold. The choice of these metal oxides was dictated by the differences in their properties. Niobium(V) oxide is a typical acidic metal oxide, zinc(II) oxide reveals amphoteric properties whereas cerium(IV) oxide exhibits redox/basic properties. Niobium and cerium oxides belong to reducible metal oxides although this property is more prominent in CeO_2. The surface properties of support influence the interaction with gold and formation of different gold active centres. Thus, the nature of the support determines the activity of gold species in the desired catalytic oxidation. We shall consider in detail the methods of gold loading, characterization of gold species formed on metal oxide supports and their catalytic activity in different oxidation processes. Direct participation of the support in the catalytic oxidation process is also examined. It is particularly important if the reaction proceeds via the Mars-van Krevelen's mechanism in which oxygen from the metal oxide support takes part in the formation of oxidized product and oxygen from the gas phase is used for the reoxidation of the catalyst. Thus, the properties of oxygen (especially its mobility) in metal oxides used as supports considered in this chapter are also discussed. The support also participates in the catalytic oxidation if the reaction proceeds via the Langmuir-Hinshelwood mechanism in which oxygen and the oxidized molecule are chemisorbed on the catalyst surface and the reaction occurs between two chemisorbed species. If this mechanism is realized in the oxidation reaction over gold/metal oxides catalysts, oxygen can be chemisorbed on gold centres or anion vacancies in the support located near Au particles, whereas the oxidized molecule can be adsorbed on metal oxide surface. It will be shown in this chapter how the different properties of metal oxide supports influence the state of gold and catalytic activity.

Gold loaded on cerium oxide

Properties of cerium oxide

Cerium dioxide (CeO_2) crystallizes in the cubic phase with a fluorite-type structure (space group Fm3m). As prepared ceria nanocrystals are isotropic in shape. However, the fluorite structure has different atomic arrangements in different facets. In the stoichiometric ceria, the coordination number of oxygen atoms is four (CN = 4) and the structure consists of OCe_4 tetrahedra, whereas the coordination number of the cerium cations is eight (CN = 8) and the crystal structure contains CeO_8 cubes (Yashima 2013). Although cerium cations occupy FCC-like sites, they are not close-packed with respect to their Ce nearest neighbours. It is characteristic that oxygen atoms are mobile compared with the Ce cations. Such properties of ceria cause the formation of defects because ceria is easily reduced to a nonstoichiometric compound, CeO_{2-x}. The reduced ceria fluorite structure contains both, Ce^{4+} and Ce^{3+} cations; the latter ones are involved in the creation of defect sites. The reduced defected ceria exhibits mixed ionic and electronic conducting (MIEC) properties (Chatzichristodoulou et al. 2013). The defects created in the form of Ce^{3+} are charge balanced by oxide ion vacancies. The theoretical calculations (Mullins 2015; Nolan et al. 2005; Paier et al. 2013) have indicated that the energy of oxygen vacancy formation depends on the facets' type. When oxygen vacancies are created, two of the Ce cations adjacent to the vacancy change from Ce^{4+} to Ce^{3+}. These vacancies play a crucial role in many catalytic oxidation processes on pure ceria and ceria doped with metals. They are also created during the catalytic oxidation. If oxygen on the surface of ceria is used in a chemical reaction (typical in the Mars-van Krevelen mechanism of oxidation), vacancies are left on the surface. During the catalytic reaction cycle, the vacancies must be filled. Such filling is dependent on oxygen conductivity in the lattice. In fact, one of the most important properties of ceria is its ability to withstand high oxygen depletion while retaining the fluorite crystal structure.

The shape and size of ceria nanocrystals are the other important factors determining catalytic properties. The Ce^{3+}/Ce^{4+} ratio can be an indicator of vacancies concentration which determines the catalytic activity in redox reactions. It has been shown that the Ce^{3+}/Ce^{4+} ratio is higher in small CeO_2 clusters and Ce^{3+} ions preferentially occupy under-coordinated sites, edge or corner sites (Kim et al. 2012; Mullins 2015); thus, increasing the concentration of defect sites leading to the enhancement of catalytic activity.

Depending on the catalytic reactions, acid-base and redox properties of ceria are employed. Both the oxygen mobility and acid-base/redox properties of ceria depend on the crystal face. In literature (Capdevila-Cortada et al. 2016; Wang et al. 2014; Yi et al. 2010; Lin et al. 2015; Piumetti et al. 2016; Mullins 2015) the low-index facets of CeO_2, that is (111), (110) and (100) are considered. The (111) surface is characterized by higher coordination of both atoms, $CN_{Ce} = 7$ and $CN_O = 3$, whereas $CN_{Ce} = 6$ in both (110) and (100), and CN_O decreases to 2 in (100), but remains 3 in (110) (Capdevila-Cortada et al. 2016). The coordination numbers mentioned determine Ce-O and O-O distances as well as the surface properties. The basicity is much higher and nearly identical for the faces (111) and (100), and lower for the (110) surface. The acidity of cerium atoms increases in the following order (100) > (110) > (111), which is inverse to their stability order. The low coordination numbers of cerium cations and oxygen ions in (100) favour the required rearrangement to incorporate the bigger Ce^{3+}, whereas in the (110) surface, the accommodation of the bigger reduced atoms is less favoured. The oxygen vacancy formation energy has been commonly accepted as a descriptor for reactivity in computational approaches. The calculation of this energy indicated that in the (111) surface it is the lowest. The energy of the formation of oxygen vacancies decreases in the same order as the surface density of atoms, that is, (111) > (100) > (110) (Yi et al. 2010).

However, not only are the as-made crystal faces important for catalytic processes but also the changes in ceria surface resulting from thermal treatment usually applied before the use of the catalyst in the reaction. Thermal treatment induces large thermal vibrations of oxygen atoms in ceria and it leads to the disorder of the oxide ions, which is responsible for ionic conduction in ceria-based materials (Yashima 2013). Ceria shows marked structure-sensitive properties which can be assessed through the shape controlled synthesis. The synthesis should be performed in such a way that would allow the achievement of a relatively large surface area and the shape of ceria particles exposing the required facet. Different syntheses of ceria-based nanomaterials have been discussed in detail and presented in many papers and books (e.g. (Li et al. 2013; Li et al. 2012; Lakshmanan et al. 2013; Zhang et al. 2013; Liu et al. 2013; Kaminski and Ziolek 2016; Kovacevic et al. 2016; Bastos et al. 2012; Wang et al. 2014; Yi et al. 2010; 2013; Piumetti et al. 2016; Mullins 2015)). As concerns pure ceria, it can be produced in a variety of shapes: (i) zero-dimensional (0D) ceria nanoparticles (Kovacevic et al. 2016; Bastos et al. 2012; Wang et al. 2014) and nanopolyhedra (e.g., nanooctahedra: (Wang et al. 2014; Mullins 2015)), (ii) one-dimensional (1D) nanorods (e.g. (Kovacevic et al. 2016; Wang et al. 2014; Yi et al. 2010; 2013; Piumetti et al. 2016; Mullins 2015)), nanotubes, nanowires and nanospindles, (iii) two-dimensional (2D) nanoplates and nanodiscs, (iv) three-dimensional mesoporous and macroporous ceria (e.g. (Li et al. 2012; Kaminski and Ziolek 2016; Piumetti et al. 2016)). Various methods have been developed to control the size and shape of ceria particles and they are described in the above indicated references. Both features are important for ceria surface properties because the size determines the number of defects, as mentioned above, and the shape determines the distribution of facets. One can discuss the differences in surface properties by comparing the catalytic performance of ceria nanorods (CeO_2-R), nanocubes (CeO_2-C) and nanooctahedra (CeO_2-O) (Kovacevic et al. 2016; Wang et al. 2014; Yi et al. 2010; 2013; Piumetti et al. 2016; Mullins 2015). Each shape exposes different planes. Wang et al. (2014) has been shown that (CeO_2-R) exposes 51% of (110) and 49% of (100) planes. The distribution of planes depends on the nanorods' size. The above-mentioned distribution was established for (10 × 20–110) nm nanorods and 80 m^2/g surface area. Bigger nanorods, for example, (10 × 50–200) nm (Yi et al. 2010) will reveal a higher excess of (110) plane. Schemes of nanostructured ceria presented in (Wang et al. 2014) with different crystal planes exposed are shown in Fig. 1.

Nearly 100% of the (CeO_2-C) exposed (100) plane, whereas ~ 100% of (CeO_2-O) in ceria described in Wang et al. (2014) exposed (111) plane. The authors of Piumetti et al. (2016) have also synthesized ceria nanocubes but they obtained material with very low surface area (4 m^2/g) and bigger size (50–200 nm) and indicated the presence of both facets (100) and (110), whereas (111) was predominantly exposed on ceria nanorods (300–350 nm). They also prepared mesoporous ceria with the use of SBA-15 as a template and in the so obtained ceria, they found the exposition of the highest number of (111) facets. This description clearly shows a significant impact of ceria size on the exposition of different planes. The properties of ceria nanorods (10 × 160 nm—from SEM), nanocubes (37 nm—from SEM) and nanoparticles (26 nm—from SEM) were compared and

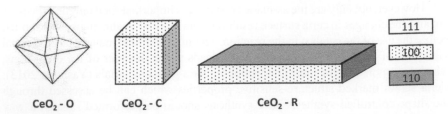

Fig. 1. Schemes of nanostructured ceria (80 m^2/g) with different crystal planes exposed: O = anooctahedra (size 50–135 nm); C = nanocubes (size 10–50 nm); R = nanorods (size (10 × 20–110) nm) (after (Wang et al. 2014)).

discussed in Kovacevic et al. (2016). The abundance of oxygen vacancies, estimated from Raman bands, decreases in the following order: nanorods > nanocubes > nanoparticles. The reducibility of ceria surfaces studied by H$_2$-TPR increased in the inverse order (nanorods < nanocubes < nanoparticles). More oxygen vacancies results in formation of less reducible surface oxygen species. The exposed ceria plane, dependent on the crystals shape, also plays a crucial role in the gold loading that is discussed below.

Properties of gold on cerium oxide

Methods of gold incorporation and dispersion of gold

Several methods of gold loading on ceria have been applied. They are described in many review papers and books (e.g. (Andreeva et al. 2013; Scirè and Liotta 2012; Barakat et al. 2013; Mullins 2015)). The technique of gold loading determines, among others, two catalytically very important parameters, that is, dispersion of metal and its oxidation state. Both parameters are dependent not only on methods of gold loading but also on the properties of ceria, especially ceria crystal facets. In this section, the focus will be on the relationship between the methods of gold loading and its dispersion, whereas in the next section, the oxidation state of gold will be described in the aspect of the interaction between gold and ceria. The methods of gold deposition can be classified in two groups. The first group is based on the deposition or adsorption of Au precursor on a preliminary prepared ceria, whereas the second produces a well-mixed material through the simultaneous formation of ceria and the gold precursor. In both groups of methods, HAuCl$_4$ is applied as the gold precursor.

Historically, the deposition-precipitation (DP) method has been used for the first time by the Haruta group (Haruta et al. 1993) with Na$_2$CO$_3$, KOH or NaOH as precipitation agents, while in the homogeneous deposition precipitation (HDP), urea was a precipitating agent. At present, the HDP method is often also denoted as DP or sometimes DPU (e.g. (Lakshmanan et al. 2013; Delannoy et al. 2010)). In this section, all deposition-precipitation methods will be denoted as DP. The DP methods have been most often used for deposition of gold on ceria (e.g. (Scirè et al. 2012; Lakshmanan et al. 2013; Delannoy et al. 2010; Scirè and Liotta 2012; Kaminski and Ziolek 2016; Qin et al. 2014; Barakat et al. 2013; Aboukaïs et al. 2012; 2016; Glisenti et al. 2010; Mullins 2015; Berrichi et al. 2015; Yang et al. 2008)), as they have been regarded as proper to give small Au particles on the support surface leading to very active catalysts. The use of urea has been shown to yield greater quantities of gold deposited on the surface in contrast to NaOH, and with urea the gold particle size is much smaller than with NaOH. Thus, the use of urea as a precipitating agent in DP is more advantageous in terms of gold loading (Barakat et al. 2013). The domination of DP methods for gold loading on ceria is clearly indicated in the review paper (Scirè and Liotta 2012). Gold deposited on ceria is often reduced to metallic species before the catalytic reaction in hydrogen flow (e.g. (Delannoy et al. 2010)) or during precipitation when various reducing agents are added to obtain metallic gold before heat treatment. Usually $NaBH_4$ is applied for this purpose (e.g. (Barakat et al. 2013)) but glycerol is also effective in this task (Lakshmanan et al. 2013) as well as hydrazine, formaldehyde or ethanol (e.g. (Barakat et al. 2013; Venugopal and Scurrell 2004)).

Dispersion of gold and its location on ceria particles in DP methods strongly depend on pH and the molarity of the gold precursor solution. It can be illustrated by the example given in Andreeva et al. (2013). Upon pH = 8.5 (close to point of zero charge for ceria), 8.1 (Kosmulski 1997) and $M = 0.2 \times 10^{-3}$ small gold particles (1.5–2 nm) homogeneously distributed on the support were obtained. The increase in M to 1×10^{-3} resulted in the formation of slightly larger particles (~ 3 nm), which were mainly distributed on grain borders. Both lower pH (7) and higher pH (10) caused the formation of large particles (15–20 nm) due to agglomeration of the smallest ones. The size of the gold particles may also affect their shape (Barakat et al. 2013). It has been pointed out that the relative amount of corner and edge atoms (important for catalytic processes) is larger on flat particles than on round particles of the same size. The temperature of gold deposition (DP method) is also important. When Au was deposited on CeO_2 (111) at 300 K, the Au particles nucleated at step edges and defect sites of the oxidized sample, whereas when the deposition temperature was 10 K, Au particles were adsorbed on terraces with no preference for edges and defects (Mullins 2015). The relation between ceria facets and gold particle size has not been clarified yet. Some authors have found (Wang et al. 2014) that the size of gold particles was almost identical on ceria nanorods, nanocubes and nanooctahedra, revealing the domination of different facets as mentioned in the previous section. On the other hand, according to Lin et al. (2015), different ceria nanostructures have impact on the size and morphology of gold particles. It has been proved by calculations (Zhu et al. 2016), that the type of ceria facet strongly determines the shape of gold particles. Au_x clusters on ceria (111) tend to be 3D semi-spherical structures for $\times$ greater than 3, except for Au_7. However, on the (110) facet, the most stable configurations are 2D structures for Au_x (x = 3–8) clusters. For bigger gold clusters, the most stable structures on CeO_2 (110) are 3D ones. Au_x (x = 1–4) on ceria (100) are prone to be monoatomically dispersed, while Au_x (x > 4) aggregate into 3D structures.

An interesting modification of DP method was used in Jia et al. (2005). The authors have deposited gold on as-precipitated cerium hydroxide, not on metal oxide. The use of this method was documented to lead to a more homogeneous dispersion of gold.

From among the second group of methods used for gold deposition on ceria, the co-precipitation (CP) one is often applied (e.g. (Barakat et al. 2013; Andreeva et al.

2013)). The CP method not only leads to lower dispersion of gold when compared with the DP method, but it also changes the ceria structure. Although both ceria samples exhibit fluorite like structure, the samples prepared by CP have a needlelike and layered bulk structure, whereas the DP samples have a uniform spherical structure. In Scirè and Liotta (2012), it has been demonstrated that the catalysts prepared by the DP method contain a higher amount of surface gold particles that are also smaller than those found in the CP samples. In the co-precipitation method due to a much lower solubility of $Au(OH)_3$ than $Ce(OH)_3$, the former appears before the latter. In this way, the aggregation of gold hydroxide in larger particles is favoured. These particles are covered by the cerium hydroxide formed subsequently. In the presence of the support (DP method), the extremely low solubility of $Au(OH)_3$ leads to the formation of a high number of nucleation centres, which interact with ceria, resulting in the appearance of small gold particles well distributed on the support surface.

Dispersion of gold is also dependent on the gold amount deposited on ceria. Andreeva et al. (2013) analysing the literature data noted that for 0.25 and 0.5 wt % loading of gold on ceria, very small gold particles are formed (1–2 and 1–3 nm, respectively). On increasing the amount of gold, the heterogeneity of gold dispersion is observed. For 1 wt % of gold loading, Au/ceria particles of 5–8 nm are present apart from smaller ones, whereas for 2.5 and 5.0 wt % of gold loading, larger gold particles of 10–20 nm and 30 nm were formed besides the smaller ones.

One of the main drawbacks of ceria based catalysts is their deactivation with time on stream. To resolve this problem, small Au nanoparticles were encapsulated inside a porous ceria layer (Andreeva et al. 2013). This methodology is based on thiolates used to protect the gold particles and on the direct self-assembly assisted precipitation of $Ce(OH)_x$ species around the preformed particles. Further calcination caused the removal of the organic layer. Good et al. (2017) have also used thiolated gold particles, that is, $Au_{38}(SR)_{24}$ (where SR= $-SCH_2CH_2Ph$) nanoclusters as the catalytic particles combined with ceria support. In their work, gold was not located inside ceria layers but was loaded on the ceria surface.

Other methods of gold deposition on ceria include photodeposition (Kominami et al. 2011) and pulsed laser ablation in the liquid phase (PLAL) (Zhang et al. 2013). In the samples prepared by PLAL, the hydroxyl groups on the gold nanoparticles resulted from the laser ablation process showed good affinity with ceria. In the photodeposition method (Kominami et al. 2011), the Au source was reduced by photogenerated electrons, and Au metal was deposited on CeO_2 particles, resulting in the formation of Au/CeO_2.

Oxidation state of gold and its interaction with the support

In the Au/CeO_2 system, ceria acts not only as a structural promoter for gold, as indicated in the previous section, but also as a chemical promoter stabilizing gold particles against sintering, thanks to the strong chemical interaction between the support and the metal loaded. Such interaction determines the oxidation state of gold, which plays an important role in catalytic oxidation processes. There is still a debate about the relation of the oxidation state of gold species on ceria and the method of gold loading as well as the ceria structure and the interaction of both components. Also, the role of the oxidation state of gold in different catalytic oxidation processes is discussed. Some examples of gold oxidation states on ceria indicated in literature are briefly mentioned below.

It has been found (e.g. (Kaminski and Ziolek 2016; Sharma et al. 2016; Abad et al. 2005; Ying et al. 2010)) that gold dispersed over CeO_2 support prepared by the DP method existed in different oxidation states (Au^0, Au^+ and Au^{3+}). In the samples obtained by the DP method used in Aboukaïs et al. (2012; 2016), 80% of gold nanoparticles were in metallic Au^0 form and 20% in cationic Au^+ species. The latter interacted directly with O^{2-} ions in ceria support. Evacuation of the sample at 673 K for 1 hr (Aboukaïs et al. 2012) caused the reduction of gold cations to metallic ones.

Within a single $Au_{38}(SR)_{24}$ nanocluster on ceria, there were differently charged gold atoms: Au^+, partial positive ($Au^{\delta+}$), neutral (Au^0) and partial negative ($Au^{\delta-}$) sites (Good et al. 2017). The samples prepared by the PLAL method (Zhang et al. 2013) comprised a relatively high content of oxidized gold species. The strong interaction between gold and ceria in these samples led to the appearance of defects on the CeO_2 surface. Gold loaded on ceria enhances the reduction of Ce^{4+} to Ce^{3+} and in this way causes an increase in the number of defects on the support surface (Liu et al. 2013). The strong plasmon resonance effect observed in the UV-Vis spectra of Au/CeO_2 prepared by DP in Ampelli et al. (2015) was attributed to the interfacial interactions between Au and CeO_2, resulting in the electron and energy transfer from the metal surface to the support. The morphology of ceria used as a support has a profound effect on the electronic structure of gold. The presence of cationic Au species has been related to the nanosized nature of the support (3.4 nm), since no Au_n–O–Ce bond has been observed on conventional ceria with very large particle size (> 17 nm) (Guan et al. 2015; Guzman et al. 2005).

It has been indicated in Delannoy et al. (2010), that ceria, especially with high surface area, used for gold loading by the DP method with urea is able to prevent extensive reduction of Au^{3+} species under calcination at 773 K. This fact has also been observed by Corma and co-workers (Concepcion et al. 2006; Carrettin et al. 2005) who have shown that nanocrystalline CeO_2 was able to stabilize Au^{3+} on its surface under CO oxidation conditions.

Scirè et al. 2012 have correlated the decrease in Au surface concentration in the AuDP sample calcined at a high temperature with an increase in the size of gold particles with the temperature increase. However, it has also been indicated that gold cations may diffuse into the ceria support during thermal treatment (Delannoy et al. 2010; Fu et al. 2003). Kaminski and Ziolek (2016) have documented that thermal treatment of gold/ceria material results in the migration of gold from the surface of the support into the bulk of ceria. It has been found that activation of Au/CeO_2 in argon flow at 573 K causes a decrease in the surface gold species from 4.9 to 1.5 at% as estimated by XPS measurements. The activation at this temperature also leads to changes in the oxidation state of gold: cationic gold disappears, whereas negatively charged gold particles are formed by the electron transfer from the support to gold particles.

Table 1. Parameters determining the oxidation state of gold in Au/CeO$_2$ catalysts.

Parameter	Description	Gold oxidation state	References
Method of Au loading	Deposition-precipitation	Domination of Au^0	(Aboukaïs et al. 2012; 2016)
	Loading of nanoclusters Au$_{38}$(SR)$_{24}$	Au$^+$, Au$^{\delta+}$, Au^0, Au$^{\delta-}$	(Good et al. 2017)
	PLAL	Domination of cationic gold	(Zhang et al. 2013)
Morphology of ceria	Small crystals (3.4 nm)	Domination of cationic gold	(Guan et al. 2015; Guzman et al. 2005)
	Very large particles (> 17 nm)	Absence of cationic gold	
Thermal treatment of Au/ceria	Calcination in air	Only Au^{3+} if ~ 1 wt% Au; for higher Au loading– partially Au^0	(Delannoy et al. 2010)
	Evacuation at 673 K	Only Au^0	(Aboukaïs et al. 2012)
	Heating in argon flow at 573 K	Au^0, Au$^{\delta-}$	(Kaminski and Ziolek 2016)

The interaction of gold with ceria can also lead to the formation of solid solution in which Au substitutes for Ce lattice site (Camellone and Fabris 2009; Venezia et al. 2005). X-ray photoemission and diffraction studies of gold/ceria catalysts prepared by DP synthesis revealed the presence of Au^{3+} ions in a cubic fluorite environment that was interpreted as a gold/ceria solid solution in which Au substitutes for Ce atoms. The formation of a mixed Au-Ce-O surface has an important impact on the catalytic properties in CO oxidation.

The above description show different parameters which determine the oxidation state of gold loaded on ceria. They are summarized in Table 1.

Catalytic application

Gold supported on ceria was applied in several catalytic reactions such as CO oxidation and preferential CO oxidation (PROX) (Scirè et al. 2012; Delannoy et al. 2010; Guan et al. 2015; Jia et al. 2005; Rodriguez et al. 2015; Good et al. 2017; Barakat et al. 2013; Camellone and Fabris 2009; Ta et al. 2012; Kim et al. 2012; Widmann et al. 2007; Wang et al. 2015), water gas shift reaction (WGS) (e.g. (Tabakova et al. 2011; Guan et al. 2015; Rodriguez 2011; Reina et al. 2015; Yi et al. 2010; Barakat et al. 2013; Shi et al. 2014; Lin et al. 2015)), oxidation of alcohols (e.g. (Lakshmanan et al. 2013; Topka and Klementová 2016; Kaminski and Ziolek 2016; Glisenti et al. 2010; Mullins 2015; Corma and Garcia 2008; Sharma et al. 2016; Mullen et al. 2017)), oxidation of volatile organic compounds (VOC) (e.g. (Li et al. 2012; Delannoy et al. 2010; Scirè and Liotta 2012; Qin et al. 2014; Jia et al. 2005; Bastos et al. 2012; Aboukaïs et al. 2016; Yang et al. 2008; Corma and

Garcia 2008)) and C-C coupling reaction (e.g. (Beaumont et al. 2010; Berrichi et al. 2015; Corma and Garcia 2008)).

For the oxidation reactions in which the Mars-van Krevelen mechanism is involved, the activity of ceria support is very important. Mullins in a review article (Mullins 2015) describes the adsorption and dissociation of O_2 with ceria surface and the desorption of O_2 from oxidized ceria depending on crystal facets. The formation of superoxo (O_2^-) and peroxo (O_2^{2-}) species on the reduced CeO_{2-x} surfaces (also on non-shape-specific reduced ceria powders) upon oxygen adsorption was observed using Raman spectroscopy. These species dissociated to lattice oxygen as follows ((g) = gas phase; (a) = adsorbed; (l) = lattice species):

$$O_2(g) \rightarrow O_2^-(a) \rightarrow O_2^{2-}(a) \rightarrow O^{2-}(l)$$

~ 20% of adsorbed oxygen desorbed as gas oxygen through the disproportionation reactions:

$$O_2^-(a) + O_2^-(a) \rightarrow O_2^{2-}(a) + O_2(g)$$

$$O_2^{2-}(a) + O_2^{2-}(a) \rightarrow 2 O^{2-}(l) + O_2(g)$$

The temperature of oxygen desorption depends on ceria particle shape and it is lower (ca. 340 K) on nanocubes than on nanorods (ca. 470 K). The thermal treatment led to the disappearance of the first superoxo and next peroxo species. The calculations for partially reduced CeO_{2-x} (111) indicated that the superoxo species were weakly bounded to the oxygen-vacancies (–0.30 to –0.38 eV), while the peroxo species were more strongly bounded (–2.80 to –3.25 eV). On partially reduced CeO_{2-x} (110) and CeO_{2-x} (100) surfaces, only the peroxo species were formed with adsorption energies of 2.05 and –2.02 eV, respectively.

Gold loading on ceria results in enhancement of the reactivity of the CeO_2 surface by capping oxygen and weakening the surface Ce–O bonds, as pointed by H_2-TPR results shown in Scirè et al. (2012). Generally, Au/CeO_2 catalysts are not stable during different catalytic reactions (Delannoy et al. 2010; Kaminski and Ziolek 2016) because, as indicated in the previous section, gold oxidation state and location in the catalyst (on the surface vs. bulk in the support) are sensitive to thermal treatment in different atmospheres (oxidative or reduction). Therefore, the control of changes in the catalyst surface is required in the study of catalytic activity of gold/ceria.

CO oxidation and PROX reaction

If gold/ceria catalysts are applied in CO oxidation both components, gold and the support can participate in the reaction mechanism. Ceria itself is active in the oxidation of CO. It has been observed in Piumetti et al. (2016) that the catalytic performance for CO oxidation over CeO_2 nanocatalysts mainly depends on the presence of highly reactive (100) and (110) surfaces. These facets, characteristic of ceria nanocubes prepared in this work (big size—50–200 nm), contain a great number of coordinative unsaturated atomic sites. On the other hand, worse CO conversion was obtained if the reaction was performed on ceria exhibiting the highest amount of stable (111) planes. It confirms the structure sensitivity of the ceria performance for CO oxidation. The activity of ceria is enhanced by the modification with gold. The type of planes in the ceria used as a support for gold is also important for the activity of Au/CeO$_2$ system in CO oxidation, as shown below.

The method for preparation of gold/ceria catalysts and the catalyst pretreatment strongly affect the catalytic activity of Au/CeO$_2$. Scirè et al. (2012) have found that the sample prepared by DP was more active than that prepared by CP. The effect of the preparation method and catalyst pretreatment accounted for both the different particle size and different surface amount of active species. The DP method led to smaller metal particles compared to those obtained in CP. The role of gold dispersion has been also stressed in Jia et al. (2005) which reported that much higher CO conversion was obtained when very small gold particles were loaded on as-precipitated cerium hydroxide than that achieved on gold/cerium oxide material. Dispersion and concentration of gold species on ceria surface are determined by the activation temperature. Temperature pretreatment of the catalyst before the CO oxidation resulted in lowering of its activity (Scirè et al. 2012) because the concentration of surface gold species decreased, as described in the previous section. A higher calcination temperature led to a lower amount of gold accessible for the reagents. This tendency resulted from the increase in the size of gold particles with temperature (Scirè et al. 2012) or migration of gold from the surface of the support into the bulk of ceria (Delannoy et al. 2010; Kaminski and Ziolek 2016). This behaviour leads to a decrease in CO conversion.

Not only the size of gold particles and concentration of surface gold species are important in the effective low temperature CO oxidation on gold/ceria system, but the morphology of ceria support and oxidation state of gold also play important roles. As concerns the influence of ceria morphology on the activity of the gold/ceria system, the particle size of the support and the exposed facets have to be considered. Carrettin et al. (2004) have prepared ceria with an average particle size of 3.3 nm and surface area of 180 m^2/g and indicated that gold nanoparticles on this material showed the activity in CO oxidation by two orders of magnitude higher as compared to those supported on conventional ceria nanoparticles of a larger size (e.g., d = 15.9 nm and S = 70 m^2/g). They explained this phenomenon by the synergistic effect of nano-Au and nano-ceria by enhancing the formation of highly reactive superoxide and peroxide species at one-electron defect sites on the support. The interaction of gold with ceria support strongly depends on the type of ceria facets which is determined by the ceria crystalline type as shown in the previous sections. Yi et al. (2009) have shown the higher activity of Au/CeO$_2$ (rod) in CO oxidation than Au/CeO$_2$ (particle) and Au/CeO$_2$ (cube) because the former support exposes the (100) and (110) faces for gold anchoring and thus determines the oxidation state of gold most favourable for CO oxidation. The fully reduced gold species on the Au/CeO$_2$ (cube) was less active than the partially reduced one. However, in Rodriguez et al. (2015), it has been indicated that the increase in the content of metallic gold on ceria surface by the reduction with hydrogen at 573 K or treatment of the catalyst at RT by the CO/O$_2$ mixture resulted in the enhancement of activity in oxidation of carbon monoxide. Thus, the effect of oxidation state of gold on the activity of the catalyst in CO oxidation is more complex and depends on the mechanism of

Table 2. Low temperature CO oxidation over Au/CeO$_2$ catalysts.

Features influencing activity	Description	References
Method of Au loading	DP – higher activity than CP	(Scirè et al. 2012)
Dispersion of Au	Higher dispersion → higher activity	(Jia et al. 2005)
Temperature of pretreatment	Higher temperature → lower amount of Au accessible for reagents	(Scirè et al. 2012)
Morphology of ceria	Smaller ceria particles and higher surface area → higher activity	(Carrettin et al. 2004)
Ceria facets	Au loaded on (100) and (110) ceria facets → higher activity than Au loaded on (111)	(Yi et al. 2009)

the reaction. The most important features which influence the effectiveness in low temperature oxidation of CO are summarized in Table 2.

The mechanism of CO oxidation has been considered on the basis of experimental results and computational calculations in several publications (e.g. (Guan et al. 2015; Good et al. 2017; Wu et al. 2014; Kung et al. 2007)). Two prominent mechanisms are generally suggested to be involved in CO oxidation catalysis: the Mars-van Krevelen (MvK) mechanism and the Langmuir-Hinshelwood (L-H) mechanism. Wu et al. (2014) have found that the Mars-van Krevelen mechanism operates about three times as fast as the Langmuir-Hinshelwood mechanism. In the first mechanism, ceria plays a double role: (i) as a support influencing the state of active gold species and (ii) a source of oxygen and oxygen defects participating in the reaction. The depleted lattice oxygen atoms are then replenished by atmospheric oxygen. In the L-H mechanism, oxygen is dissociatively chemisorbed on surface gold species ($\frac{1}{2}O_2 + (Au)_{surf} \rightarrow O \cdots (Au)_{surf}$) and CO is associatively chemisorbed ($CO + (Au)_{surf} \rightarrow (Au)_{surf} \cdots CO$). Next, the reaction occurs between both chemisorbed species towards CO$_2$ ($(Au)_{surf} \cdots CO + O \cdots (Au)_{surf} \rightarrow (Au)_{surf} \cdots CO_2$) which finally desorbs (Nikolaev et al. 2015). It has been found by Guan et al. (2015) that the MvK mechanism does not occur for gold nanoclusters loaded on the CeO$_2$ (111) facet. However, small particles of Au dispersed on CeO$_2$ (111) displayed high catalytic activity in CO oxidation (Rodriguez et al. 2015), although the reaction does not proceed according to MvK mechanism. In order to generate different concentrations of ionic (AuO$_x$) and metallic (Au) gold species, the authors of Rodriguez et al. (2015) have performed low-temperature (283–343 K) oxidation in oxygen rich mixture (O$_2$/CO = 4) on Au/CeO$_2$ (111) after pre-oxidation (20% O$_2$/He, 573 K) and pre-reduction (5% H$_2$/He, 573 K) treatments. It was indicated that both oxidized and reduced gold nanostructures were able to catalyse the oxygen-rich low-temperature CO oxidation, but the ones in metallic form were more active.

In contrast, the MvK mechanism occurs if the Au/CeO$_2$ (110) catalyst is applied (e.g. (Guan et al. 2015; Good et al. 2017)). This mechanism can be discussed on the basis of the scheme presented in Fig. 2.

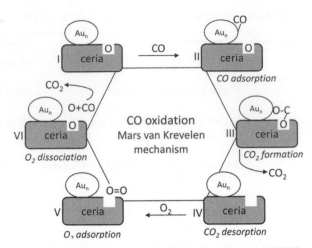

Fig. 2. CO oxidation on Au_n/CeO_2 (110) (after (Guan et al. 2015)).

According to the description in the studies by Guan et al. (2015) and Song and Hensen (2013a), for gold nanoclusters loaded on CeO_2 (110), CO is adsorbed on top of the Au atom and interacts with a neighbouring O atom of the ceria support, which results in adsorption of CO_2. The removal of one surface O atom from ceria results in reduction of one of the surface Ce^{4+} cations to Ce^{3+}. The product CO_2 is only weakly adsorbed to the interface between the Au nanorods and is easily desorbed leading to the formation of oxygen vacancy in the ceria support. In this vacancy, molecular oxygen from the gas phase is adsorbed. The bond distance in the adsorbed O_2 molecule becomes slightly elongated and the electrons are transferred from the ceria support to the antibonding orbital of the adsorbed O_2. In accordance with this, one Ce^{3+} is oxidized to Ce^{4+} and the adsorbed oxygen molecule is dissociated. The generated atomic O is then easily reduced by CO. For the CeO_2 (111) surface supported Au nanocluster, the desorption energy of CO_2 is high and therefore, the MvK mechanism does not apply and the L-H mechanism may occur (in which both the carbon monoxide and the atmospheric oxygen are coordinated by a gold nanocluster). One of the most prominent functions of the gold nanocluster is to coordinate the CO. Gold nanoclusters contain differently charged gold atoms: Au^+, partial positive ($Au^{\delta+}$), neutral (Au^0) and partial negative ($Au^{\delta-}$) sites. It has been confirmed that the CO adsorbed at the partial positive site is most readily oxidized to CO_2 (Wu et al. 2014). In the dominant mechanism for CO oxidation on gold/ceria catalysts, that is, the MvK mechanism, the rate limiting step is the coordination of lattice oxygen to carbon monoxide by the Ce^{4+} site (Good et al. 2017). The role of ceria support is crucial as it is able to both release oxygen from the lattice and incorporate atmospheric oxygen to replenish the supply of lattice oxygen. The Au atoms at the gold-ceria perimeters are the active sites for the recombination of adsorbed CO molecules and the active oxygen species on ceria to form CO_2 molecules (Ta et al. 2012).

The effect of oxygen vacancies in Au/CeO_2 (111) on CO oxidation has been discussed by Camellone and Fabris (2009) on the basis of DFT calculations. The authors have identified two different reaction pathways catalysed by Au^+ adatoms supported on the stoichiometric (111) CeO_2 surface and by Au^{3+} atoms dispersed into (111) $Au_xCe_{1-x}O_{2-\delta}$ surface structures. The results showed that both supported and substitutional Au atoms promote oxidation via oxygen vacancy formation. These vacancies can be efficiently filled by molecular oxygen only when the Au atoms are substitutional for Ce ions, and a full catalytic cycle can be established in this case. In the case of Au^+ adatoms supported on the (111) CeO_2, the O vacancy formed during the CO oxidation attracts the supported Au^+ adatoms during the reaction, turning them into $Au^{\delta-}$ species that prevent further CO adsorption and thus, deactivate the catalyst. Deactivation of the gold/ceria catalyst can also occur by the reaction products, water and CO_2. It has been found by Scirè et al. (2012) that for Au/CeO_2 (DP), a moderate deactivation occurs in the presence of H_2O and to a higher extent in the presence of CO_2 (reaction performed at 333 K). Deactivation with water is reversible (i.e., the activity is fully recovered by switching to a water free feed), whereas no sensible recovery has been observed upon switching off the CO_2 from the feed.

Similar to CO oxidation, preferential oxidation of CO (PROX reaction) in the presence of hydrogen depends on both the amount and size of the surface gold species (Scirè et al. 2012). It has been found that PROX occurs through the MvK reaction mechanism which involves active lattice oxygen from ceria support reacting with CO adsorbed and activated on gold particles. It has been suggested that the CO activation on gold affects the PROX reaction more than the reactivity of ceria surface oxygen. The greater role of nanogold particles than the ceria morphology has also been pointed out by Lai et al. (2006). The DFT calculations for Au (111) face showed a lower energy barrier for CO oxidation than for hydrogen oxidation (Andreeva et al. 2013; Scirè and Liotta 2012; Barakat et al. 2013; Mullins 2015).

Water gas shift reaction (WGS)

WGS reaction (the reaction between CO and H_2O towards the formation of CO_2 and hydrogen) has been widely studied over gold/ceria catalysts by several authors (Andreeva et al. 2013; Scirè and Liotta 2012; Barakat et al. 2013; Mullins 2015; Tabakova et al. 2011; Guan et al. 2015; Rodriguez 2011; Yi et al. 2010; Shi et al. 2014; Lin et al. 2015). The water-gas shift reaction is considered as a critical step in fuel processors for preliminary CO clean up and additional hydrogen generation prior to the CO preferential oxidation step. This reaction has a long historical application as an industrially important process for hydrogen production.

Similar to what happens in CO oxidation and PROX reaction performed over Au/CeO_2, in the WGS reaction, the role of gold (atomic vs. clusters/particles) and ceria (morphology) species have to be discussed. It is obvious that the method for the Au/CeO_2 preparation strongly influences WGS activity because it determines gold particle size and availability of active gold site at the surface. Different mechanisms of WGS reaction on gold/ceria catalysts have been proposed. At present, these mechanisms have been categorized into two groups: one involving the O-H bond cleavage (redox mechanism and formate mechanism) and the other one involving COOH as an intermediate (carboxyl mechanism). Alternatively, OH reacts with CO to form COOH (Guan et al. 2015). A simplified description of the reaction steps can be as follows: the adsorption and dissociation of water takes place on ceria, CO is adsorbed on the gold nanoparticles and all subsequent reaction steps occur at the gold-ceria interface (Rodriguez 2011). Generally, WGS reaction occurs at the boundary between the small metallic gold particles and the ceria, where CO adsorption on gold and H_2O dissociation on oxygen-vacancy defects in the ceria takes place (Andreeva et al. 2013). However, the course of the reaction is more complicated and depends on many factors, especially on ceria facets (depending on crystal shape and size) and the state of gold (atomic gold or clusters).

The morphology effect of ceria has been first indicated in (Si and Flytzani-Stephanopoulos 2008), in which a strong shape/crystal plane effect of ceria on the gold/ceria activity for the WGS reaction has been identified. Higher activity of gold loaded on rod-like ceria (containing (110) and (100) planes) than on cube CeO_2 was evidenced. Defects and oxygen vacancies of ceria surfaces play a crucial role in H_2O dissociative adsorption and activation (Guan et al. 2015; Mullins et al. 2012; Chen et al. 2013). Water adsorption is the rate determining step. If single gold atoms are loaded on the (110) facet of ceria, H_2O is adsorbed at the interface between Au and the defective ceria surface towards the production of H atom on the Au atom and the OH group attached to Ce^{3+} site from the ceria support (Song and Hensen 2014). The redox mechanism for H_2 formation involves the reaction of OH groups on the ceria surface with Au-bound H. Thus, the energy of O-H bond is important in this reaction step. As on Au/CeO_2 (110), this energy is high, the redox mechanism will only be operative at relatively high temperatures because of the strong O-H bond. The same applies to the formate mechanism in which CO interacts with the surface OH groups to form the HCOO intermediate (e.g. (Jacobs et al. 2003; 2004)). Some authors (Meunier et al. 2007), on the basis of operando diffuse reflectance IR spectroscopy, argued that formate does not play the role of intermediate in the reaction pathway but is a spectator species. As an alternative, the carboxyl mechanism has been proposed.

As was mentioned above, ceria support is active in the adsorption and dissociation of water molecules, whereas CO can be chemisorbed on gold species. Such a chemisorption strongly depends on the nuclearity of the gold phase (single Au atom vs. clustered Au atoms). On the Au (single atom)/CeO_2 (110) catalyst, carbon monoxide cannot be chemisorbed on single gold species because it is negatively charged and therefore, the carboxyl mechanism is not operative in the low temperature WGS reaction for this single Au atom model. As a consequence, oxygen defects on the ceria surface, important for H2O adsorption and dissociation, cannot be formed by the interaction of CO chemisorbed with ceria oxygen. Therefore, as an alternative way for the removal of surface oxygen, the Eley-Rideal mechanism has been proposed (Guan et al. 2015; Song and Hensen 2014). In this mechanism, CO from the gas phase directly reacts with the surface OH group to form COOH.

Au (single atom)/CeO_2 (111) is also an unsatisfactory catalyst for the low temperature CO oxidation because CO adsorption may occur as long as the gold is positively charged. As reported in (Song and Hensen 2013b), upon CO_2 formation and desorption, the Au ion migrates to the resulting oxygen vacancy site and blocks the sites for water adsorption and dissociation. Moreover, the Au atom becomes negatively charged in this position, preventing further CO adsorption. In contrast, the initiating step of CO oxidation proceeds much more easily on Au (clusters)/CeO_2 (110), because CO can be adsorbed on the gold clusters and react with a negligible barrier with the ceria surface oxygen atom to form CO_2. Thus for the oxygen vacancy formation step, the gold clusters are more active than a single gold atom. As oxygen vacancies take part in H_2O adsorption and dissociation, which is the rate determining step of the WGS reaction, one can conclude that gold clusters loaded on ceria are more effective catalysts for this reaction than single gold atoms supported on ceria.

Summarizing the consideration of various mechanisms for the WGS reaction, it is important to point out that the dominant mechanism depends on the reaction conditions, in particular the temperature and H_2O/CO ratio. The redox mechanism is expected to dominate at higher temperatures, whereas at low temperatures and depending on the concentration of water in the reaction mixture, carbonate or formate decomposition steps are more important (Andreeva et al. 2013).

Oxidation of volatile organic compounds (VOCs)

Supported gold catalysts exhibit superior catalytic activity at low temperatures for the oxidation of various volatile organic compounds (VOCs) and therefore, they have attracted considerable research interests and have been the subject of several reviews (e.g. (Scirè and Liotta 2012; Corma and Garcia 2008; Scirè et al. 2003; Good et al. 2017; Wang et al. 2014; Ampelli et al. 2015)). From among different gold supported catalysts, the gold/ceria systems are particularly attention-grabbing. Different volatile organic compounds (e.g., a mixture of 2-propanol, methanol and toluene in Scirè et al. (2003); formaldehyde in Li et al. (2012), Qin et al. (2014) and Jia et al. (2005); propene in Delannoy et al. (2010) and; Aboukaïs et al. (2016); chlorinated VOCs (dichloromethane, o-dichlorobenzene, o-chlorobenzene) in Scirè and Liotta (2012); ethyl acetate, ethanol and toluene in Bastos et al. (2012); benzene in Yang et al. (2008)) have been totally oxidized over gold/ceria catalysts in order to reduce the environmental impact of VOCs.

As with the oxidation of carbon monoxide, the activity of gold/ceria catalysts in the oxidation of VOCs was dependent on the preparation method, which enormously affected the size of the gold particles and their oxidation states (Venezia et al. 2013). If gold is loaded on ceria prepared by the DP method (leading to high gold dispersion), the activity of the catalysts in propene total oxidation is higher than that of the samples prepared by CP (e.g. (Delannoy et al. 2010; Venezia et al. 2013)) or impregnation (Aboukaïs et al. 2016). The DP method not only allows obtaining well dispersed gold particles but also increases the reducibility of ceria surface, thus enhancing the mobility of the surface lattice oxygen involved in the VOC oxidation through the MvK mechanism (Scirè et al. 2003).

It has been proposed that small gold nanoparticles weaken the surface Ce-O bonds adjacent to Au atoms facilitating the mobility of surface oxygen. In Aboukaïs et al. (2016), it has been indicated that the total oxidation of VOC is dependent not only on the crystallites of gold but also on the size of ceria crystallites. Smaller crystallites of both gold and ceria lead to higher activity in VOC oxidation. Porosity of ceria is another important factor which has to be considered in VOC oxidation (Scirè and Liotta 2012). It has been evidenced that macroporous ceria is effective support for gold to obtain a catalyst highly active in formaldehyde total oxidation (Li et al. 2012; Scirè and Liotta 2012; Zhang et al. 2009). The authors have found that the catalysts with 80 nm pores displayed the highest catalytic activity due to the highest surface area and the optimal pore size leading to the uniform distribution of small gold nanoparticles. The role of macropores in ceria support for gold in formaldehyde total oxidation was also stressed in Li et al. (2012). It was possible to achieve the effective removal of formaldehyde on the gold/ceria catalyst at 353–373 K (Jia et al. 2005). Shen et al. (2008) indicated that Au/CeO$_2$ catalysts containing 0.85 wt % Au exhibited very high activity in formaldehyde total oxidation at temperature below 373 K. In Jia et al. (2005), 100% oxidation of formaldehyde was obtained at 353 K. If ceria with 80 nm pores was used as gold support, 100% formaldehyde conversion was achieved at 348 K. The importance of CeO$_2$ properties in formaldehyde total oxidation on gold/ceria catalysts is well illustrated by the data in Table 3.

The amount of gold dispersed on ceria surface also an important role in VOC oxidation. In Qin et al. (2014), the authors indicated that with a gold content of only 1.83 wt %, the total combustion of formaldehyde was achieved at 393 K. The gold loading effect was also observed for 2-propanol total oxidation (Venezia et al. 2013; Liu and Yang 2008). Comparison of the effect of 0.3, 1.6 and 2.1 wt % gold loading showed that the maximum activity was achieved for 1.6 wt % gold loaded on ceria. The authors explained this behaviour by the optimal gold particle size (below 5 nm) and the proper number of Au$^+$ species.

The effect of gold oxidation state on the activity in VOC oxidation should also be considered. It has been found that ceria stabilizes cationic gold (Au^{3+}) up to the limiting loading (Delannoy et al. 2010). In the low loaded samples (~ 1 wt % of Au) after calcination, all gold remained in the initial cationic state, whereas if 4 wt % of Au was loaded on ceria, gold was partially reduced upon calcination. The authors have also found that the activation of gold/ceria catalysts under hydrogen flow at 573 K gave rise to a higher activity in the total combustion of propene than activation in O$_2$/He at 773 K. Thus, for propene total oxidation, gold is more active when it is in metallic form than when it is unreduced. In contrast, Zhang et al. (2009) emphasized the importance of coexistence of mixed valence states of gold, cationic (Au^{3+}) and metallic, for formaldehyde combustion.

To sum up, there is no doubt that low temperature VOC total oxidation over Au/ceria catalysts is governed by both the support and the gold properties, which amplify their effects due to a synergistic effect.

Table 3. Characteristic of Au/CeO$_2$ catalysts and their activity in formaldehyde total combustion.

Au content, wt%	Ceria properties	Temperature of 100% formaldehyde combustion, K	References
1.83	n.d.	393	(Qin et al. 2014)
1.63	macroporous CeO$_2$; pores ~ 80 nm	393	(Li et al. 2012)
1.06	CeO$_2$ – ordinary powder	503	(Li et al. 2012)
2.00	CeO$_2$ S = 47 m^2/g	373	(Jia et al. 2005)
2.00	Ce – as precipitated hydroxide; S = 122 m^2/g	353	(Jia et al. 2005)

Gold loaded on zinc oxide

Properties of zinc oxide

Thanks to its unique and attractive properties, ZnO belongs to the metal oxides that are widely used in the industry, especially in catalysis and photocatalysis. Besides the use of ZnO as an active phase, it is also an interesting candidate as a support. One of the most important of its properties is strong metal-support interaction (SMSI), which allows modification of electronic and catalytic properties of metals, and thus permits tuning their activity (Liu et al. 2012; Liu et al. 2013; Wu et al. 2011).

Zinc oxide is an n-type semiconductor, which crystallizes in three different crystal structures of wurtzite, zinc blende and rocksalt (Fig. 3) (Morkoc and Ozgur 2009).

From among these structures, only wurtzite is thermodynamically stable under ambient conditions. Wurtzite has a hexagonal unit cell with two lattice parameters a and c in the ratio of $c/a = 1.633$ (in the ideal wurtzite structure) and belongs to the space group C_{6v}^4 in the Schoenflies notation (Morkoc and Ozgur 2009). The wurtzite structure of ZnO is characterized by two interconnecting sublattices of Zn^{2+} and O^{2-}, so that each Zn ion is surrounded by a tetrahedron of O ions, and vice-versa. This tetrahedral coordination gives rise to polar symmetry along the hexagonal axis (Coleman and Jagadish 2006). The four most common face terminations of wurtzite ZnO are: the polar Zn-terminated (0001) and O-terminated (000$\bar{1}$) faces (c-axis oriented) and the non-polar (11$\bar{2}$0) (a-axis) and (10$\bar{1}$0) faces which both contain the same number of Zn and O atoms (Coleman and Jagadish 2006).

The great advantage of ZnO in terms of catalysis is the possibility to tune its morphology by changing the synthesis conditions. Literature data show that so far numerous one- (1D) (nano-rods, -needles, -helixes, -rings, -belts, -wires, etc.), two- (2D) (nano-plate, -sheet, -pellets) and three-dimensional (3D) (flower, dandelion, snowflakes, etc.) structures of zinc oxide have been successfully synthesized and their surface areas varied in the range from ca. 1 to ca. 200 m^2/g (Kołodziejczak-Radzimska and Jesionowski 2014). Application of ZnO with different sizes and shapes in catalysis allowed the observation that the properties of this metal oxide and its activity in different processes are strongly affected by the type of faces, which are exposed on the surface of the catalyst (Sun et al. 2016; Kołodziejczak-Radzimska and Jesionowski 2014). Morphology of ZnO nanoparticles alter not only the type

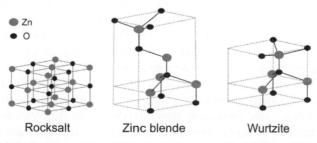

Fig. 3. Schematic representation of ZnO crystal structures (after (Morkoc and Ozgur 2009)).

and concentration of defects, acid-base properties of surface and the ability to adsorb reagents but also the nature of interaction between the support and metals, stability of active components as well as their electronic and magnetic properties (Sun et al. 2016; Kołodziejczak-Radzimska and Jesionowski 2014). Polar faces of ZnO in comparison to non-polar ones were found to be more active in adsorption of oxygen and exhibited enhanced ability to form oxygen vacancies (Sun et al. 2016; Li et al. 2008).

The properties of ZnO were found to be affected not only by the exposition of polar or nonpolar planes; the type of element exposed on the faces was also of importance (Liu and Zhang 2011; Sun et al. 2016). For example, Liu et al. (2011) have reported that the exposition of facets which contained O^{2-} (strong base), that is, (0002), ($10\bar{1}0$), ($\bar{1}100$) and ($01\bar{1}0$), was responsible for the enhancement of the catalytic activity of ZnO catalyst during bio-diesel production. The composition of planes also affected the sorption properties of ZnO. It was found that Zn-terminated surfaces of (0001) planes exhibited higher ability to chemisorb oxygen than other non-polar planes, which contained the same amount of Zn and O atoms (Alenezi et al. 2013). A similar phenomenon has been also observed by Kaneti et al. (2014). These authors synthesized ZnO nanoplates and nanorods, with exposed Zn-terminated polar (0001) planes and non-polar ($10\bar{1}0$) crystal planes, respectively. They reported that ethanol was stronger adsorbed on (0001) planes of ZnO than on ($10\bar{1}0$), and this better affected the sensing properties of ZnO nanoplates in ethanol detection. All these reports clearly show that the type and composition of exposed planes affected the surface properties of catalysts.

Another very important feature of ZnO in terms of catalysis is its ability to induce, upon absorption of light, the chemical transformation of different reagents (Sun et al. 2016; Li and Haneda 2003). ZnO is characterized by a wide band gap of ca. 3.4 eV (Janotti and Van de Walle 2009; Sun et al. 2016; Chen et al. 2014). Consequently, the electrons localized in the valence band of ZnO can be excited by the absorption of light with energy equal to or higher than the band gap value leading to the formation of excited electrons in the conduction band and positively charged holes in the valence band. The photogenerated charge carriers (electrons in the conduction band and holes in the valence band) can then migrate to the surface of ZnO, where they can undergo different redox reactions, for example, positively charged holes can capture electrons from organic compounds leading to their oxidation (Mills and Le Hunte 1997). High electron mobility, which is typical of ZnO, allows the high efficiency of photogenerated charge carries separation, and thus promotes high photocatalytic activity of this material.

Similarly like ceria, zinc oxide exhibits the ability to form crystal lattice defects due to the loss of oxygen and the formation of non-stoichiometric ZnO_{1-x} upon high thermal treatment. However, these defects are unstable and are easily refilled by oxygen after temperature decrease (Drouilly et al. 2012; Peng et al. 2014; Zheng et al. 2007). Many authors have reported that various stable defects in the lattice of ZnO (e.g., zinc vacancies, zinc interstitials, oxygen vacancies, oxygen interstitials, etc.) can be easily formed for example by doping of ZnO with other elements (Yang et al. 2017; Layek et al. 2016; Satheesan et al. 2017), treatment of ZnO with $NaBH_4$ (Wang et al. 2015), and thermal decomposition of ZnO_2 (Renaud et al. 2015). Literature data show that the presence of defects in zinc oxide lattice leads to important changes in the optical properties of ZnO and enhancement of its catalytic activity (e.g. (Peng et al. 2014; Zheng et al. 2007; Zhang et al. 2014; Satheesan et al. 2017; Chen et al. 2014)). This indicates that the ability of ZnO to form different stable defects is another very important feature of this metal oxide, which increases its attractiveness as a support for catalysts addressed to oxidation reactions.

Properties of gold loaded on zinc oxide

Methods of gold incorporation and dispersion of gold

One of the most widely used techniques for the preparation of Au/ZnO catalysts is co-precipitation (CP). This preparation approach allows production of materials with large surface area of 50–90 m^2/g and relatively small gold nanoparticles in the range from 3 to 8 nm at relatively high gold loading (up to 5 at%) (Manzoli et al. 2004; Sakurai and Haruta 1996; Gabriel and Scurrell 2016; Wang et al. 2002). From among different precipitation agents ((NH_4)$_2CO_3$, NaOH and NH_4OH, Na_2CO_3) the most widely used is Na_2CO_3, which leads to materials with the highest surface area and the smallest gold particle size (Wang et al. 2003). A very important factor affecting the properties of Au/ZnO catalyst prepared by CP is the temperature of calcination (Wang et al. 2003). It was found that thermal treatment at temperatures lower than 513 K led to the formation of hydrozincite instead of ZnO, and gold species supported on this material were less active and stable than those supported on ZnO. In order to avoid the formation of hydrozincite, the as-prepared catalysts should be calcined at temperatures equal to or higher than 513 K, at which hydrozincite is fully transformed into the crystalline ZnO phase.

The other methods for the synthesis of Au/ZnO by the *in situ* growth of ZnO nanostructures in the presence of colloidal gold nanoparticles are the hydrothermal one (She et al. 2017; Sun et al. 2011) and colloidal method described in Misra et al. (2014). Hydrothermal synthesis allows the obtainment of ZnO nanorods decorated with Au NPs with large size of 20–40 nm, while the colloidal method leads to the formation of core-shell structures, in which the gold core is decorated by ZnO nanoparticles (shell). Interesting synthesis approach was proposed by Pawinrat et al. (2009). The authors used flame spray pyrolysis (FSP) as an effective, one step method for the production of Au/ZnO catalysts. This method provided obtainment of small gold nanoparticles with the size of 4.0 to 6.7 nm for 1 and 3 wt % of gold loading, respectively. The authors stressed that an important advantage of FSP is the production of highly pure particles without additional treatment like drying and calcination.

From amongst all the post synthesis methods of gold loading on ZnO, the most widely used is deposition-precipitation (DP). Na_2CO_3 is often applied as a precipitation agent. The size of gold nanoparticles prepared by this method is ca. 5 nm for gold loading equal to 1 at% (Naknam et al. 2009). One of the most important parameters, which influence the size of gold nanoparticles prepared by DP is the reaction temperature. Castillejos et al. (2012) have reported that gold nanoparticles precipitated at 283 K exhibited significantly smaller size than those prepared at 313 K (3.6 vs. 7.6 nm, respectively).

Interesting results have been reported in Liu et al. (2016). The authors have synthesized the Au/ZnO catalyst (4.1 wt % of Au) supported on highly deficient ZnO mesocrystals using a simple DP method. In this catalyst, Au NPs were both supported on and doped in the lattice of ZnO. The introduction of gold into the lattice of ZnO was possible because of a very high concentration of vacancy defects (especially Zn vacancies) in the structure of zinc oxide. The "heavy doping" of ZnO by Au was confirmed by the lack of metallic gold nanoparticles on the surface of as-prepared material. A small amount of ultra-fine gold nanoparticles with the size of ca. 2 nm appeared on the surface of the catalyst after heat treatment at 473 K in an oxygen atmosphere. It was evidenced that this phenomenon resulted from the migration of gold atoms from the ZnO lattice to the surface of the metal oxide. Very small gold nanoparticles of ca. 3.8 nm were also observed for the catalyst in which gold loading was equal to 18 wt %. Simple estimation on the basis of the particle sizes indicated that the Au NPs on surface could not account for the large 18% loading. Based on this observation, the authors concluded that substantial amount of gold in the prepared catalyst was incorporated into the structure of ZnO. This novel synthesis approach led to a new catalytic system, where Au/ZnO can function as a reducible metal oxide because of the presence of Au–O–Zn within the support.

Another synthesis approach for the preparation of Au/ZnO catalysts is deposition-reduction with the use of $NaBH_4$ as a reducing agent. Literature data show that this method allowed the obtainment gold nanoparticles with the size of 5–8 nm for gold loading ranging from 1.37 to 5.40 wt % (Xie et al. 2017; Kaneti et al. 2016; Arunkumar et al. 2017). The other reducing agent which was also used is sodium citrate (Nagajyothi et al. 2016; Khan et al. 2016), but Au NPs prepared using this reagent instead of $NaBH_4$ led to the formation of much larger particles (~ 12 nm) (Khan et al. 2016).

An interesting modification of the deposition-reduction method has been proposed in Massola et al. (2015). The authors had applied layered zinc hydroxide (ZHL) as a support for gold. It was shown that gold nanoparticles were rapidly formed when the ZHL-layered was added to the ethanol chloroauric acid solution at room temperature. The double role of ethyl alcohol as a solvent for the gold precursor and a reducing agent was pointed out. Au(III) ions were mainly reduced via a redox reaction between the metallic precursor and the solvent. Thermal decomposition of this catalyst at 473 K led to formation of Au/ZnO. The size of gold nanoparticles prepared by this synthesis route was strongly dependent on the temperature of calcination and was found to be 19, 23 and 28 nm for the samples treated at 473, 523 and 573 K, respectively.

One of the most effective methods for the deposition of small gold nanoparticles on the surface of ZnO is the anchoring of Au NPs functionalized with different organic molecules. From among them, much attention was paid to glutathione, which strongly interacts with the surface of ZnO and thus leads to the fast and effective deposition of functionalized Au NPs on the surface of the metal oxide (Udawatte et al. 2011). The application of this functionalization agent allowed the obtainment of very small (ca. 3.4 nm) monodispersed gold nanoparticles, whose size and loading could be simply tuned during the synthesis (Udawatte et al. 2011).

Very small gold nanoparticles on the surface of ZnO can also be achieved by the colloidal deposition method, in which gold sol is formed by the reduction of $HAuCl_4$ in the presence of polyvinyl alcohol. ZnO is added to as-prepared gold sol and the mixture is stirred until the total adsorption of Au NPs takes place. This method is time consuming, but allows for the synthesis of very small gold nanoparticles with the size of ca. 2.7–3.4 nm depending on the gold loading (Strunk et al. 2009). In Wu et al. (2011), the authors have reported that this method allows obtainment of ultra-fine and homogeneously dispersed gold nanoparticles with the size of ca. 3.1 nm, even for the catalyst which was calcined at 623 K.

In recent years, much attention has been paid to application of the photodeposition method as a simple and environmentally friendly approach for the synthesis of the Au/ZnO catalyst. Many authors have reported that small gold nanoparticles can be efficiently deposited on the surface of ZnO using this synthesis route. For instance, gold nanoparticles with the size in the range from 0.77 to 2.30 nm has been successfully synthesized by He et al. (2014). Similar results were also obtained by Dulnee et al. (2014) who acquired Au NPs with the size of ca. 2.4 nm. On the other hand, Ismail et al. (2016) produced much larger Au NPs (4–10 nm) when the irradiation time was significantly longer than in the previous methods. The factors influencing the formation of Au NPs on the surface of ZnO during light irradiation have been precisely investigated by Fernando et al. (2016). The authors have revealed an important role of the solvent properties in the determination of the size of Au NPs. Au NPs formed in the reaction in which ethanol was used as solvent were significantly larger than those obtained in the reaction with the use of toluene (14 vs. 6 nm). The other important parameters were pre-irradiation of the $HAuCl_4$ solution, type of light source and concentration of dissolved oxygen.

The sizes of Au NPs on the surface of catalysts prepared by above described methods are summarized in Table 4. These data clearly show the remarkable influence of the synthesis route applied on the gold particle size.

Oxidation state of gold and its interaction with the support

Similarly to Au/CeO_2, the oxidation state of gold in Au/ZnO catalysts is strongly affected by synthesis conditions, especially by the deposition method and calcination temperature.

The oxidation state of gold in the catalysts prepared by DP method was studied *inter alia* by Liu et al. (2012). In the as-prepared material (sample not pretreated at high temperature), gold was present both in the metallic form (more than half of the gold species), and in the form of Au^+ and Au^{3+}. The authors have also observed the relationship between the oxidation state of gold in Au/ZnO and the calcination temperature. The higher the temperature of heat treatment, the higher the content of metallic gold and the smaller the content of the cationic gold species.

The results reported in (Liu et al. 2016) have indicated that the oxidation state of gold is not only affected by the method of gold deposition, but also by the properties of support. In this work, Au/ZnO catalysts were prepared by the same method as above (DP) but with the use of zinc oxide containing a high concentration of defect

Table 4: The influence of the synthesis route and calcination temperature on the size of supported gold nanoparticles.

Preparation method	Gold loading, at% or wt%	Calcination temperature, K	Average particle size, nm	References
Co-precipitation	5.0 at%	673	5.0	(Manzoli et al. 2004)
Co-precipitation	5.0 at%	673	3.5	(Sakurai and Haruta 1996)
Co-precipitation	33.0 at%	673	19.5	
Impregnation	5.0 at%	673	33.9	
Deposition-precipitation	4.1 wt %	473	1.9	(Liu et al. 2016)
	18.0 wt %	473	3.8	
Deposition-precipitation	1.0 at%	773	4.9	(Naknam et al. 2009)
Deposition-precipitation	1.7 wt %	As-prepared	3.6	(Castillejos et al. 2012)
Anchoring of Au NPs functionalized with glutathione	0.4 wt %	As-prepared	3.3	(Udawatte et al. 2011)
	1.0 wt %	As-prepared	3.3	
	2.0 wt %	As-prepared	3.4	
	2.0 wt %	523	3.6	
Spray pyrolysis	1.0 wt %	As-prepared	4.0	(Pawinrat et al. 2009)
	3.0 wt %	As-prepared	6.7	
Colloidal deposition	1.0 wt %	n.d.	5.5	(Noei et al. 2012)
Colloidal deposition	1.25 wt %	623	3.1	(Wu et al. 2011)
Photodeposition	n.d.	As-prepared		(Fernando et al. 2016)
solvent – cyclohexane			12.0	
solvent – toluene			6.0	
solvent – methanol			18.0	
solvent – acetonitrile			22.0	
Photodeposition	molar Au/ZnO = 0.02	As-prepared	0.8	(He et al. 2014)
	molar Au/ZnO = 0.04	As-prepared	1.4	
	molar Au/ZnO = 0.10	As-prepared	2.3	

sites (zinc and oxygen vacancies). It was found that in as-prepared material, gold was present mainly in the cationic form (Au^+–27.4% and Au^{3+}–70.4%), and was located in the lattice of ZnO. Heat treatment of such a material led to a significant increase in the content of metallic gold on the surface of the catalyst. XPS analysis revealed that after the heating of Au/ZnO catalyst at 473 K in oxygen atmosphere, the concentration of metallic gold increased from 7.3 to 73.6%, and a small amount of cationic gold species was still observed (Au^{3+}–8.7% and Au^+–17.7%). The defected ZnO enhances SMSI and leads to the formation of mixed oxide of Au/ZnO as indicated in Liu et al. (2016).

Au/ZnO catalysts prepared with the use of other methods like co-precipitation (Manzoli et al. 2004) and deposition-reduction (Arunkumar et al. 2017), all contained both metallic gold species and cationic one ($Au^{\delta+}$). In Manzoli et al.

(2004), the authors have suggested that the presence of oxidized form of gold in the Au/ZnO catalyst resulted from the strong interaction of Au with the support, possibly thought the support defects. The influence of defects on the oxidation state of gold was studied in detail in Liu et al. (2012) with the use of electron paramagnetic resonance (EPR) spectroscopy. The authors pointed out the important role of ionized oxygen vacancies in the electron transfer form Au to ZnO. They have documented that the above-mentioned oxygen vacancies were able to gain electron from gold nanoparticles under high temperature treatment in oxygen atmosphere, and convert them back after treatment under hydrogen atmosphere.

Behl and Jain (2014) revealed that electron transfer from oxygen vacancies of ZnO to gold nanoparticles significantly influence the surface properties of ZnO. Ionization of the defect sites by electron transfer to Au NPs led to a significant increase in the acidity of the catalyst surface, probably due to the formation of electron deficient Zn sites, which can play a role of Lewis acid sites.

The influence of the above described gold deposition methods and conditions of catalyst pretreatment on oxidation state of gold is summarized in Table 5.

Strong interaction between the Au and ZnO has been reported by many authors. An evidence for this phenomenon was the hemispherical shape of Au NPs observed in Liu et al. (2012) and Wu et al. (2011), and the presence of a well distinguished contact perimeter or metal-oxide interface between the Au and ZnO, such as the one reported in Wu et al. (2011). Another evidence for the presence of SMSI in Au/ZnO systems has been given by Misra et al. (2014), who investigated the optical properties of

Table 5. The influence of synthesis conditions and heat pretreatment on the oxidation state of gold in Au/ZnO catalysts.

Preparation method	Heat treatment	Oxidation state of gold	References
Co-precipitation	O_2, 573 K	Au^{0}[b], Au^{6+}[b]	(Manzoli et al. 2004)
Co-precipitation	Air, 423 K	Au^{0}[c], Au^{3+}[c]	(Zhang et al. 2003)
	Air, 513 K	Au^{0}[c], Au^{3+}[c]	
	Air, 543 K	Au^{0}[c]	
Deposition-precipitation	As-prepared	Au^{0}[a], Au^{+}[a], Au^{3+}[a]	(Liu et al. 2016)
	O_2, 473 K	Au^{0}[a], Au^{+}[a], Au^{3+}[a]	
Deposition-precipitation	Air, 773 K	Au^{0}[c], Au^{3+}[d]	(Naknam et al. 2009)
Deposition-precipitation	O_2, 473 K	Au^{0}[a], Au^{+}[a], Au^{3+}[a], Au^{6+}[b]	(Liu et al. 2012)
	O_2, 573 K	Au^{0}[a], Au^{+}[a], Au^{3+}[a], $Au^{\delta+}$[b]	
	H_2, 573 K	Au^{0}[a], Au^{+}[a], $Au^{\delta-}$[b]	
Deposition-reduction	Dried at 373 K	Au^{0}[a]	(Castillejos et al. 2012)
Deposition-reduction	Air, 723 K	Au^{0}[a]	(Xie et al. 2017)
Colloidal deposition	Vacuum, 680 K	$Au^{\delta-}$[b]	(Noei et al. 2012)
	O_2, 450 K, then vacuum, 680 K	Au^{0}[b]	

From: [a] XPS, [b] CO adsorption, [c] XRD, [d] TPR, [e] UV-Vis.

Au-core-ZnO-shell structures. They observed that in such structures, the absorption spectrum of ZnO was red-shifted in comparison to that of ZnO nanoparticles, while the SPR band of the gold core was blue-shifted in comparison to that of unmodified Au NPs. This phenomenon resulted from the electronic interaction between Au and ZnO, which led to the modification of the optical properties of both Au and ZnO.

The electron transfer from ZnO to Au in Au/ZnO catalysts has also been observed with the use of UV-Vis spectroscopy by many researches (e.g. (Xie et al. 2017; Rahman and Ghosh 2016)). The evidence for such electron transfer has been reported by Kaneti et al. (2016). The authors used X-ray photoelectron spectroscopy (XPS). They revealed that deposition of Au on the surface of ZnO led to an increase in the binding energy (BE) of $Zn2p_{3/2}$. The increase in the binding energy of Zn2p3/2 was associated with a decrease in the binding energy of Au $4f_{7/2}$ from 84.0 eV, which is characteristic of Au NPs, to 83.5 eV. The authors concluded that the decrease in binding energies of gold species was caused by the larger electronegativity of Au compared to that of Zn atoms in ZnO, which promotes electron transfer from ZnO to Au. A similar phenomenon has been also noted by other authors (e.g. (Arunkumar et al. 2017) – BE Au $4f_{7/2}$ = 83.6 eV, (Xie et al. 2017) – BE Au $4f_{7/2}$ =83.4 eV and (Behl and Jain 2014) – BE Au $4f_{7/2}$ = 83.3 eV)).

Strunk et al. (2009) have stressed the enhancement of reducibility of ZnO after the gold deposition, which followed from the increased number of removable oxygen atoms in the ZnO structure and formation of more exposed oxygen vacancies. The same feature has been reported by Dulnee et al. (2014). They reported that reducibility of ZnO was higher in the Au/ZnO catalyst which was calcined at high temperatures.

The SMSI in Au/ZnO catalysts was supported by reversible mass transport phenomenon indicated in Liu et al. (2012). The authors have observed that during the heat treatment of the catalyst at 573 K in an oxygen atmosphere, zinc oxide migrated on the top surface of Au NPs, while the gold particle size remained almost unchanged (ca. 3.4 nm). On the other hand, the heat treatment in hydrogen atmosphere resulted in the return of ZnO to the original positions. Another important factor in SMSI is the electronic interaction. This phenomenon was investigated by the use of *in situ* diffuse reflectance infrared Fourier transform spectroscopy (DRIFTS) combined with CO adsorption, and XPS studies. For the sample dried at 333 K (without any pretreatment), the band at 2101 cm^{-1} characteristic of CO adsorbed on metallic Au NPs was observed. Upon increasing temperature, this band was blue-shifted and its intensity decreased. For the catalyst pretreated at 573 K in an oxygen atmosphere, the CO adsorption band was shifted from 2101 to 2113 cm^{-1}, which indicated that the surfaces of the gold nanoparticles were slightly positively charged. The following treatment at 573 K in the hydrogen atmosphere led to the reappearance of the band at 2108 cm^{-1}, characteristic of CO adsorbed on metallic gold and the formation of a new band at 2048 cm^{-1}, typical of CO adsorbed on negatively charged Au NPs (Au$^{\delta-}$). In this experiment, the authors clearly confirmed that depending on the pretreatment conditions, the reversible transport of electron from ZnO to Au NPs occurred, and thus indicating the presence of another very important characteristic of the SMSI in Au/ZnO systems.

Catalytic application

Literature data shows that Au/ZnO catalyst have been successfully used in many different catalytic processes, like the synthesis of methanol from syngas (Strunk et al. 2009), hydrogenation of CO_2 (Sakurai and Haruta 1996), selective hydrogenation of cinnamaldehyde (Castillejos et al. 2012), dehydrogenation of propane (Tóth et al. 2016), preferential CO oxidation (PROX) (Naknam et al. 2009; Zhang et al. 2003; Dulnee et al. 2014) and total benzene oxidation (Wu et al. 2011). Au/ZnO catalysts have also been found to be interesting candidates for the detection of different organic and inorganic molecules like: ethyl acetate (Xie et al. 2017), hydrazine (Ismail et al. 2016), NO (Gogurla et al. 2014), n-butylamine (Kaneti et al. 2016) and CO (Arunkumar et al. 2017). From among these numerous processes, one of the most investigated was the low-temperature CO oxidation (Al-Sayari et al. 2007; Manzoli et al. 2004; Noei et al. 2012; Wang et al. 2003; Liu et al. 2012; Wang et al. 2002). Similar to Au/CeO$_2$, the activity of Au/ZnO in this process was strongly affected by the synthesis conditions. Al-Sayari et al. (2007) have reported that the activity of Au/ZnO catalyst prepared by co-precipitation is influenced by numerous factors like: pH of reaction mixture, temperature of precipitation, aging time and thermal treatment of the catalyst. It has been established that even a slight modification of synthesis conditions may remarkably affect the properties and catalytic activity of the materials. The authors highlighted that the catalyst prepared without calcination usually exhibited low catalytic activity at the beginning of the reaction, and its activity increased with time on stream (TOS). They postulated that this phenomenon may result from the reduction of cationic gold species to metallic gold caused by the interaction of the catalyst surface with CO. Another explanation given by the authors assumes that the above-mentioned increase in the catalyst activity comes from non-optimal concentration of hydroxyl groups on the catalyst surface at the beginning of the reaction.

The impact of the calcination step on the activity of catalysts prepared by co-precipitation has also been indicated by Wang et al. (2003). They reported that the most effective calcination temperature which led to the catalysts of the highest activity was 513 K. On the other hand, Liu et al. (2012) have reported that for the Au/ZnO catalyst prepared by deposition-precipitation, the optimal temperature of heat treatment in the oxygen atmosphere was lower and estimated as 473 K. They indicated that calcination at higher temperatures (T = 573 K) led to a significant decrease in the catalytic activity of Au/ZnO due to the encapsulation of gold nanoparticles by ZnO, which reduced the accessibility of active sites to CO oxidation.

Another very important factor affecting the activity of Au/ZnO catalyst is the gold particle size. It is known that small gold particles not only provide more active sites for the reversible adsorption of CO but also appreciably increase the amount of oxygen adsorbed on the oxide support (Wang et al. 2003). Liu et al. (2016) have shown that the Au/ZnO catalyst with a very small size of Au NPs (ca. 2 nm) exhibited higher catalytic activity than two of the most active catalysts in this process, that is, Au/TiO$_2$ and Au/FeTiO$_2$. Noei et al. (2012) have revealed that another important factor, which can enhance the activity of Au/ZnO at low temperature CO oxidation is the presence of reduced $Au^{\delta-}$ nanoparticles. The authors found that this additional negative charge on Au NPs led to the easier bond breaking of chemisorbed O_2 molecules, and thus enhanced the efficiency of the oxidation process. The enhancement of catalytic activity in the presence of $Au^{\delta-}$ also resulted from a lower stability of adsorbed CO species in comparison to those adsorbed on metallic Au NPs.

The mechanism of CO oxidation over Au/ZnO catalysts has been studied by Noei et al. (2012) and Manzoli et al. (2004). The authors have found that the first step in CO oxidation was the activation of both CO and O_2 molecules on small gold nanoparticles, possibly near an oxygen vacancy (Manzoli et al. 2004). Then, the oxidation of CO proceeded by the direct reaction of Au-bonded CO with co-adsorbed oxygen at the Au/ZnO interface. However, Noei et al. (2012) reported that the oxidation of CO may also proceed by another pathway, that is, adsorption of CO on the surface of ZnO followed by its spillover to the Au/ZnO interface, at which the reaction with activated oxygen may lead to the formation of CO_2. Noei et al. (2012) have also reported that the activation of molecular oxygen on Au/ZnO is promoted by the preadsorbed CO molecules. DFT studies performed by these authors indicated that the activation of oxygen can occur via two different pathways: (i) direct dissociation at the Au/ZnO interface yielding atomic species or (ii) by bimolecular reaction forming an $OC-O_2$ intermediate complex. This complex is then activated by charge transfer from the Au/ZnO substrate to O_2 in the complex which weakens the O-O bond, leading to easier dissociation of the molecular oxygen and to the formation of CO_2. The remaining atomic oxygen is bound to nearby Au atoms and can further react with the second CO molecule.

Much attention has also been paid to the deactivation of Au/ZnO catalysts during CO oxidation. It was found that the main factor affecting the decrease in catalytic activity of Au/ZnO was the formation of carbonate-like species, which cover the active interface sites and inhibit the back-spillover of oxygen species activated on the support to the gold (Manzoli et al. 2004; Noei et al. 2012). Wang et al. (2002) have reported that the formation of carbonate-like species was strongly affected by gold loading and the presence of water. The optimum Au/Zn atomic ratio for CO oxidation at 298 K was found to be 5/100. Using this catalyst, CO was completely transformed into CO_2 in 2100 hr (in the reaction in the presence of water, this time was shorter—1600 hr). It was also found that deactivation of the catalyst by the formation of carbonate-like species may be significantly reduced in different ways (e.g., by increasing the concentration of CO in the reaction mixture (Wang et al. 2002), presence of Na^+ in Au/ZnO catalysts (Wang et al. 2003) or co-feeding with H_2 (Manzoli et al. 2004)).

Au/ZnO catalysts have also attracted much attention as effective photocatalysts for waste water remediation. As follows from literature data, Au/ZnO can be successfully used for the degradation of numerous organic compounds like: rhodamine 6G (Udawatte et al. 2011), thionine (Udawatte et al. 2011), methyl orange (Misra et al. 2014), methylene blue (He et al. 2014; Pawinrat et al. 2009), salicylic acid (He et al. 2014), rhodamine B (She et al. 2017), Evans blue (Rahman and Ghosh 2016), phenol (Silva et al. 2014) and benzene (Yu et al. 2012). It was proved that under both UV and visible light irradiation, the presence of gold nanoparticles significantly enhanced the activity of ZnO, leading to the obtainment of materials whose activity was comparable or even higher than that of TiO$_2$ (e.g. (Pawinrat et al. 2009)). Many authors have reported that the role of Au NPs in photocatalytic degradation with the use of Au/ZnO catalysts depends on the type of light source (Fig. 4).

Under UV irradiation, photogenerated electrons from the conduction band of ZnO were transferred to Au NPs. This electron transfer increased the efficiency of the photogenerated charge carriers separation and allowed the increase in photocatalytic activity of the composite (Udawatte et al. 2011; Nagajyothi et al. 2016). Thus, under UV irradiation, Au NPs acted as electron sink. Lee et al. (2011) reported that the efficiency of electron transfer from ZnO to Au NPs was affected by the size of Au NPs. From among the catalyst with different particle sizes (1.1, 1.6 and 2.6 nm), the highest activity was observed for Au NPs with the largest particles of 2.6 nm. When visible light was used, ZnO could not be excited due its relatively high band gap value of ca. 3.2–3.4 eV. It was found that in these conditions, the electrons from Au NPs were excited and transferred from gold to the conduction band of ZnO, where they can react with oxygen leading to the formation of superoxide anions (Misra et al. 2014; Yu et al. 2012; Silva et al. 2014; She et al. 2017). At the same time, positively charged holes in Au NPs can react with H$_2$O or OH$^-$ leading to formation of strongly oxidative hydroxyl radicals (Yu et al. 2012; Silva et al. 2014; She et al. 2017). However, not all Au/ZnO catalysts were active upon visible light irradiation. Silva et al. (2014) have reported that the Au/ZnO catalyst with very small Au NPs (2.9 nm) was not activated during irradiation with visible light (λ = 546 nm). This resulted from the fact that the gold nanoparticles with diameter below 5 nm do not show surface plasmon resonance due to quantum-size limitations, and thus hot electrons cannot be generated in this system. The relationship between the size of

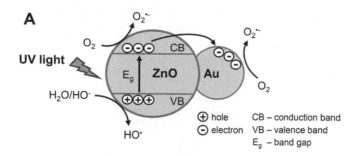

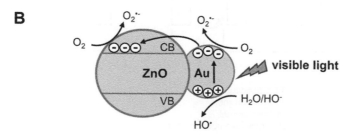

Fig. 4. The role of gold in photocatalytic processes under UV (A) and visible light (B) irradiation (after (Yu et al. 2012) and (She et al. 2017)).

supported Au NPs and their photocatalytic activity upon visible light irradiation was investigated by Rahman and Ghosh (2016). The authors synthesized hybrid ZnO-Au nanostructures with different sizes of Au NPs (8, 10, 13, 16, and 25 nm). They noted the highest efficiency in photocatalytic degradation of Evans blue for the catalyst containing gold nanoparticles with the diameter of 8 nm. Based on the above-desribed results, one can conclude that the optimum size of supported gold nanoparticles for the obtainment of high efficiency in degradation of dyes under visible light irradiation is in the range from 5 to 8 nm. It is also important to stress that unsupported Au NPs have been found inactive in the degradation of organic compounds under visible light irradiation (Sun et al. 2016).

Gold loaded on niobium oxide

Properties of niobium oxide

Niobium pentoxide (Nb_2O_5) is the most thermodynamically stable oxide from among the niobium-oxygen systems (Nico et al. 2016). It is a white, air-stable and water-insoluble solid material, which is difficult to reduce. For the reduction of bulk Nb_2O_5 to bulk Nb_2O_4, the temperature of 1573 K is needed (Wachs et al. 2000; Ziolek 2003). Nb_2O_5 has a relatively complicated structure that displays extensive polymorphism (Nowak and Ziolek 1999; Rani et al. 2014; Nico et al. 2016). The most common crystal phases are pseudohexagonal (TT-Nb_2O_5), orthorhombic (T-Nb_2O_5) and monoclinic (H-Nb_2O_5). Synthesized amorphous Nb_2O_5 crystallizes at 773 K into TT or T phases, at 1073 K it undergoes transformation into the M phase (tetragonal), and above 1273 K into the H phase (Ko and Weissman 1990) (Fig. 5).

Niobium oxide is characterized by acidic surface properties. Amorphous hydrated niobium oxide (niobic acid—$Nb_2O_5 \cdot nH_2O$), especially, is known as a

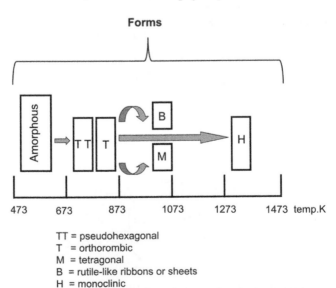

Forms

473   673   873   1073   1273   1473 temp.K

TT = pseudohexagonal
T = orthorombic
M = tetragonal
B = rutile-like ribbons or sheets
H = monoclinic

Fig. 5. Polymorphism of Nb_2O_5 (after (Ko and Weissman 1990)).

solid acid that retains high acidic strength (Ho ≤ −5.6) on the surface (Ushikubo et al. 1996). Lewis and Brønsted acid sites were found on the surface of niobic acid (Prasetyoko et al. 2005). NbO_6 sites (octahedrally coordinated) act as Lewis acid sites, whereas the slightly distorted surface NbO_6 sites, as well as NbO_7 and NbO_8 sites are associated with Brønsted acid centres (Jehng and Wachs 1990). However, the strong acidic properties (BAS) of hydrated Nb_2O_5 disappear after calcination at temperatures higher than 800 K. Niobia is a typical SMSI (strong metal-support interaction) oxide. These interactions, present when Nb_2O_5 is used as support for various metals, are the most relevant to catalysis (Nowak and Ziolek 1999; Ziolek 2003).

Properties of gold loaded on niobium oxide

Up to now, Nb_2O_5 oxide has been rarely used as the support for gold. There are only a few publications concerning the introduction of gold on the surface of niobia, its characterization and catalytic applications.

Methods of gold incorporation and dispersion of gold

Although the most popular methods for modification of metal oxide with gold in order to obtain highly dispersed nanoparticles on the support surface are deposition-precipitation (metal hydroxide is supposed to precipitate on the support) and deposition-reduction (with the use of a reducing agent), it is difficult to effectively use these methods (and to generate small gold particles) for the metal oxides with the isoelectric point below pH = 5 (for Nb_2O_5 the isoelectric point was estimated to be 2.8 (Vallar et al. 1999; Murayama and Haruta 2016). One of the first reports concerning the Au/Nb_2O_5 catalyst describes the samples prepared by strong electrostatic adsorption (SEA) (Miller et al. 2006). $HAuCl_4$ solution as a source of gold was adsorbed on the support (160 m²/g) without washing, for the removal of residual Cl⁻ ions. After adsorption, the samples were dried at 373 K and reduced with H_2 at 423 or 523 K. Au/Nb_2O_5-1 and Au/Nb_2O_5-2 samples with 2.1 and 2.5 wt % of gold and 1.3 and 2.2 wt % of Cl were obtained, respectively. For Au/Nb_2O_5-1, a much higher dispersion (based on EXAFS studies) was obtained than for Au/Nb_2O_5-2, with a higher content of Cl (85 and 42%, respectively). Chloride ions promoted agglomeration of gold species during thermal treatment. It was found that with the increasing dispersion of gold particles, the Au–Au bond length decreased. Moreover, with decreasing particle size, the d-electron density in small particles increased. Campo et al. (2008) have used the direct anionic exchange (DAE) method with the addition of ammonia (developed by Ivanova (Ivanova et al. 2004)) for the preparation of the Au/Nb_2O_5 catalyst. Niobia (calcined at 1073 K, 52 m²/g) was applied as the support and chloroauric acid as the source of gold. The prepared catalyst was washed with hot water in order to eliminate Cl⁻ and NH_4^+ ions, dried at 373 K and calcined under air at 573 K. The real gold loading in the prepared Au/Nb_2O_5 sample was 1.5 wt % and it was lower than assumed (2 wt %). It is important to stress that the amount of chlorine species was lower than 200 ppm, indicating that the NH_4OH treatment effectively eliminates the precursor ligands. In the XRD pattern of Au/Nb_2O_5, the reflexes corresponding to metallic gold were not observed, indicating that Au crystallites on support were smaller than 5 nm.

The conventional deposition-precipitation method with NaOH (pH = 9) for gold introduction on crystalline Nb_2O_5 was used by Tong et al. (2016). The source of gold was $HAuCl_4$, the obtained Au/Nb_2O_5 catalyst was calcined at 673 K. The gold loading was low upon the preparation conditions applied by the authors. The particle size distribution obtained on the basis of TEM images showed the domination of gold particles in the range 9–10 nm (ca. 60%). Sobczak et al. (2010) have described Au/Nb_2O_5 catalysts prepared by two different methods: gold-sol with tetrakis (hydroxymethyl) phosphonium chloride (THPC) as a reducing agent (GS) and deposition-precipitation with urea (DPU). Gold from $HAuCl_4$ was deposited on two kinds of niobia supports: crystalline (cr; 7 m^2/g) and amorphous ones (am; $Nb_2O_5 \cdot nH_2O$, 30 m^2/g). Au/Nb_2O_5 (cr) and Au/Nb_2O_5 (am) containing 1 wt % of gold were washed until the filtrate was free of chloride, dried at 373 K and calcined at 623 K. A significant effect of the preparation method on the size of gold particles was observed. The size of gold crystallites on Nb_2O_5 prepared by the DPU method was much larger than that in the samples modified by the GS method. The average particle size estimated from TEM images for Au/Nb_2O_5 (DPU) was about 25 nm, whereas in the samples prepared by the GS method it was 6 and 10 nm for Au/Nb_2O_5 (cr) and Au/Nb_2O_5 (am), respectively. However, it is important to add that the average gold particle size on Au/Nb_2O_5 (cr) (GS) calculated from the XPS spectra (depth of analysis 5–8 nm) was much smaller – 2.2 nm (Musialska et al. 2010). The explanation was that the THPC used for the GS method stabilizes the colloid gold solutions and that is why gold particles are smaller and their dispersion is greater (Sobczak et al. 2010). Five years later, the same research group (Sobczak and Wolski 2015) applied another method developed by Mou's (Liu et al. 2009) for the modification of Nb_2O_5 ($Nb_2O_5 \cdot nH_2O$, calcined at 673 K). Before gold introduction, niobia was grafted with 3-aminopropyl-trimethoxysilane (APMS) in order to functionalize the support and to generate active sites at which the gold precursor could be adsorbed. The functionalized NH_2-Nb_2O_5 was stirred in an aqua solution of chloroauric acid (2 wt % of Au as assumed) and then the gold in recovered solid was reduced with $NaBH_4$. The samples obtained were dried at 373 K and calcined at 773 K. It was indicated that the interaction of gold (in the form of $AuCl_4^-$ ions (Liu et al. 2009)) with NH_2 groups (in the form of NH_3^+ ions (Liu et al. 2009)) was effective for gold incorporation into Nb_2O_5. The average gold particle size for Au/Nb_2O_5 was 4.9 nm (with ca. 50% of particles between 4–5 nm).

The above described methods of gold loading on Nb_2O_5 and their influence on the properties of gold loaded on niobia are summarized in Table 6.

Very recently, more detailed studies concerning Au/Nb_2O_5 catalysts based on different types of niobia and modified with gold by various methods were performed by Haruta et al. (Murayama and Haruta 2016; Murayama et al. 2016; Table 7). The novelty of this study was the use of support of crystalline layered structure type (deformed orthorhombic) niobium oxide with a high surface area of 208 m^2/g, synthesized by a hydrothermal method from ammonium niobium oxalate (denoted Nb_2O_5 (HT)). For comparison, commercial Nb_2O_5 (5.8 m^2/g) and $Nb_2O_5 \cdot nH_2O$ (19 m^2/g) calcined at 673 K were used as the supports for gold (Murayama and Haruta 2016). Gold was

Table 6. The methods used for Au/Nb$_2$O$_5$ catalysts preparation and the characterization of gold.

Method of Au loading (HAuCl$_4$ as source of metal)	Loading of gold, wt%	Heat treatment	Average particle size, nm	Oxidation state of Au	References
Strong electrostatic adsorption	2.1 2.5	Air-373 K and H$_2$-423 K Air-373 K and H$_2$-523 K	n.d. n.d.	n.d. n.d.	(Miller et al. 2006)
Direct anionic exchange	1.5	Air-573 K	< 5	n.d.	(Campo et al. 2008)
Deposition-precipitation (NaOH)	very low	Air-673 K	9–10	Au^0	(Tong et al. 2016)
Deposition-precipitation (urea)	1.0	Air-623 K	25	Au^0	(Sobczak et al. 2010)
Gold-sol	1.0	Air-623 K	6–10	Au^0	(Sobczak et al. 2010)
Grafting	2.0	Air-773 K	4.9	(Au^0)$^{\delta-}$	(Sobczak and Wolski 2015)

Table 7. The loading and properties of gold supported on different types of niobia (Murayama and Haruta 2016; Murayama et al. 2016).

Type of niobia	Method of Au loading (source)	Loading of gold, wt%	The average size of Au crystallites, nm
Nb$_2$O$_5$ (HT)	Deposition-precipitation (Au(en)$_2$Cl$_3$)	0.81	5.0
Nb$_2$O$_5$ (HT)	Deposition-precipitation (HAuCl$_4$)	0.18	7.5
Nb$_2$O$_5$	Deposition-precipitation (Au(en)$_2$Cl$_3$)	0.10	23.0
Nb$_2$O$_5$·nH$_2$O	Deposition-precipitation (Au(en)$_2$Cl$_3$)	0.88	6.0
Nb$_2$O$_5$ (HT)	Deposition-reduction (Au(en)$_2$Cl$_3$)	0.93	4.9
Nb$_2$O$_5$ (HT)	Deposition-reduction (HAuCl$_4$)	0.26	4.3
Nb$_2$O$_5$	Deposition-reduction (Au(en)$_2$Cl$_3$)	1.00	14.0
Nb$_2$O$_5$·nH$_2$O	Deposition-reduction (Au(en)$_2$Cl$_3$)	2.00	13.0
Nb$_2$O$_5$ (HT)	Solid grinding (Au(en)$_2$Cl$_3$)	1.00	4.1
TT-Nb$_2$O$_5$	Sol immobilization (HAuCl$_4$)	1.07	2.6
T-Nb$_2$O$_5$	Sol immobilization (HAuCl$_4$)	0.79	2.7
Nb$_2$O$_5$ (HT)	Sol immobilization (HAuCl$_4$)	1.01	2.7
Nb$_2$O$_5$-P	Sol immobilization (HAuCl$_4$)	0.76	6.1

introduced by deposition-precipitation (DP) method with NaOH. Additionally, for Nb$_2$O$_5$ (HT), deposition reduction with NaBH$_4$ (DR) and solid grinding (SG) methods were used. In all the synthesis routes the cationic source of gold (Au(en)$_2$Cl$_3$; en = ethylenediamine) was used and the catalysts prepared were calcined at 573 K. It was found that using the DP method, the morphology and large surface area of Nb_2O_5 (HT) are important factors for the formation of small gold particles dispersed on Nb_2O_5. The smallest average gold particle size (measured from TEM images) was found for layered Au/Nb_2O_5 (HT) (5 nm). The gold diameters for Au/$Nb_2O_5 \cdot nH_2O$ and Au/Nb_2O_5 were bigger, independent of the method, 6 and 23 nm for DP (Murayama and Haruta 2016) and 13 and 14 nm for DR (with wide particle size distribution between 2–50 nm), respectively (Murayama et al. 2016). The DR and SG methods used for Nb_2O_5 (HT) modification also gave small gold crystallites on the oxide surface, of 4.9 and 4.1 nm (narrow particle size distribution), respectively. Moreover, the efficiency of gold incorporation was different. The loading of gold was similar for Au/Nb_2O_5 (HT) (DP), Au/Nb_2O_5 (HT) (DR) and Au/$Nb_2O_5 \cdot nH_2O$ (DP) (0.81, 0.93, 0.88 wt %), whereas on Au/Nb_2O_5 (DP) the metal loading was very low (0.1 wt %). The results suggested that the strength of electrostatic interactions between the Au(en)$_2^{3+}$ species and the surface of Nb_2O_5, which shows a negative electrical charge in water, was different for different types of Nb_2O_5. Moreover, the Au/Nb_2O_5 (HT) (DP) sample was prepared with $HAuCl_4$ as a gold precursor. In these conditions the gold introduction was also ineffective (0.18 wt % of Au loading, the average gold particle diameter 7.5 nm). Au(OH)$_4^-$ species formed at pH = 9 (Tsubota et al. 1995) do not interact with Nb_2O_5 because of its low isoelectric point. The application of Au(en)$_2$Cl$_3$, which forms Au(en)$_2^{3+}$ cation, increases the gold loading. Using the DR method and $HAuCl_4$, the gold loading on Nb_2O_5 (HT) was low (0.26 wt %), but the size of the gold particles was much smaller (4.3 nm). The explanation was that the structure of Nb_2O_5 (HT) affected the interactions of Au(OH)$_4^-$ on $Nb_2O_5 \cdot Nb_2O_5$ (HT) has a layered-type structure, and consists of NbO_6 octahedra, NbO_7 and micropore channels in its structure. TEM images indicated that on Au/Nb_2O_5 (HT) prepared with $HAuCl_4$, gold particles were deposited on the basal planes of the rod-type Nb_2O_5 particles (Fig. 6). Au^{3+} cations were preferentially reduced with $NaBH_4$ on the basal planes of the rod.

In another paper of Haruta (Murayama et al. 2016), the studies of Au/Nb_2O_5 have been extended to gold catalysts supported on Nb_2O_5 oxides with different crystalline structures: pseudohexagonal (TT-Nb_2O_5, 43 m²/g), orthorhombic (T-Nb_2O_5, 24 m²/g) and pyrochlore (denoted as Nb_2O_5-P, 30 m²/g) niobium oxides. These supports were

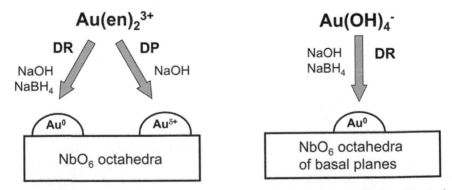

Fig. 6. Schematic localization of gold species on octahedral based layered-type Nb_2O_5 depending on the deposition method (after (Murayama and Haruta 2016; Murayama et al. 2016)).

modified with gold (from $HAuCl_4$) by the sol immobilization method (SI) (1 wt % of Au as assumed) and calcined at 573 K. Using the same method, Au/Nb_2O_5 (HT) (SI) was also prepared. The average gold particle diameter was almost the same for $Au/TT-Nb_2O_5$ (SI) (2.6 nm), $Au/T-Nb_2O_5$ (SI) (2.7 nm) and Au/Nb_2O_5 (HT) (SI) (2.7 nm) with a narrow particle size distribution (between 1–7 nm). The exception was Au/Nb_2O_5-P, for which a much wider particle size distribution was obtained (between 1–20 nm; the average gold diameter was 6.1 nm).

Oxidation state of gold and its interaction with the support

The abovementioned strong metal support interactions (SMSI) which characterize niobium oxide are expected to influence the oxidation state of gold loaded on this support, the dispersion of metal and the activity of the catalysts.

Tong et al. (2016) and Sobczak et al. (2010) have used UV-Vis technique for the determination of the state of gold loaded on Nb_2O_5 by deposition-precipitation (Tong et al. 2016; Sobczak et al. 2010) and gold-sol methods (Sobczak et al. 2010). The catalysts prepared contained metallic Au particles detected on the basis of the intense and well-resolved band at 550–560 nm in their UV-Vis spectra. TEM images confirmed the presence of metallic gold crystallites. The better gold dispersion on crystalline niobia than on amorphous ones was a result of the strong interaction of gold particles with unsaturated niobium species. Niobium in the oxide crystal phase occupied positions at the corners and edges of the crystals in which Nb species are unsaturated. Au/Nb_2O_5 (cr) (GS) were also studied by XP spectroscopy. The binding energies of 83.7 and 87.4 eV, measured for $Au4f_{7/2}$ and $Au4f_{5/2}$, were typical of metallic supported gold (Moroz et al. 2009). Moreover, it was found that the atomic ratio Nb/O for Au/Nb_2O_5 (cr) (GS) was lower than 0.4 (the value stoichiometric for Nb_2O_5), suggesting the storage of oxygen on niobia support, most probably on the interface with gold. On the other hand, negatively charged gold particles $(Au^0)^{\delta-}$ were identified by XPS on the surface of Au/Nb_2O_5 prepared by grafting with APMS and gold adsorption on $NH_2-Nb_2O_5$ (Sobczak and Wolski 2015). The binding energy of $Au4f_{7/2}$ was shifted relative to that bulk Au^0 from 84.0 to 83.2 eV. The presence of the $(Au^0)^{\delta-}$ species was a result of the preparation method. In this procedure, the negatively charged $AuCl_4^-$ ions are adsorbed on NH_3^+ ions from organosilane and after the reduction with $NaBH_4$, the negatively charged gold particles are produced. The niobia-gold interactions were identified by the observed shift of UV-Vis band from octahedral Nb species from 272 nm (Nb_2O_5) to 283 nm (Au/Nb_2O_5). XRD and UV-Vis techniques confirmed the presence of metallic gold on the surface of the Au/Nb_2O_5 sample as the reflections characteristic of metallic gold at $2\theta = 38.2°$ and $44.4°$ and a band ca. 540 nm were observed in the XRD patterns and UV-Vis spectra, respectively. The oxidation state of gold in Nb_2O_5 catalysts prepared by the sol immobilization method was studied using FTIR spectroscopy (DRIFT-IR) with the adsorption of CO and XPS (Murayama et al. 2016). The IR spectra of Au/Nb_2O_5 (HT) (SI), $Au/TT-Nb_2O_5$ (SI) and $Au/T-Nb_2O_5$ (SI) showed two bands at 2130 and 2116 cm^{-1}. They were assigned to CO adsorbed on Au^0 sites. No FTIR bands were observed after CO adsorption on $Au/P-Nb_2O_5$ (SI), indicating the modification of the surface of Au particles by strong metal-support interactions (SMSI). The XP spectra showed the presence of negatively charged gold particles on the sample surface (BE measured for Au4f$_{7/2}$ were between 83.0 and 83.5 eV). Interestingly, the position of the bands observed was dependent on the structure of Nb$_2$O$_5$ catalyst. Moreover, the higher BE of Au4f$_{7/2}$ resulted in lower BE of Nb3d$_{5/2}$. This effect indicates the gold-support interactions.

Catalytic application

The activity of gold catalysts is determined mainly by the oxidation state of metal and the size of gold particles. Au/Nb$_2$O$_5$ were mostly applied in oxidation (carbon monoxide and alcohols oxidation) and hydrogenation reactions.

Oxidation reactions

CO Oxidation. CO oxidation at low temperature is the most commonly studied process in the field of catalysis by gold. This reaction is structure sensitive; to obtain high activity of a catalyst in CO oxidation, the diameter of gold particles should be smaller than 5 nm, which increases the surface boundary between the metal nanoparticle and the support (Takei et al. 2011). Haruta et al. (Murayama et al. 2016; Murayama and Haruta 2016) have studied gold catalysts in CO oxidation based on different types of niobia and prepared by various methods. The activities of Nb$_2$O$_5$ (HT) characterized with layered structure and high surface area, commercial Nb$_2$O$_5$ and Nb$_2$O$_5$·nH$_2$O (described above) modified with gold by the DP method (with Au(en)$_2$Cl$_3$) were compared and the highest was found for Au/Nb$_2$O$_5$ (HT) (T$_{1/2}$ = 358 K) (Murayama and Haruta 2016). The catalytic activities of Au/Nb$_2$O$_5$·nH$_2$O and Au/Nb$_2$O$_5$ were much lower (T$_{1/2}$ = 444 and 523 K, respectively). The reason for the low activity of these catalysts was the low surface area of the supports giving the low gold dispersion and the low loading of gold in Au/Nb$_2$O$_5$ (0.1 wt %). The effect of different preparation methods on the activity of Au/Nb$_2$O$_5$ (HT) was also studied (Murayama and Haruta 2016; Murayama et al. 2016). The catalysts activity decreased in the order: SI (T$_{1/2}$ = 284 K) > SG (T$_{1/2}$ = 325 K) > DR (T$_{1/2}$ = 346 K) > DP (T$_{1/2}$ = 358 K). This sequence was in agreement with the decrease in gold loading and the increase in the size of gold particles. A decrease in the catalysts activity with the reaction time was also observed (at 373 K), but on the basis of TEM measurements, it was concluded that the deactivation process was not due to the aggregation of gold particles. The activity was recovered by calcination of the samples at 523 K, which resulted in retrieving active lattice oxygen within the perimeter of gold nanoparticles. The application of HAuCl$_4$ instead of Au(en)$_2$Cl$_3$ as a source of gold in the DP and DR methods gave Au/Nb$_2$O$_5$ (HT) catalysts with low metal loading and lower activity in CO oxidation (T$_{1/2}$ = 535 K for Au/Nb$_2$O$_5$ (HT) (DP) and T$_{1/2}$ = 383 K for Au/Nb$_2$O$_5$ (HT) (DR)). The higher activity of Au/Nb$_2$O$_5$ (HT) (DR) was a result of the presence of smaller gold particles on its surface. The results obtained clearly show that smaller Au particles on the Nb$_2$O$_5$ surface give higher catalytic activity in CO oxidation. The active sites are localized in the perimeter of the Au particles. That is why, it was suggested that a much higher activity of the Au/Nb$_2$O$_5$ catalyst can be achieved for catalysts with tiny cluster size particles (smaller than 2 nm in diameter).

To study the effect of the crystalline structure of niobia support (Nb_2O_5 (HT), TT-Nb_2O_5, T-Nb_2O_5) on the activity of gold catalysts in CO oxidation, the samples (prepared by the SI method) with similar loading of metal and almost the same average gold particle sizes (2.6–2.7 nm) were compared (Murayama et al. 2016). The observed differences in catalytic activity were dependent on the support structure. Au/Nb_2O_5 (HT) (SI) had the highest activity ($T_{1/2}$ = 284 K), whereas the activities of Au/TT-Nb_2O_5 (SI) and Au/T-Nb_2O_5 (SI) were significantly lower ($T_{1/2}$ = 352 K and $T_{1/2}$ = 332 K, respectively). It was concluded that the crystalline form of the support affected the reactivity even when the Au particles with the same diameter were present. The experiment indicates that the reactivities of the active sites formed around the perimeters of gold particles are different despite the same diameter.

Alcohols Oxidation. Gold catalysts supported on crystalline and amorphous niobia were found to be active in the oxidation of alcohols: glycerol in the liquid phase and methanol in the gas phase (Sobczak et al. 2010; Musialska et al. 2010; Sobczak and Wolski 2015). The activity of catalysts in both reactions strongly depended on the nature of niobia (i.e., whether it was amorphous or crystalline) (Ziolek et al. 2011).

Glycerol is a co-product of triglyceride transesterification in the production of fatty acid esters employed as biodiesel (Centi and Van Santen 2007). One of the ways to transform it into valuable chemicals is the oxidation process. Not only the crystallinity but also the method of modification with gold influenced the activity of catalysts in glycerol oxidation at 333 K. Amorphous Nb_2O_5 modified with gold by the GS method was less active (31% of conversion) than its crystalline form (67% of conversion). Moreover, the Au/Nb_2O_5 (cr) (GS) activity was higher than that of gold catalysts based on Al_2O_3 (Sobczak et al. 2010). It was related to a higher ionicity of the Nb-O bond than the Al-O one (concluded from XPS study) and a consequent better dispersing ability of niobia than alumina. Furthermore, the enhancement of activity was caused by the generation of active oxygen on the interface between gold particles and the niobia support. The role of gold-niobium interaction in Au/Nb_2O_5 catalysts was also emphasized. The stronger metal-support (Au–Nb) interaction in the crystalline Nb_2O_5 than in amorphous Nb_2O_5 resulted in the higher activity of the crystalline catalyst. Both types of niobia catalysts were highly selective to glyceric acid (ca. 50%). In contrast, the activity of Au/Nb_2O_5 (cr) (DPU) and Au/Nb_2O_5 (am) (DPU), containing larger gold particles, was much lower (11 and 10%, respectively). This observation indicated the effect of the gold particles size on the activity.

Similarly, the best oxidative properties was achieved on Au/Nb_2O_5 (cr) (GS) studied in the methanol oxidation process used in chemical industry for the synthesis of various valuable products (e.g., formaldehyde) (Ziolek et al. 2011). Gold loaded on amorphous Nb_2O_5 (Au/Nb_2O_5 (am) (GS)) was highly active in MeOH oxidation (81% of MeOH conversion was obtained at 523 K), but the catalyst activated mainly the total oxidation of methanol (88% of selectivity to CO_2). Gold on crystalline Nb_2O_5 (Au/Nb_2O_5 (cr) (GS)) appeared to be a selective oxidation catalyst although its activity was much lower (12% of MeOH conversion, 44% of selectivity to HCHO and 48% to $HCOOCH_3$ at 523 K).

The result indicated that the amorphous state of niobia support does not play an important role in the catalytic properties of gold modified catalysts in selective alcohols' oxidation with gas oxygen. The crystalline Au/Nb_2O_5 (cr) (GS) is much more active than gold loaded on amorphous niobia in the oxidative dehydrogenation to aldehyde which is the first step in the oxidation of both methanol and glycerol, irrespective of the reaction phase (gas or liquid). The unsaturated niobium species in Nb_2O_5 (cr) interact more strongly with metallic gold leading to electron transfer to oxygen, making oxygen species more basic. This phenomenon increases the ease of abstraction of proton from alcohol, which occurs at the basic sites of the catalyst.

Hydrogenation reactions

The selective hydrogenation of α,β-unsaturated aldehydes to the corresponding unsaturated alcohols is an important process in fine chemistry, because of the products' significance. Gold has a low ability to chemisorb H_2 in comparison to the platinum group metals, and that is why it is less active for the reactions involving dissociation of hydrogen (McEwan et al. 2010). Despite this, it has been reported that high selectivity to unsaturated alcohol can be obtained using gold catalysts supported on, for example, TiO_2, Fe_2O_3 and ZnO (Zanella et al. 2004; Milone et al. 2005; Bailie et al. 2001). However, the influence of the support nature on the activity and selectivity is important. It was found that the higher the reducibility of the support, the higher the catalytic performance of gold (Milone et al. 2004).

Au/Nb$_2$O$_5$ catalysts prepared by deposition-precipitation were studied in the selective hydrogenation of crotonaldehyde towards crotyl alcohol. The reaction was performed both in liquid (4 atm, 333 K, solvent: isopropanol) and in gas (393 K, atmospheric pressure) phases (Campo et al. 2008). The activity and selectivity of Au/Nb_2O_5 was compared with those of Au/CeO_2. It was concluded that niobia, contrary to ceria, is an unfavourable support for gold when this metal is used as a catalyst for selective hydrogenation. In the gas phase, the conversion of crotonaldehyde on Au/Nb_2O_5 was lower than that on Au/CeO_2. Moreover, the niobia supported catalyst was unselective towards crotyl alcohol (the selectivity of 78% was obtained for Au/CeO_2). Taking into account that the nature of the support strongly influences the morphology of gold particles, it was considered that niobia favours the presence of gold crystallites with a low concentration of edges (which are active in C=O hydrogenation), and consequently, the catalyst becomes unselective. Similar results were obtained using Au/Nb_2O_5 in the reaction performed in the liquid phase.

Condensation

Oxidative condensation of renewable furfural (one of the most common chemicals derived from lignocellulose) and ethanol to produce furan-2-acrolein in the presence of oxygen was performed on Au/Nb_2O_5 (prepared by DP method) with the addition of K_2CO_3 as promoter (Tong et al. 2016). The obtained conversion of furfural was 39.6% at 84.6% selectivity to furan-2-acrolein at 403 K after 4 hr of the reaction. Much higher activity under the same reactions conditions showed $Au/CeO_2+K_2CO_3$ (97.3% of conversion); however, its selectivity was much lower (54.9%). It was concluded that the high activity of Au catalyst supported on CeO_2 was attributed to the large surface area (94 m^2/g) and smaller Au nanoparticles (7–8 nm). Moreover, weak acid sites on Au/CeO_2 played a key role in the condensation reaction. The

Au/Nb_2O_5 studied was characterized by very low surface area (0.9 m^2/g), relatively large gold particles (9–10 nm) and had mainly moderately strong and strong acid sites (measured by NH_3-TPD).

Summing up, the acidity of niobia support for gold catalysts was not useful for many catalytic reactions such as hydrogenation or oxidative dehydrogenation. The catalysts described can be applied in the future in the reactions that require the presence of both acid and redox sites (Murayama and Haruta 2016; Murayama et al. 2016).

Summary

Gold particle size and the catalytic performance of gold supported on metal oxides are strongly dependent on the gold-support interaction, which is related to the reducibility of metal in oxides (for example, ceria and niobia), being the main requisite for improving gold dispersion and achieving high oxidation activity. Au-support interaction is stronger if ceria (basic/redox oxide) instead of niobia (acidic oxide) is used as the support. Interestingly, although ZnO is not an easy reducible metal oxide, the Au-ZnO interaction is strong, due to the small differences in lattice parameters between Au (111) and ZnO (101) atomic planes. The three metal oxide supports described in this chapter differ not only in their reducible properties and acid-base properties but also in their point of zero charge (PZC: $Nb_2O_5 \ll CeO_2 <$ ZnO), which influences the amount of deposited gold (the lower the PZC, the higher the gold loading).

Gold loaded on CeO_2 and ZnO occurs in different oxidation states (Au^+, Au^{3+}, $Au^{\delta+}$, Au^0, $Au^{\delta-}$), whereas only Au^0 and $(Au^0)^{\delta-}$ are observed on Nb_2O_5. Charges of gold particles depend on several parameters discussed in this chapter. They, of course, strongly influence the activity of gold catalysts.

As concerns the catalytic application, it is clear that in oxidation processes, the base support for gold (ceria) and amphoteric one (ZnO) are more effective for the achievement of high activity than niobia (acid oxide). The basicity of oxygen and its mobility characteristic of basic and amphoteric metal oxide, are the evident advantages leading to the enhancement of gold species activity. Gold loaded on ceria exhibits very high catalytic performance in several oxidation processes, not all of them have been discussed in this chapter. In contrast, the acidic properties of niobia seem to decrease the selectivity of oxidation reactions because of the activity in the side reactions caused by the support acidity. Therefore, niobia as a support for gold can be recommended for the catalytic processes requiring bifunctional catalyst surfaces (redox and acidic). Moreover, gold loaded on niobia and ZnO are potential photocatalysts and the development of these systems seems to be addressed to photocatalytic reactions.

Acknowledgements

National Science Centre in Poland (Grants No. 2014/15/B/ST5/00167 and 2016/21/N/ST5/00533) is acknowledged for the financial support. Anna Wojtaszek-Gurdak (from Adam Mickiewicz University) is acknowledged for the revision of this chapter.

Reference cited

Abad, A., P. Concepcion, A. Corma and H. Garcia. 2005. A collaborative effect between gold and a support induces the selective oxidation of alcohols. Angew. Chem. Int. Ed. 44: 4066–4069.

Aboukaïs, A., S. Aouad, H. El-Ayadi, M. Skaf, M. Labaki, R. Cousin and E. Abi-AAd. 2012. Physicochemical characterization of Au/CeO$_2$ solid. Part 1: The deposition-precipitation preparation method. Mater. Chem. Phys. 137: 34–41.

Aboukaïs, A., M. Skaf, S. Hany, R. Cousin, S. Aouad, M. Labaki and E. Abi-AAd. 2016. A comparative study of Cu, Ag and Au doped CeO$_2$ in the total oxidation of volatile organic compounds (VOCs). Mater. Chem. Phys. 177: 570–576.

Al-Sayari, S., A.F. Carley, S.H. Taylor and G.J. Hutchings. 2007. Au/ZnO and Au/Fe$_2$O$_3$ catalysts for CO oxidation at ambient temperature: comments on the effect of synthesis conditions on the preparation of high activity catalysts prepared by co-precipitation. Top. Catal. 44: 123–128.

Alenezi, M.R., A.S. Alshammari, K.D.G.I. Jayawardena, M.J. Beliatis, S.J. Henley and S.R.P. Silva. 2013. Role of the exposed polar facets in the performance of thermally and UV activated ZnO nanostructured gas sensors. J. Phys. Chem. C 117: 17850–17858.

Ampelli, C., S.G. Leonardi, A. Bonavita, C. Genovese, G. Papanikolaou, S. Perathoner, G. Centi and G. Neri. 2015. Electrochemical H$_2$O$_2$ sensors based on Au/CeO$_2$ nanoparticles for industrial applications. Chem. Eng. Trans. 43: 733–738.

Andreeva, D., T. Tabakova and L. Ilieva. 2013. Ceria-based gold catalysts: synthesis, properties, and catalytic performance for the WGS and PROX processes. pp. 497–564. *In:* Trovarelli, A. and P. Fornasiero (eds.). Catalysis by Ceria and Related Materials. Imperial College Press, London, UK.

Arunkumar, S., T. Hou, Y.-B. Kim, B. Choi, S.H. Park, S. Jung and D.-W. Lee. 2017. Au decorated ZnO hierarchical architectures: Facile synthesis, tunable morphology and enhanced CO detection at room temperature. Sensors Actuators B Chem. 243: 990–1001.

Bailie, J.E., H.A. Abdullah, J.A. Anderson, C.H. Rochester, N.V. Richardson, N. Hodge, J.-G. Zhang, A. Burrows, C.J. Kiely and G.J. Hutchings. 2001. Hydrogenation of but-2-enal over supported Au/ZnO catalysts. Phys. Chem. Chem. Phys. 3: 4113–4121.

Barakat, T., J.C. Rooke, E. Genty, R. Cousin, S. Siffert and B.-L. Su. 2013. Gold catalysts in environmental remediation and water-gas shift technologies. Energy Environ. Sci. 6: 371–391.

Bastos, S.S.T., S.A.C. Carabineiro, J.J.M. Órfão, M.F.R. Pereira, J.J. Delgado and J.L. Figueiredo. 2012. Total oxidation of ethyl acetate, ethanol and toluene catalyzed by exotemplated manganese and cerium oxides loaded with gold. Catal. Today 180: 148–154.

Beaumont, S.K., G. Kyriakou and R.M. Lambert. 2010. Identity of the active site in gold nanoparticle-catalyzed sonogashira coupling of phenylacetylene and iodobenzene. J. Am. Chem. Soc. 132: 12246–12248.

Behl, M. and P.K. Jain. 2014. Catalytic activation of a solid oxide in electronic contact with gold nanoparticles. Angew. Chem. Int. Ed. 53: 1–7.

Berrichi, A., R. Bachir, M. Benabdallah and N. Choukchou-Braham. 2015. Supported nano gold catalyzed three-component coupling reactions of amines, dichloromethane and terminal alkynes (AHA). Tetrahedron Lett. 56: 1302–1306.

Camellone, M.F. and S. Fabris. 2009. Reaction mechanisms for the CO oxidation on Au/CeO$_2$ catalysts: Activity of substitutional Au^{3+}/Au$^+$ cations and deactivation of supported Au$^+$ adatoms. J. Am. Chem. Soc. 131: 10473–10483.

Campo, B.C., S. Ivanova, C. Gigola, C. Petit and M.A. Volpe. 2008. Crotonaldehyde hydrogenation on supported gold catalysts. Catal. Today 133-135: 661–666.

Capdevila-Cortada, M., G. Vilé, D. Teschner, J. Pérez-Ramírez and N. López. 2016. Reactivity descriptors for ceria in catalysis. Appl. Catal. B Environ. 197: 299–312.

Carrettin, S., P. Concepcion, A. Corma, J.M.L. Nieto and V.F. Puntes. 2004. Nanocrystalline CeO$_2$ increases the activity of Au for CO oxidation by two orders of magnitude. Angew. Chem. Int. Ed. 43: 2538–2540.

Carrettin, S., A. Corma, M. Iglesias and F. Sanchez. 2005. Stabilization of Au(III) on heterogeneous catalysts and their catalytic similarities with homogeneous Au(III) metal organic complexes. Appl. Catal. A Gen. 291: 247–252.

Castillejos, E., E. Gallegos-Suarez, B. Bachiller-Baeza, R. Bacsa, P. Serp, A. Guerrero-Ruiz and I. Rodríguez-Ramos. 2012. Deposition of gold nanoparticles on ZnO and their catalytic activity for hydrogenation applications. Catal. Commun. 22: 79–82.

Centi, G. and R.A. Van Santen. 2007. Catalysis for Renewables—from Feedstock to Energy Production. Wiley-VCH Verlag GmbH & Co. KGaA, Weinheim, Germany.

Chatzichristodoulou, Ch., P.T. Blennoow, M. Sogaard, P.V Hendriksen and M.B. Mogensen. 2013. Ceria and its use in solid oxide cells and oxygen membranes. pp. 623–782. *In*: Trovarelli, A. and P. Fornasiero (eds.). Catalysis by Ceria and Related Materials. Imperial College Press, London, UK.

Chen, B., Y. Ma, L. Ding, L. Xu, Z. Wu and Q. Yuan. 2013. Reactivity of hydroxyls and water on a CeO_2 (111) thin film surface: the role of oxygen vacancy. J. Phys. Chem. C 117: 5800–5810.

Chen, D., Z. Wang, T. Ren, H. Ding, W. Yao, R. Zong and Y. Zhu. 2014. Influence of defects on the photocatalytic activity of ZnO. J. Phys. Chem. C 118: 15300–15307.

Chen, Y., D. Zeng, K. Zhang, A. Lu, L. Wang and D.-L. Peng. 2014. Au-ZnO hybrid nanoflowers, nanomultipods and nanopyramids: one-pot reaction synthesis and photocatalytic properties. Nanoscale 6: 874–881.

Coleman, V.A. and Ch. Jagadish. 2006. Basic properties and applications of ZnO. pp. 1–20. *In*: Jagadish, Ch. and S. Pearton (eds.). Zinc Oxide Bulk, Thin Films and Nanostructures. Elsevier Science.

Concepcion, P., S. Carrettin and A. Corma. 2006. Stabilization of cationic gold species on Au/CeO_2 catalysts under working conditions. Appl. Catal. A Gen. 307: 42–45.

Corma, A. and H. Garcia. 2008. Supported gold nanoparticles as catalysts for organic reactions. Chem. Soc. Rev. 37: 2096–2126.

Delannoy, L., K. Fajerwerg, P. Lakshmanan, C. Potvin, C. Méthivier and C. Louis. 2010. Supported gold catalysts for the decomposition of VOC: Total oxidation of propene in low concentration as model reaction. Appl. Catal. B Environ. 94: 117–124.

Drouilly, C., J.-M. Krafft, F. Averseng, S. Casale, D. Bazer-Bachi, C. Chizallet, V. Lecocq, H. Vezin, H. Lauron-Pernot and G. Costentin. 2012. ZnO oxygen vacancies formation and filling followed by *in situ* photoluminescence and *in situ* EPR. J. Phys. Chem. C 116: 21297–21307.

Dulnee, S., A. Luengnaruemitchai and R. Wanchanthuek. 2014. Activity of Au/ZnO catalysts prepared by photo-deposition for the preferential CO oxidation in a H_2-rich gas. Int. J. Hydrogen Energy 39: 6443–6453.

Fernando, J.F.S., M.P. Shortell, C.J. Noble, J.R. Harmer, E.A. Jaatinen and E.R. Waclawik. 2016. Controlling Au photodeposition on large ZnO nanoparticles. ACS Appl. Mater. Interfaces 8: 14271–14283.

Fu, Q., H. Saltsburg and M. Flytzani-Stephanopoulos. 2003. Active nonmetallic Au and Pt species on ceria-based water-gas shift catalysts. Science 301: 935–938.

Gabriel, S. and M.S. Scurrell. 2016. The conversion of methanol into higher hydrocarbons catalyzed by gold. ChemCatChem. 8: 1–4.

Glisenti, A., A. Frasson, A. Galenda and M.M. Natile. 2010. Au/CeO_2 supported nanocatalysts: Interaction with methanol. Nanosci. Nanotechnol. Lett. 2: 213–219.

Gogurla, N., A.K. Sinha, S. Santra, S. Manna and S.K. Ray. 2014. Multifunctional Au-ZnO plasmonic nanostructures for enhanced UV photodetector and room temperature NO sensing devices. Sci. Rep. 4: 6483.

Good, J., P.N. Duchesne, P. Zhang, W. Koshut, M. Zhou and R. Jin. 2017. On the functional role of the cerium oxide support in the $Au_{38}(SR)_{24}/CeO_2$ catalyst for CO oxidation. Catal. Today 280: 239–245.

Guan, Y., W. Song and E.J.M. Hensen. 2015. Gold clusters and nanoparticles stabilized by nanoshaped ceria in catalysis. pp. 99–132. *In*: Wu, Z. and S.H. Overbury (eds.). Catalysis by Materials with Well-Defined Structures. Elsevier Science.

Guzman, J., S. Carrettin, J.C. Fierro-Gonzalez, Y. Hao, B.C. Gates and A. Corma. 2005. CO oxidation catalyzed by supported gold: cooperation between gold and nanocrystalline rare-earth supports forms reactive surface superoxide and peroxide species. Angew. Chem. Int. Ed. 44: 4778–4781.

Haruta, M., T. Kobayashi, H. Sano and N. Yamada. 1987. Novel gold catalysts for the oxidation of carbon monoxide at a temperature far below 0°C. Chem. Lett. 16: 405–408.

Haruta, M., N. Yamada, T. Kobayashi and S. Iijima. 1989. Gold catalysts prepared by coprecipitation for low-temperature oxidation of hydrogen and of carbon monoxide. J. Catal. 115: 301–309.

Haruta, M., S. Tsubota, T. Kobayashi, H. Kageyama, M.J. Genet and B. Delmon. 1993. Low-temperature oxidation of CO over gold supported on TiO_2, α-Fe_2O_3, and Co_3O_4. J. Catal. 144: 175–192.

He, W., H.-K. Kim, W.G. Warner, D. Melka, J.H. Callahan and J.-J. Yin. 2014. Photogenerated charge carriers and reactive oxygen species in ZnO/Au hybrid nanostructures with enhanced photocatalytic and antibacterial activity. J. Am. Chem. Soc. 136: 750–757.

Ismail, A.A., F.A. Harraz, M. Faisal, A.M. El-Toni, A. Al-Hajry and M.S. Al-Assiri. 2016. A sensitive and selective amperometric hydrazine sensor based on mesoporous Au/ZnO nanocomposites. Mater. Des. 109: 530–538.

Ivanova, S., C. Petit and V. Pitchon. 2004. A new preparation method for the formation of gold nanoparticles on an oxide support. Appl. Catal. A Gen. 267: 191–201.

Jacobs, G., L. Williams, U. Graham, G.A. Thomas, D.E. Sparks and B.H. Davis. 2003. Low temperature water-gas shift: *in situ* DRIFTS-reaction study of ceria surface area on the evolution of formates on Pt/CeO_2 fuel processing catalysts for fuel cell applications. Appl. Catal. A Gen. 252: 107–118.

Jacobs, G., E. Chenu, P.M. Patterson, L. Williams, D. Sparks, G. Thomas and B.H. Davis. 2004. Water-gas shift: comparative screening of metal promoters for metal/ceria systems and role of the metal. Appl. Catal. A Gen. 258: 203–214.

Janotti, A. and C.G. Van de Walle. 2009. Fundamentals of zinc oxide as a semiconductor. Rep. Prog. Phys. 72: 126501.

Jehng, J.-M. and I.E. Wachs. 1990. The molecular structures and reactivity of supported niobium oxide catalysts. Catal. Today 8: 37–55.

Jia, M., Y. Shen, C. Li, Z. Bao and S. Sheng. 2005. Effect of supports on the gold catalyst activity for catalytic combustion of CO and HCHO. Catal. Letters 99: 235–239.

Kaminski, P. and M. Ziolek. 2016. Mobility of gold, copper and cerium species in Au, Cu/Ce, Zr-oxides and its impact on total oxidation of methanol. Appl. Catal. B Environ. 187: 328–341.

Kaneti, Y.V., Z. Zhang, J. Yue, Q.M.D. Zakaria, C. Chen, X. Jiang and A. Yu. 2014. Crystal plane-dependent gas-sensing properties of zinc oxide nanostructures: experimental and theoretical studies. Phys. Chem. Chem. Phys. 16: 11471–11480.

Kaneti, Y.V., X. Zhang, M. Liu, D. Yu, Y. Yuan, L. Aldous and X. Jiang. 2016. Experimental and theoretical studies of gold nanoparticle decorated zinc oxide nanoflakes with exposed {10$\bar{1}$0} facets for butylamine sensing. Sensors Actuators B Chem. 230: 581–591.

Khan, R., J.-H. Yun, K.-B. Bae and I.-H. Lee. 2016. Enhanced photoluminescence of ZnO nanorods via coupling with localized surface plasmon of Au nanoparticles. J. Alloys Compd. 682: 643–646.

Kim, H.Y., H.M. Lee and G. Henkelman. 2012. CO oxidation mechanism on CeO_2-supported Au nanoparticles. J. Am. Chem. Soc. 134: 1560–1570.

Ko, E.I. and J.G. Weissman. 1990. Structures of niobium pentoxide and their implications on chemical behavior. Catal. Today 8: 27–36.

Kołodziejczak-Radzimska, A. and T. Jesionowski. 2014. Zinc oxide-from synthesis to application: A review. Materials 7: 2833–2881.

Kominami, H., A. Tanaka and K. Hashimoto. 2011. Gold nanoparticles supported on cerium(IV) oxide powder for mineralization of organic acids in aqueous suspensions under irradiation of visible light of λ = 530 nm. Appl. Catal. A Gen. 397: 121–126.

Kosmulski, M. 1997. Attempt to determine pristine points of zero charge of Nb_2O_5, Ta_2O_5, and HfO_2. Langmuir 13: 6315–6320.

Kovacevic, M., B.L. Mojet, J.G. Van Ommen and L. Lefferts. 2016. Effects of morphology of cerium oxide catalysts for reverse water gas shift reaction. Catal. Letters 146: 770–777.

Kung, M.C., R.J. Davis and H.H. Kung. 2007. Understanding Au-catalyzed low-temperature CO oxidation. J. Phys. Chem. C 111: 11767–11775.

Lai, S.-Y., Y. Qiu and S. Wang. 2006. Effects of the structure of ceria on the activity of gold/ceria catalysts for the oxidation of carbon monoxide and benzene. J. Catal. 237: 303–313.

Lakshmanan, P., P.P. Upare, N.-T. Le, Y.K. Hwang, D.W. Hwang, U.-H. Lee and H.R. Kim and J.-S. Chang. 2013. Facile synthesis of CeO_2-supported gold nanoparticle catalysts for selective oxidation of glycerol into lactic acid. Appl. Catal. A Gen. 468: 260–268.

Layek, A., S. Banerjee, B. Manna and A. Chowdhury. 2016. Synthesis of rare-earth doped ZnO nanorods and their defect–dopant correlated enhanced visible-orange luminescence. RSC Adv. 6: 35892–35900.

Lee, J., H.S. Shim, M. Lee, J.K. Song and D. Lee. 2011. Size-controlled electron transfer and photocatalytic activity of ZnO-Au nanoparticle composites. J. Phys. Chem. Lett. 2: 2840–2845.

Li, C., S. Chen, Z. Zhang and J. Zhang. 2012. Gold catalysts supported on the macroporous nanoparticles composited of cerium oxide for oxidation of formaldehyde. Adv. Mater. Res. 347-353: 2117–2120.

Li, D. and H. Haneda. 2003. Morphologies of zinc oxide particles and their effects on photocatalysis. Chemosphere 51: 129–137.

Li, G.R., T. Hu, G.L. Pan, T.Y. Yan, X.P. Gao and H.Y. Zhu. 2008. Morphology-function relationship of ZnO: Polar planes, oxygen vacancies, and activity. J. Phys. Chem. C 112: 11859–11864.

Li, Z.-X., W. Feng, C. Zhang, L.-D. Sun, Y.-W. Zhang and C.-H. Yan. 2013. Two-dimensional and three-dimensional ceria-based nanoarchitectures. pp. 295–360. In: Trovarelli, A. and P. Fornasiero (eds.). Catalysis by Ceria and Related Materials. Imperial College Press, London, UK.

Lin, Y., Z. Wu, J. Wen, K. Ding, X. Yang, K.R. Poeppelmeier and L.D. Marks. 2015. Adhesion and atomic structures of gold on ceria nanostructures: the role of surface structure and oxidation state of ceria supports. Nano Lett. 15: 5375–5381.

Liu, F. and Y. Zhang. 2011. Controllable growth of "multi-level tower" ZnO for biodiesel production. Ceram. Int. 37: 3193–3202.

Liu, M.-H., Y.-W. Chen, X. Liu, J.-L. Kuo, M.-W. Chu and C.-Y. Mou. 2016. Defect-mediated gold substitution doping in ZnO mesocrystals and catalysis in CO oxidation. ACS Catal. 6: 115–122.

Liu, S.Y. and S.M. Yang. 2008. Complete oxidation of 2-propanol over gold-based catalysts supported on metal oxides. Appl. Catal. A Gen. 334: 92–99.

Liu, X., A. Wang, X. Yang, T. Zhang, C.-Y. Mou, D.-S. Su and J. Li. 2009. Synthesis of thermally stable and highly active bimetallic Au–Ag nanoparticles on inert supports. Chem. Mater. 21: 410–418.

Liu, X., M.-H. Liu, Y.-C. Luo, C.-Y. Mou, S.D. Lin, H. Cheng, J.-M. Chen, J.-F. Lee and T.-S. Lin. 2012. Strong metal–support interactions between gold nanoparticles and ZnO nanorods in CO oxidation. J. Am. Chem. Soc. 134: 10251–10258.

Liu, X.Y., A. Wang, T. Zhang and C.-Y. Mou. 2013. Catalysis by gold: New insights into the support effect. Nano Today 8: 403–416.

Liu, Y., B. Liu, Y. Liu, Q. Wang, W. Hu, P. Jing, L. Liu, S. Yu and J. Zhang. 2013. Improvement of catalytic performance of preferential oxidation of CO in H_2-rich gases on three-dimensionally ordered macro- and meso-porous Pt-Au/CeO_2 catalysts. Appl. Catal. B Environ. 142-143: 615–625.

Manzoli, M., A. Chiorino and F. Boccuzzi. 2004. Interface species and effect of hydrogen on their amount in the CO oxidation on Au/ZnO. Appl. Catal. B Environ. 52: 259–266.

Massola, B.C.P., N.M. Pereira de Souza, F.F.F. Stachack, E.W.R. da S. Oliveira, J.C. Germino, A.J. Terezo and F.J. Quites. 2015. Au-ZnO prepared by simple in situ reduction and spontaneous of gold nanoparticles on the surface of the layered zinc hydroxide using a novel one-pot method. Mater. Chem. Phys. 167: 152–159.

McEwan, L., M. Julius, S. Roberts and J.C.Q. Fletcher. 2010. A review of the use of gold catalysts in selective hydrogenation reactions. Gold Bull. 43: 298–306.

Meunier, F.C., D. Reid, A. Goguet, S. Shekhtman, C. Hardacre, R. Burch, W. Deng and M. Flytzani-Stephanopoulos. 2007. Quantitative analysis of the reactivity of formate species seen by DRIFTS over a Au/Ce(La)O_2 water-gas shift catalyst: First unambiguous evidence of the minority role of formates as reaction intermediates. J. Catal. 247: 277–287.

Miller, J.T., A.J. Kropf, Y. Zha, J.R. Regalbuto, L. Delannoy, C. Louis, E. Bus and J.A. van Bokhoven. 2006. The effect of gold particle size on Au–Au bond length and reactivity toward oxygen in supported catalysts. J. Catal. 240: 222–234.

Mills, A. and S. Le Hunte. 1997. An overview of semiconductor photocatalysis. J. Photochem. Photobiol. A Chem. 108: 1–35.

Milone, C., R. Ingoglia, A. Pistone, G. Neri, F. Frusteri and S. Galvagno. 2004. Selective hydrogenation of α, β-unsaturated ketons to α, β-unsaturated alcohols on gold-supported catalysts. J. Catal. 222: 348–356.

Milone, C., R. Ingoglia, L. Schipilliti, C. Crisafulli, G. Neri and S. Galvagno. 2005. Selective hydrogenation of α,β-unsaturated ketone to α,β-unsaturated alcohol on gold-supported iron oxide catalysts: Role of the support. J. Catal. 236: 80–90.

Misra, M., P. Kapur and M.L. Singla. 2014. Surface plasmon quenched of near band edge emission and enhanced visible photocatalytic activity of Au@ZnO core-shell nanostructure. Appl. Catal. B Environ. 150-151: 605–611.

Morkoc, H. and U. Ozgur. 2009. General properties of ZnO. pp. 1–71. *In*: Morkoc, H. and U. Ozgur (eds.). Zinc Oxide: Fundamentals, Materials and Device Technology. Wiley-VCH Verlag GmbH & Co. KGaA, Weinheim, Germany.

Moroz, B.L., P.A. Pyrjaev, V.I. Zaikovskii and V.I. Bukhtiyarov. 2009. Nanodispersed Au/Al$_2$O$_3$ catalysts for low-temperature CO oxidation: Results of research activity at the Boreskov Institute of Catalysis. Catal. Today 144: 292–305.

Mullen, G.M., E.J. Evans, I. Sabzevari, B.E. Long, K. Alhazmi, B.D. Chandler and C.B. Mullins. 2017. Water influences the activity and selectivity of ceria-supported gold catalysts for oxidative dehydrogenation and esterification of ethanol. ACS Catal. 7: 1216–1226.

Mullins, D.R., P.M. Albrecht, T. Chen, F.C. Calaza, M.D. Biegalski, H.M. Christen and S.H. Overbury. 2012. Water dissociation on CeO$_2$ (100) and CeO$_2$ (111) thin films. J. Phys. Chem. C 116: 19419–19428.

Mullins, D.R. 2015. The surface chemistry of cerium oxide. Surf. Sci. Rep. 70: 42–85.

Murayama, T. and M. Haruta. 2016. Preparation of gold nanoparticles supported on Nb$_2$O$_5$ by deposition precipitation and deposition reduction methods and their catalytic activity for CO oxidation. Chinese J. Catal. 37: 1694–1701.

Murayama, T., W. Ueda and M. Haruta. 2016. Deposition of gold nanoparticles on niobium pentoxide with different crystal structures for room-temperature carbon monoxide oxidation. ChemCatChem. 8: 1–6.

Musialska, K., E. Finocchio, I. Sobczak, G. Busca, R. Wojcieszak, E. Gaigneaux and M. Ziolek. 2010. Characterization of alumina- and niobia-supported gold catalysts used for oxidation of glycerol. Appl. Catal. A Gen. 384: 70–77.

Nagajyothi, P.C., H. Lim, J. Shim and S.B. Rawal. 2016. Au nanoparticles supported nanoporous ZnO sphere for enhanced photocatalytic activity under UV-light irradiation. J. Clust. Sci. 27: 1159–1170.

Naknam, P., A. Luengnaruemitchai and S. Wongkasemjit. 2009. Au/ZnO and Au/ZnO-Fe$_2$O$_3$ prepared by deposition-precipitation and their activity in the preferential oxidation of CO. Energy and Fuels 23: 5084–5091.

Nico, C., T. Monteiro and M.P.F. Graça. 2016. Niobium oxides and niobates physical properties: Review and prospects. Prog. Mater. Sci. 80: 1–37.

Nikolaev, S.A., E.V. Golubina, I.N. Krotova, M.I. Shilina, A.V. Chistyakov and V.V. Kriventsov. 2015. The effect of metal deposition order on the synergistic activity of Au-Cu and Au-Ce metal oxide catalysts for CO oxidation. Appl. Catal. B Environ. 168-169: 303–312.

Noei, H., A. Birkner, K. Merz, M. Muhler and Y. Wang. 2012. Probing the mechanism of low-temperature CO oxidation on Au/ZnO catalysts by vibrational spectroscopy. J. Phys. Chem. C 116: 11181–11188.

Nolan, M., S.C. Parker and G.W. Watson. 2005. The electronic structure of oxygen vacancy defects at the low index surfaces of ceria. Surf. Sci. 595: 223–232.

Nowak, I. and M. Ziolek. 1999. Niobium compounds: preparation, characterization, and application in heterogeneous catalysis. Chem. Rev. 99: 3603–3624.

Paier, J., C. Penschke and J. Sauer. 2013. Oxygen defects and surface chemistry of ceria: quantum chemical studies compared to experiment. Chem. Rev. 113: 3949–3985.

Pawinrat, P., O. Mekasuwandumrong and J. Panpranot. 2009. Synthesis of Au-ZnO and Pt-ZnO nanocomposites by one-step flame spray pyrolysis and its application for photocatalytic degradation of dyes. Catal. Commun. 10: 1380–1385.

Peng, Y., Y. Wang, Q.-G. Chen, Q. Zhu and A.W. Xu. 2014. Stable yellow ZnO mesocrystals with efficient visible-light photocatalytic activity. CrystEngComm. 16: 7906–7913.

Piumetti, M., T. Andana, S. Bensaid, N. Russo, D. Fino and R. Pirone. 2016. Study on the CO oxidation over ceria-based nanocatalysts. Nanoscale Res. Lett. 11: 165.

Prasetyoko, D., Z. Ramli, S. Endud and H. Nur. 2005. Preparation and characterization of bifunctional oxidative and acidic catalysts Nb$_2$O$_5$/TS-1 for synthesis of diols. Mater. Chem. Phys. 93: 443–449.

Qin, Y., H. Ye, F. Li, X. Kong and J. Mei. 2014. Nano-gold particles on cerium oxide for catalytic combustion of formaldehyde. Appl. Mech. Mater. 525: 150–153.

Rahman, D.S. and S.K. Ghosh. 2016. Manipulating electron transfer in hybrid ZnO–Au nanostructures: size of gold matters. J. Phys. Chem. C 120: 14906–14917.

Rani, R.A., A.S. Zoolfakar, A.P. O'Mullane, M.W. Austin and K. Kalantar-Zadeh. 2014. Thin films and nanostructures of niobium pentoxide: fundamental properties, synthesis methods and applications. J. Mater. Chem. A 2: 15683–15703.

Reina, T.R., S. Ivanova, M.A. Centeno and J.A. Odriozola. 2015. Boosting the activity of a Au/CeO$_2$/Al$_2$O$_3$ catalyst for the WGS reaction. Catal. Today 253: 149–154.

Renaud, A., L. Cario, X. Rocquelfelte, P. Deniard, E. Gautron, E. Faulques, T. Das, F. Cheviré, F. Tessier and S. Jobic. 2015. Unravelling the origin of the giant Zn deficiency in wurtzite type ZnO nanoparticles. Sci. Rep. 5: 12914.

Rodriguez, J.A. 2011. Gold-based catalysts for the water-gas shift reaction: Active sites and reaction mechanism. Catal. Today 160: 3–10.

Rodriguez, J.A., R. Si, J. Evans, W. Xu, J.C. Hanson, J. Tao and Y. Zhu. 2015. Active gold-ceria and gold-ceria/titania catalysts for CO oxidation: From single-crystal model catalysts to powder catalysts. Catal. Today 240: 229–235.

Sakurai, H. and M. Haruta. 1996. Synergism in methanol synthesis from carbon dioxide over gold catalysts supported on metal oxides. Catal. Today 29: 361–365.

Satheesan, M.K., K.V. Baiju and V. Kumar. 2017. Influence of defects on the photocatalytic activity of Niobium-doped ZnO nanoparticles. J. Mater. Sci. Mater. Electron. 28: 4719–4724.

Scirè, S., S. Minicò, C. Crisafulli, C. Satriano and A. Pistone. 2003. Catalytic combustion of volatile organic compounds on gold/cerium oxide catalysts. Appl. Catal. B Environ. 40: 43–49.

Scirè, S., C. Crisafulli, P.M. Riccobene, G. Patanè and A. Pistone. 2012. Selective oxidation of CO in H$_2$-rich stream over Au/CeO$_2$ and Cu/CeO$_2$ catalysts: An insight on the effect of preparation method and catalyst pretreatment. Appl. Catal. A Gen. 417-418: 66–75.

Scirè, S. and L.F. Liotta. 2012. Supported gold catalysts for the total oxidation of volatile organic compounds. Appl. Catal. B Environ. 125: 222–246.

Sharma, A.S., H. Kaur and D. Shah. 2016. Selective oxidation of alcohols by supported gold nanoparticles: recent advances. RSC Adv. 6: 28688–28727.

She, P., K. Xu, Q. He, S. Zeng, H. Sun and Z. Liu. 2017. Controlled preparation and visible light photocatalytic activities of corn cob-like Au-ZnO nanorods. J. Mater. Sci. 52: 3478–3489.

Shen, Y., X. Yang, Y. Wang, Y. Zhang, H. Zhu, L. Gao and M. Jia. 2008. The states of gold species in CeO$_2$ supported gold catalyst for formaldehyde oxidation. Appl. Catal. B Environ. 79: 142–148.

Shi, J., A. Schaefer, A. Wichmann, M.M. Murshed, T.M. Gesing, A. Wittstock and M. Bäumer. 2014. Nanoporous gold-supported ceria for the water–gas shift reaction: UHV inspired design for applied catalysis. J. Phys. Chem. C 118: 29270–29277.

Si, R. and M. Flytzani-Stephanopoulos. 2008. Shape and crystal-plane effects of nanoscale ceria on the activity of Au-CeO$_2$ catalysts for the water-gas shift reaction. Angew. Chem. Int. Ed. 47: 2884–2887.

Silva, C.G., M.J. Sampaio, S.A.C. Carabineiro, J.W.L. Oliveira, D.L. Baptista, R. Bacsa, B.F. Machado, P. Serp, J.L. Figueiredo, A.M.T. Silva and J.L. Faria. 2014. Developing highly active photocatalysts: Gold-loaded ZnO for solar phenol oxidation. J. Catal. 316: 182–190.

Sobczak, I., K. Jagodzinska and M. Ziolek. 2010. Glycerol oxidation on gold catalysts supported on group five metal oxides—A comparative study with other metal oxides and carbon based catalysts. Catal. Today 158: 121–129.

Sobczak, I. and Ł. Wolski. 2015. Au–Cu on Nb$_2$O$_5$ and Nb/MCF supports—Surface properties and catalytic activity in glycerol and methanol oxidation. Catal. Today 254: 72–82.

Song, W. and E.J.M. Hensen. 2013a. A computional DFT study of CO oxidation on a Au nanorod supported on CeO$_2$ (110): on the role of the support termination. Catal. Sci. Technol. 3: 3020–3029.

Song, W. and E.J.M. Hensen. 2013b. Structure sensitivity in CO oxidation by a single Au atom supported on ceria. J. Phys. Chem. C 117: 7721–7726.

Song, W. and E.J.M. Hensen. 2014. Mechanistic aspects of the water-gas shift reaction on isolated and clustered Au atoms on CeO$_2$ (110): A density functional theory study. ACS Catal. 4: 1885–1892.

Strunk, J., K. Kahler, X. Xia, M. Comotti, F. Schuth, T. Reinecke and M. Muhler. 2009. Au/ZnO as catalyst for methanol synthesis: The role of oxygen vacancies. Appl. Catal. A Gen. 359: 121–128.

Sun, L., D. Zhao, Z. Song, C. Shan, Z. Zhang, B. L and D. Shen . 2011. Gold nanoparticles modified ZnO nanorods with improved photocatalytic activity. J. Colloid Interface Sci. 363: 175–181.

Sun, Y., L. Chen, Y. Bao, Y. Zhang, J. Wang, M. Fu, J. Wu and D. Ye. 2016. The applications of morphology controlled ZnO in catalysis. Catalysts 6: 188.

Sun, Y., Y. Sun, T. Zhang, G. Chen, F. Zhang, D. Liu, W. Cai, Y. Li, X. Yang and C. Li. 2016. Complete Au@ZnO core–shell nanoparticles with enhanced plasmonic absorption enabling significantly improved photocatalysis. Nanoscale 8: 10774–10782.

Ta, N., J. Liu, S. Chenna, P.A. Crozier, Y. Li, A. Chen and W. Shen. 2012. Stabilized gold nanoparticles on ceria nanorods by strong interfacial anchoring. J. Am. Chem. Soc. 134: 20585–20588.

Tabakova, T., M. Manzoli, D. Paneva, F. Boccuzzi, V. Idakiev and I. Mitov. 2011. CO-free hydrogen production over $Au/CeO_2-Fe_2O_3$ catalysts: Part 2. Impact of the support composition on the performance in the water-gas shift reaction. Appl. Catal. B Environ. 101: 266–274.

Takei, T., N. Iguchi and M. Haruta. 2011. Support effect in the gas phase oxidation of ethanol over nanoparticulate gold catalysts. New J. Chem. 35: 2227–2233.

Tong, X., Z. Liu, J. Hu and S. Liao. 2016. Au-catalyzed oxidative condensation of renewable furfural and ethanol to produce furan-2-acrolein in the presence of molecular oxygen. Appl. Catal. A Gen. 510: 196–203.

Topka, P. and M. Klementová. 2016. Total oxidation of ethanol over $Au/Ce_{0.5}Zr_{0.5}O_2$ cordierite monolithic catalysts. Appl. Catal. A Gen. 522: 130–137.

Tóth, A., G. Halasi, T. Bánsági and F. Solymosi. 2016. Reactions of propane with CO_2 over Au catalysts. J. Catal. 337: 57–64.

Tsubota, S., D.A.H. Cunningham, Y. Bando and M. Haruta. 1995. Preparation of nanometer gold strongly interacted with TiO_2 and the structure sensitivity in low-temperature oxidation of CO. Stud. Surf. Sci. Catal. 91: 227–235.

Udawatte, N., M. Lee, J. Kim and D. Lee. 2011. Well-defined Au/ZnO nanoparticle composites exhibiting enhanced photocatalytic activities. ACS Appl. Mater. Interfaces 3: 4531–4538.

Ushikubo, T., Y. Koike, K. Wada, L. Xie, D. Wang and X. Guo. 1996. Study of the structure of niobium oxide by X-ray absorption fine structure and surface science techniques. Catal. Tod. 28: 59–69.

Vallar, S., D. Houivet, J. El Fallah, D. Kervadec and J.-M. Haussonne. 1999. Oxide slurries stability and powders dispersion: Optimization with zeta potential and rheological measurements. J. Eur. Ceram. Soc. 19: 1017–1021.

Venezia, A.M., G. Pantaleo, A. Longo, G. Di Carlo, M.P. Casaletto, F.L. Liotta and G. Deganello. 2005. Relationship between structure and CO oxidation activity of ceria-supported gold catalysts. J. Phys. Chem. B 109: 2821–2827.

Venezia, A.M., L.F. Liotta, G. Pantaleo and A. Longo. 2013. Ceria-based catalysts for air pollution abatement. pp. 813–880. *In*: Trovarelli, A. and P. Fornasiero (eds.). Catalysis by Ceria and Related Materials. Imperial College Press, London, UK.

Venugopal, A. and M.S. Scurrell. 2004. Low temperature reductive pretreatment of Au/Fe_2O_3 catalysts, TPR/TPO studies and behaviour in the water–gas shift reaction. Appl. Catal. A Gen. 258: 241–249.

Wachs, I.E., L.E. Briand, J.-M. Jehng, L. Burcham and X. Gao. 2000. Molecular structure and reactivity of the group V metal oxides. Catal. Today 57: 323–330.

Wang, C., D. Wu, P. Wang, Y. Ao, J. Hou and J. Qian. 2015. Effect of oxygen vacancy on enhanced photocatalytic activity of reduced ZnO nanorod arrays. Appl. Surf. Sci. 325: 112–116.

Wang, G., W. Zhang, H. Lian, Q. Liu, D. Jiang and T. Wu. 2002. Effect of Au loading, H_2O and CO concentration on the stability of Au/ZnO catalysts for room-temperature CO oxidation. React. Kinet. Catal. Lett. 75: 343–351.

Wang, G.Y., W.X. Zhang, H.L. Lian, D.Z. Jiang and T.H. Wu. 2003. Effect of calcination temperatures and precipitant on the catalytic performance of Au/ZnO catalysts for CO oxidation at ambient temperature and in humid circumstances. Appl. Catal. A Gen. 239: 1–10.

Wang, H., Y. Li, Y. Zhang, J. Chen, G. Chu and L. Shao. 2015. Preparation of CeO_2 nano-support in a novel rotor-stator reactor and its use in Au-based catalyst for CO oxidation. Powder Technol. 273: 191–196.

Wang, M., F. Wang, J. Ma, M. Li, Z. Zhang, Y. Wang, X. Zhang and J. Xu. 2014. Investigations on the crystal plane effect of ceria on gold catalysis in the oxidative dehydrogenation of alcohols and amines in the liquid phase. Chem. Commun. 50: 292–294.

Widmann, D., R. Leppelt and R.J. Behm. 2007. Activation of a Au/CeO_2 catalyst for the CO oxidation reaction by surface oxygen removal/oxygen vacancy formation. J. Catal. 251: 437–442.

Wu, H., L. Wang, J. Zhang, Z. Shen and J. Zhao. 2011. Catalytic oxidation of benzene, toluene and p-xylene over colloidal gold supported on zinc oxide catalyst. Catal. Commun. 12: 859–865.

Wu, Z., D. Jiang, A.K.P. Mann, D.R. Mullins, Z.-A. Qiao, L.F. Allard, C. Zeng, R. Jin and S.H. Overbury. 2014. Thiolate ligands as a double-edged sword for CO oxidation on CeO_2 supported $Au_{25}(SCH_2CH_2Ph)_{18}$ nanoclusters. J. Am. Chem. Soc. 136: 6111–6122.

Xie, X., X. Wang, J. Tian, J. Liu, H. Cao, X. Song, N. Wei and H. Cui. 2017. Facile synthesis and superior ethyl acetate sensing performance of Au decorated ZnO flower-like architectures. Ceram. Int. 43: 5053–5060.

Yang, S.M., D.M. Liu and S.Y. Liu. 2008. Catalytic combustion of benzene over Au supported on ceria and vanadia promoted ceria. Top. Catal. 47: 101–108.

Yang, W., B. Zhang, Q. Zhang, L. Wang, B. Song, Y. Ding and C.P. Wong. 2017. Adjusting the band structure and defects of ZnO quantum dots via tin doping. RSC Adv. 7: 11345–11354.

Yashima, M. 2013. Crystal and electronic structures, structural disorder, phase transformation, and phase diagram of ceria-zirconia and ceria-based materials. pp. 1–46. *In*: Trovarelli, A. and P. Fornasiero (eds.). Catalysis by Ceria and Related Materials. Imperial College Press, London, UK.

Yi, G., Z. Xu, G. Guo, K. Tanaka and Y. Yuan. 2009. Morphology effects of nanocrystalline CeO_2 on the preferential CO oxidation in H_2-rich gas over Au/CeO_2 catalyst. Chem. Phys. Lett. 479: 128–132.

Yi, N., R. Si, H. Saltsburg and M. Flytzani-Stephanopoulos. 2010. Active gold species on cerium oxide nanoshapes for methanol steam reforming and the water gas shift reactions. Energy Environ. Sci. 3: 831–837.

Yi, N., H. Saltsburg and M. Flytzani-Stephanopoulos. 2013. Hydrogen production by dehydrogenation of formic acid on atomically dispersed gold on ceria. ChemSusChem. 6: 816–819.

Ying, F., S. Wang, C.-T. Au and S.-Y. Lai. 2010. Effect of the oxidation state of gold on the complete oxidation of isobutane on Au/CeO_2 catalysts. Gold Bull. 43: 241–251.

Yu, H., H. Ming, H. Zhang, H. Li, K. Pan, Y. Liu, F. Wang, J. Gong and Z. Kang. 2012. Au/ZnO nanocomposites: Facile fabrication and enhanced photocatalytic activity for degradation of benzene. Mater. Chem. Phys. 137: 113–117.

Zanella, R., C. Louis, S. Giorgio and R. Touroude. 2004. Crotonaldehyde hydrogenation by gold supported on TiO_2: structure sensitivity and mechanism. J. Catal. 223: 328–339.

Zhang, J., X. Wang, B. Chen, C. Li and D. Wu. 2003. Selective oxidation of CO in hydrogen rich gas over platinum–gold catalyst supported on zinc oxide for potential application in fuel cell. Energy Convers. Manag. 44: 1805–1815.

Zhang, J., Y. Jin, C. Li, Y. Shen, L. Han, Z. Hu, X. Di and Z. Liu. 2009. Creation of three-dimensionally ordered macroporous Au/CeO_2 catalysts with controlled pore sizes and their enhanced catalytic performance for formaldehyde oxidation. Appl. Catal. B Environ. 91: 11 20.

Zhang, J., G. Chen, M. Chaker, F. Rosei and D. Ma. 2013. Gold nanoparticle decorated ceria nanotubes with significantly high catalytic activity for the reduction of nitrophenol and mechanism study. Appl. Catal. B Environ. 132-133: 107–115.

Zhang, X., J. Qin, Y. Xue, P. Yu, B. Zhang, L. Wang and R. Liu. 2014. Effect of aspect ratio and surface defects on the photocatalytic activity of ZnO nanorods. Sci. Rep. 4: 4596.

Zheng, Y., C. Chen, Y. Zhan, X. Lin, Q. Zheng, K. Wei, J. Zhu and Y. Zhu. 2007. Luminescence and photocatalytic activity of ZnO nanocrystals: correlation between structure and property. Inorg. Chem. 46: 6675–6682.

Zhu, K.J., Y.J. Yang, J.J. Lang, B.T. Teng, F.M. Wu, S.Y. Du and X.-D. Wen. 2016. Substrate-dependent Au_x cluster: A new insight into Au_x/CeO_2. Appl. Surf. Sci. 387: 557–568.

Ziolek, M. 2003. Niobium-containing catalysts—the state of the art. Catal. Today 78: 47–64.

Ziolek, M., P. Decyk, I. Sobczak, M. Trejda, J. Florek, H. Golinska, W. Klimas and A. Wojtaszek. 2011. Catalytic performance of niobium species in crystalline and amorphous solids—Gas and liquid phase oxidation. Appl. Catal. A Gen. 391: 194–204.

Index

About the Editors

Vanesa Calvino-Casilda, Madrid (Spain), 1978, obtained her Ph.D. degree in Chemistry at UNED (Madrid, Spain) in 2008. Later, she worked in Catalytic Spectroscopy Laboratory as postdoctoral researcher at the Institute of Catalysis and Petrochemistry (CSIC, Madrid) (2008–2013). After that, she worked at the Faculty of Science (UNED) as postdoctoral researcher (2014–2017). Since 2017, she is an Assistant Professor at School of Engineers (UNED) in the Electrical, Electronics, Control and Telematics Engineering and Chemistry Applied to Engineering Department (DIEECTQAI). Her research focuses on Green Chemistry in the intensification of heterogeneous catalytic processes for fine chemicals synthesis and real-time reaction monitoring.
ORCID: http://orcid.org/0000-0002-2756-2164

Antonio José López-Peinado, Granada (Spain), 1959, obtained his Ph.D. degree in Chemistry at University of Granada (Spain) in 1984. He completed his formation at the University of Amsterdam (Holland) 1987, at the Penn State University (USA) 1991 and NIRE-Tsukuba (Japan) 1992. He worked at the University of Extremadura (Spain) and presently, since 1989, at UNED (Madrid, Spain). Since 2011, he is Full Professor in Inorganic Chemistry. His research focuses on the use of carbon and zeolitic materials as catalysts for processes for fine chemicals synthesis.
ORCID: https://orcid.org/0000-0002-6162-3151

Rosa María Martín-Aranda, Toledo (Spain), 1964, PhD 1992. Full Professor of Inorganic Chemistry, at the Universidad Nacional de Educación a Distancia (Madrid, UNED). She has a multidisciplinary academic background, degree in Organic Chemistry and PhD in Inorganic Chemistry from the Autónoma University of Madrid. Her research activity is focused on Sustainable Chemistry and Catalysis, it covers more than 120 scientific papers published in international journals, 8 patents and active participation in international conferences. From her PhD, she combines Organic with Inorganic Chemistry to develop new solid materials (zeolites, carbons, clays) to be used as catalysts in the preparation of fine chemicals and drugs. She is a member of different societies and national and international scientific committees. She is a member of the UNESCO Chair in Environmental Education and Sustainable Development. She actively collaborates with companies related Chemicals & Environment. She is a person convinced of the importance of popularizing science and activities for the dissemination of science.
ORCID: http://orcid.org/0000-0001-5628-8144

Elena Pérez-Mayoral, Madrid (Spain), 1969, obtained her Ph.D. degree in Chemical Sciences working in Organic Photochemistry, at the University Complutense of Madrid (Madrid, Spain), in 1999. In this year she joined the Department of Organic Chemistry and Biology, in the Science Faculty, at the National University of Distance Education (Madrid, UNED) (1999–2007) developing new organometallic complexes for Magnetic Resonance Imaging. More recently, she works in the Department of Inorganic Chemistry and Technical Chemistry, on the development of new methodologies, with reduced environmental impact, for the synthesis of biologically relevant heterocyclic compounds, promoted by porous catalytic systems, in the framework of sustainable environmental chemistry and catalysis. Since 2011 she is Professor of Inorganic Chemistry, at the UNED. Her research focuses on the development of new methodologies based on catalytic technologies for the synthesis of valuable compounds as well as the understanding of the reaction mechanisms. ORCID: http://orcid.org/0000-0003-4554-141X

Printed and bound by CPI Group (UK) Ltd, Croydon, CR0 4YY

24/10/2024

01778307-0010